D0114131

UNDERLAND

By the same author

Mountains of the Mind

The Wild Places

The Old Ways

Landmarks

The Gifts of Reading

The Lost Words (*with Jackie Morris*)

UNDERLAND

A Deep Time Journey

ROBERT MACFARLANE

W. W. NORTON & COMPANY
Independent Publishers Since 1923
New York | London

ST. JOHN THE BAPTIST PARISH LIBRARY
2920 NEW HIGHWAY 51
LAPLACE, LOUISIANA 70068

Copyright © 2019 by Robert Macfarlane
First American Edition 2019

All rights reserved
Printed in the United States of America

For information about permission to reproduce selections from this book, write to Permissions, W. W. Norton & Company, Inc., 500 Fifth Avenue, New York, NY 10110

For information about special discounts for bulk purchases, please contact W. W. Norton Special Sales at specialsales@wwnorton.com or 800-233-4830

Manufacturing by LSC Communications, Harrisonburg

Library of Congress Cataloging-in-Publication Data

Names: Macfarlane, Robert, 1976– author.
Title: Underland : a deep time journey / Robert Macfarlane.
Description: First American edition. | New York : W.W. Norton & Company, 2019. | Includes bibliographical references and index.
Identifiers: LCCN 2019000216 | ISBN 9780393242140 (hardcover)
Subjects: LCSH: Civilization, Subterranean. | Underground areas—History. | Voyages and travels. | Geology.
Classification: LCC GN755 .M295 2019 | DDC 551.44/7—dc23
LC record available at https://lccn.loc.gov/2019000216

W. W. Norton & Company, Inc., 500 Fifth Avenue, New York, N.Y. 10110
www.wwnorton.com

W. W. Norton & Company Ltd., 15 Carlisle Street, London W1D 3BS

1 2 3 4 5 6 7 8 9 0

Is it dark down there
Where the grass grows through the hair?
Is it dark in the under-land of Null?

Helen Adam, 'Down There in the Dark', 1952

The void migrates to the surface . . .

Advances in Geophysics, 2016

Contents

PART III HAUNTING (THE NORTH)

FIRST CHAMBER

The way into the underland is through the riven trunk of an old ash tree.

Late-summer heatwave, heavy air. Bees browsing drowsy over meadow grass. Gold of standing corn, green of fresh hay-rows, black of rooks on stubble fields. Somewhere down on lower ground an unseen fire is burning, its smoke a column. A child drops stones one by one into a metal bucket, *ting*, *ting*, *ting*.

Follow a path through fields, past a hill to the east that is marked by a line of nine round burial barrows, nubbing the land like the bones of a spine. Three horses in a glinting cloud of flies, stock-still but for the swish of a tail, the twitch of a head.

Over a stile in a limestone wall and along a stream to a thicketed dip from which grows the ancient ash. Its crown flourishes skywards into weather. Its long boughs lean low around. Its roots reach far underground.

Swallows curve and dart, feathers flashing. Martins criss-cross the middle air. A swan flies high and south on creaking wings. This upper world is very beautiful.

Near the ash's base its trunk splits into a rough rift, just wide enough that a person might slip into the tree's hollow heart – and there drop into the dark space that opens below. The rift's edges are smoothed to a shine by those who have gone this way before, passing through the old ash to enter the underland.

Beneath the ash tree, a labyrinth unfurls.

Down between roots to a passage of stone that deepens steeply into the earth. Colour depletes to greys, browns, black. Cold air pushes past. Above is solid rock, utter matter. The surface is scarcely thinkable.

The passage is taken; the maze builds. Side-rifts curl off. Direction is difficult to keep. Space is behaving strangely – and so too is time. Time moves differently here in the underland. It thickens, pools, flows, rushes, slows.

The passage turns, turns again, narrows – and leads into surprising space. A chamber is entered. Sound now booms, resonates. The walls of the chamber appear bare at first, but then something extraordinary happens. Scenes from the underland start to show themselves on the stone, distant from one another in history, but joined by echoes.

In a cave within a scarp of karst, a figure inhales a mouthful of red ochre dust, places its left hand against the cave wall – fingers spread, thumb out, palm cold on the rock – and then blows the ochre hard against the hand's back. There is an explosion of dust – and when the hand is lifted its ghostly print remains, the stone around having taken the red of the ochre. The hand is shifted, more dust is blown and another pale outline is left. Calcite will run over these prints, sealing them in. The prints will survive for more than 35,000 years. Signs of what? Of joy? Of warning? Of art? Of life in the darkness?

In the shallow sandy soil of northern Europe, some 6,000 years ago, the body of a young woman – dead in childbirth along with her son – is lowered gently into a grave. Next to her is laid the white wing of a swan. Then onto the wing is placed the body of her son, so that the baby is doubly cradled in death – by the swan's feathers and his mother's arms. A round mound of earth is raised to mark their burial place: the woman, the child and the white swan's wing.

On an island in the Mediterranean 300 years before the founding of the Roman Empire, a metalworker completes the design of a silver coin. The coin's face shows a square labyrinth with a single entrance on its upper edge and a complex path to its centre. The walls of the labyrinth – like the rim of the coin – are slightly raised and polished to a sheen. Tooled into the labyrinth's centre is the figure of a creature with the head of a bull and the legs of a man: the Minotaur, waiting in darkness for whatever comes next.

Six hundred years later, a young woman sits for a portrait painter in Egypt. She has dressed most handsomely for the sitting. She has strong dark eyebrows and wide dark eyes, almost black. Her hair is pulled back from her forehead by a metal band topped with a gold bead, and she wears a golden scarf and brooch. The painter works with hot beeswax, gold leaf and coloured pigments, layering them onto wood. He is creating the young woman's death image. When she dies it will be wrapped into the bands of cloth used to mummify her corpse such that it takes the place of her real face. As her body decays beneath its swaddling, the portrait will remain un-aged. It is well to do such things early, when one looks most glowing. Her body will be placed in a necropolis – a city of the dead built at the entrance to a sunken depression of desert, in a buried chamber lined with limestone and covered with quartzite slabs to deter grave robbers, close to vaults that hold the mummified corpses of more than a million ibises.

Beneath a plateau in southern Africa, late in the nineteenth century, miners crawl through miles of narrow tunnel – cut deeper underground here than anywhere else on Earth at this time – lugging ore from a sunken reef of gold. Some of these men, who have migrated to the area in their thousands to work, will die soon in rockfalls and accidents. More will die slowly of silicosis from breathing the rock dust down there in the killing dark, year after year. Here the human body is largely disposable in the view of the corporations that own the mine and the markets that drive it: a small, unskilled tool of extraction to be replaced when it fails or wears out. The ore the men bring up is crushed and smelted, and the wealth it yields lines the pockets of shareholders in distant countries.

In a cave in the foothills of the Indian Himalayas not long after Partition, a young woman meditates sixteen hours a day, for seventy-five days. She sits stone-still while meditating, save for her mouth, which moves as she murmurs mantras. She emerges most often at night; when it is cloudless the Milky Way can be seen spilling across the sky above the peaks. She lives on water drunk with cupped hands from a sacred river, and on foraged wild berries and fruits. The mantras, the solitude and the darkness bring perceptions that are new to her, and she experiences a profound change in her vision. When at last she completes her retreat she feels vast as the skies, old as the mountains, formless as starlight.

Thirty years ago a boy and his father use the claw of a hammer to prise up a floorboard in a house they are soon to leave. They have made a jam-jar time capsule. Into the jar the boy has placed objects and messages. The die-cast metal model of a bomber aeroplane. The outline of his left hand traced in red ink on plain paper. A self-description for whoever finds the jar – *Quite tall for my age, very blonde hair, almost white. Biggest fear, nuclear war* – written in pencil on a notebook page. A stopped watch with luminous hands and dial, around which he likes to cup his hands to see the numbers glow. He pours a handful of rice into the jar to absorb moisture, screws the jar's brass lid tightly shut, puts it in its hiding place and nails the floorboard back down.

Deep in an extinct volcano a tunnel network has been bored above a crustal fault known as Ghost Dance. Access drifts incline through tilted strata to level out in a repository zone, organized into emplacement corridors. The intent is to inter high-level nuclear waste in these corridors: radioactive uranium pellets encased in iron, then encased in copper, then buried above the Ghost Dance fault to pulse out their half-lives for millions of years to come. The timescale of the hazard is such that those responsible for entombing this waste must now face the question of how to communicate its danger to the distant future. This is a risk that will outlast not only the life of its makers but perhaps also the species of its makers. How to mark this site? How to tell whatever beings will come to this desert place that what is kept in this rock sarcophagus is desperately harmful, is *not* of value, *must never be disturbed*?

And on a muddy ledge, two and a half miles into the cave system of a mountain in which they have become trapped by flood waters, twelve boys and their football coach sit in utter blackness, conserving the batteries of their phones, waiting day after day to see if the waters will rise or fall – or if by miracle someone will come to rescue them. With each passing hour the oxygen in their chamber is reduced by their breathing, and carbon dioxide levels increase. Above the mountain the monsoon clouds build, threatening more rain. Outside the mountain thousands of rescuers from six countries gather. At first they do not know if the boys are alive. Then they find handprints in mud on the walls of a chamber two miles into the system. Hope is given. Divers push further and further along the flooded passageways. Nine days after entering the mountain, the boys hear sounds coming from the river that flows past their ledge. Then they see lights glowing in the water. Bubbles seethe up. The lights rise. A man breaks the surface. The boys and their coach blink in the beam of his head-torch. One of the boys raises a hand in greeting, and the diver raises his in reply. 'How many of you?' asks the diver. 'Thirteen,' one replies. 'Many people are coming,' says the diver.

So these scenes from the underland unfold along the walls of this impossible chamber, down in the labyrinth beneath the riven ash. The same three tasks recur across cultures and epochs: to shelter what is precious, to yield what is valuable, and to dispose of what is harmful.

Shelter (memories, precious matter, messages, fragile lives).

Yield (information, wealth, metaphors, minerals, visions).

Dispose (waste, trauma, poison, secrets).

Into the underland we have long placed that which we fear and wish to lose, and that which we love and wish to save.

I
Descending

We know so little of the worlds beneath our feet. Look up on a cloudless night and you might see the light from a star thousands of trillions of miles away, or pick out the craters left by asteroid strikes on the moon's face. Look down and your sight stops at topsoil, tarmac, toe. I have rarely felt as far from the human realm as when only ten yards below it, caught in the shining jaws of a limestone bedding plane first formed on the floor of an ancient sea.

The underland keeps its secrets well. Only in the last twenty years have ecologists succeeded in tracing the fungal networks that lace woodland soil, joining individual trees into intercommunicating forests – as fungi have been doing for hundreds of millions of years. In China's Chongqing province, a cave network explored in 2013 was found to possess its own weather system: ladders of stacked mist that build in a huge central hall, cold fog that drifts in giant cloud chambers far from the reach of the sun. A thousand feet underground in northern Italy, I abseiled into an immense rotunda of stone, cut by a buried river and filled with dunes of black sand. Traversing those dunes on foot was like trudging through a windless desert on a lightless planet.

Why go low? It is a counter-intuitive action, running against the grain of sense and the gradient of the spirit. Deliberately to place something in the underland is almost always a strategy to shield it from easy view. Actively to retrieve something from the underland almost always requires effortful work. The underland's difficulty of access has long made it a means of symbolizing what cannot openly be said or seen: loss, grief, the mind's obscured depths, and what Elaine Scarry calls the 'deep subterranean fact' of physical pain.

A long cultural history of abhorrence exists around underground spaces, associating them with 'the awful darkness inside the world', in Cormac McCarthy's phrase. Fear and disgust are the usual responses to such environments; dirt, mortality and brutal labour the dominant connotations. Claustrophobia is surely the sharpest of all common phobias. I have often noticed how claustrophobia – much more so than vertigo – retains its disturbing power even when being experienced indirectly as narrative or description. Hearing stories of confinement below ground, people shift uneasily, step away, look to the light – as if words alone could wall them in.

I still remember as a ten-year-old reading the account, in Alan Garner's novel *The Weirdstone of Brisingamen*, of two children escaping danger by descending the mining tunnels that riddle the sandstone outcrop of Alderley Edge in Cheshire. Deep inside the Edge, the embrace of the stone becomes so tight that it threatens to trap them:

> They lay full length, walls, floor and roof fitting them like a second skin. Their heads were turned to one side, for in any other position the roof pressed their mouths into the sand and they could not breathe. The only way to advance was to pull with the fingertips and to push with the toes, since it was impossible to flex their legs at all, and any bending of the elbows threatened to jam the arms helplessly under the body. [Then Colin's] heels jammed against the roof: he could move neither up nor down and the rock lip dug into his shins until he cried out with the pain. But he could not move . . .

Those passages took cold grip of my heart, emptied my lungs of air. Rereading them now, I feel the same sensations. But the situation also exerted a powerful narrative traction upon me – and still does. Colin could not move and I could not stop reading.

An aversion to the underland is buried in language. In many of the metaphors we live by, height is celebrated but depth is despised. To be 'uplifted' is preferable to being 'depressed' or 'pulled down'. 'Catastrophe' literally means a 'downwards turn', 'cataclysm' a 'downwards violence'. A bias against depth also runs through mainstream conventions of observation and representation. In his book *Vertical*, Stephen Graham describes the dominance of what he calls the 'flat tradition' of geography and cartography, and the 'largely horizontal worldview' that has resulted. We find it hard to escape the 'resolutely flat perspectives' to which we have become habituated, Graham argues – and he finds this to be a political failure as well as a perceptual one, for it disinclines us to attend to the sunken networks of extraction, exploitation and disposal that support the surface world.

Yes, for many reasons we tend to turn away from what lies beneath. But now more than ever we need to understand the underland. 'Force yourself to see more flatly,' orders Georges Perec in *Species of Spaces*. 'Force yourself to see more deeply,' I would counter. The underland is vital to the material structures of contemporary existence, as well as to our memories, myths and metaphors. It is a terrain with which we daily reckon and by which we are daily shaped. Yet we are disinclined to recognize the underland's presence in our lives, or to admit its disturbing forms to our imaginations. Our 'flat perspectives' feel increasingly inadequate to the deep worlds we inhabit, and to the deep time legacies we are leaving.

We are presently living through the Anthropocene, an epoch of immense and often frightening change at a planetary scale, in which 'crisis' exists not as an ever-deferred future apocalypse but rather as an ongoing occurrence experienced most severely by the most vulnerable. Time is profoundly out of joint – and so is place. Things that should have stayed buried are rising up unbidden. When confronted by such surfacings it can be hard to look away, seized by the obscenity of the intrusion.

In the Arctic, ancient methane deposits are leaking through 'windows' in the earth opened by melting permafrost. Anthrax spores are being released from reindeer corpses buried in once-frozen soil, now exposed by erosion and warmth. In the forests of Eastern Siberia a crater is yawning in the softening ground, swallowing tens of thousands of trees and revealing 200,000-year-old strata: local Yakutian people refer to it as a 'doorway to the underworld'. Retreating Alpine and Himalayan glaciers are yielding the bodies of those engulfed by their ice decades before. Across Britain, recent heatwaves have caused the imprints of ancient structures – Roman watchtowers, Neolithic enclosures – to shimmer into view as crop-marks visible from above: aridity as X-ray, the land's submerged past rising up in parched visitation. Where the River Elbe flows through the Czech Republic, summer water levels have recently dropped so far that 'hunger stones' have been uncovered – carved boulders used for centuries to commemorate droughts and warn of their consequences. One of the hunger stones bears the inscription '*Wenn du mich siehst, dann weine*': 'If you see me, weep.' In north-west Greenland an American Cold War missile base, sealed under the ice cap fifty years ago and containing hundreds of thousands of gallons of chemical contaminants, has begun to move towards the light. 'The problem,' writes the archaeologist Þóra Pétursdóttir, 'is not that things become buried deep in strata – but that they endure, outlive us, and come back at us with a force we didn't realise they had . . . a dark force of "sleeping giants"', roused from their deep time slumber.

'Deep time' is the chronology of the underland. Deep time is the dizzying expanses of Earth history that stretch away from the present moment. Deep time is measured in units that humble the human instant: epochs and aeons, instead of minutes and years. Deep time is kept by stone, ice, stalactites, seabed sediments and the drift of tectonic plates. Deep time opens into the future as well as the past. The Earth will fall dark when the sun exhausts its fuel in around 5 billion years. We stand with our toes, as well as our heels, on a brink.

There is dangerous comfort to be drawn from deep time. An ethical lotus-eating beckons. What does our behaviour matter, when *Homo sapiens* will have disappeared from the Earth in the blink of a geological eye? Viewed from the perspective of a desert or an ocean, human morality looks absurd – crushed to irrelevance. Assertions of value seem futile. A flat ontology entices: all life is equally insignificant in the face of eventual ruin. The extinction of a species or an ecosystem scarcely matters in the context of the planet's cycles of erosion and repair.

We should resist such inertial thinking; indeed, we should urge its opposite – deep time as a radical perspective, provoking us to action not apathy. For to think in deep time can be a means not of escaping our troubled present, but rather of re-imagining it; countermanding its quick greeds and furies with older, slower stories of making and unmaking. At its best, a deep time awareness might help us see ourselves as part of a web of gift, inheritance and legacy stretching over millions of years past and millions to come, bringing us to consider what we are leaving behind for the epochs and beings that will follow us.

When viewed in deep time, things come alive that seemed inert.

New responsibilities declare themselves. A conviviality of being leaps to mind and eye. The world becomes eerily various and vibrant again. Ice breathes. Rock has tides. Mountains ebb and flow. Stone pulses. We live on a restless Earth.

~

The oldest of underland stories concerns a hazardous descent into darkness in order to reach someone or something consigned to the realm of the dead. A variant to the *Epic of Gilgamesh* – written around 2100 BC in Sumeria – tells of such a descent, made by Gilgamesh's servant Enki to the 'netherworld' on behalf of his master to retrieve a lost object. Enki sails through storms of hailstones that strike him like 'hammers', his boat trembles from the impact of waves that attack it like 'butting turtles' and 'lions', but still he reaches the netherworld. There, however, he is promptly imprisoned – only to be freed when the young warrior Utu opens a hole to the surface and carries Enki back out on a lofting breeze. Up in the sunlight Enki and Gilgamesh embrace, kiss, and talk for hours. Enki has not retrieved the lost object, but he has brought back precious news of vanished people. 'Did you see my little stillborn children who never knew existence?' asks Gilgamesh desperately. 'I saw them,' answers Enki.

Similar stories recur throughout world myth. Classical literature records numerous instances of what in Greek were known as the *katabasis* (a descent to the underland) and the *nekyia* (a questioning of ghosts, gods or the dead about the earthly future), among them Orpheus' attempt to retrieve his beloved Eurydice from Hades, and Aeneas' voyage – led by the Sibyl, protected by the Golden Bough – to seek counsel with the shade of his father. The recent rescue of the

Thai footballers from their lonely chamber far inside a mountain was a modern *katabasis*: the story seized global attention in part because it possessed the power of myth.

What these narratives all suggest is something seemingly paradoxical: that darkness might be a medium of vision, and that descent may be a movement towards revelation rather than deprivation. Our common verb 'to understand' itself bears an old sense of passing beneath something in order fully to comprehend it. 'To discover' is 'to reveal by excavation', 'to descend and bring to the light', 'to fetch up from depth'. These are ancient associations. The earliest-known works of cave art in Europe – taking the form of painted ladders, dots and hand stencils on the walls of Spanish caves – have been dated to around 65,000 years ago, some 20,000 years before *Homo sapiens* are believed to have first arrived in Europe from Africa. Neanderthal artists left these images. Long before anatomically modern humans reached what is now Spain, writes one of the archaeologists responsible for the dating of this art, 'People were making journeys into the darkness.'

Underland is a story of journeys into darkness, and of descents made in search of knowledge. It moves over its course from the dark matter formed at the universe's birth to the nuclear futures of an Anthropocene-to-come. During the deep time voyage undertaken between those two remote points, the line about which the telling folds is the ever-moving present. Across its chapters, in keeping with its subject, extends a subsurface network of echoes, patterns and connections.

For more than fifteen years now I have been writing about the relationships between landscape and the human heart. What began as a wish to solve a personal mystery – why I was so drawn to mountains as a young man that I was, at times, ready to die for love of them – has unfolded into a project of deep-mapping carried out over five books and around 2,000 pages. From the icy summits of the world's highest peaks, I have followed a downwards trajectory to what must surely be a terminus, exploring the storeys of place that lie beneath the surface. 'The descent beckons / as the ascent beckoned,' wrote William Carlos Williams in a late poem. It has taken me until the second half of my life to understand something of what Williams meant. In the underland I have seen things I hope I will never forget – and things I wish I had never witnessed. What I thought would be my least human book has become, to my surprise, my most communal. If the image at the centre of much that I have written before is that of the walker's placed and lifted foot, the image at the heart of these pages is that of the opened hand, extended in greeting, compassion or the making of a mark.

I have for some time now been haunted by the Saami vision of the underland as a perfect inversion of the human realm, with the ground always the mirror-line, such that 'the feet of the dead, who must walk upside down, touch those of the living, who stand upright'. The intimacy of that posture is moving to me – the dead and the living standing sole to sole. Seeing photographs of the early hand-marks left on the cave walls of Maltravieso, Lascaux or Sulawesi, I imagine laying my own palm precisely against the outline left by those unknown makers. I imagine, too, feeling a warm hand pressing through from within the cold rock, meeting mine fingertip to fingertip in open-handed encounter across time.

~

Shortly before beginning the journeys recounted here, I was given two objects. Each came with a request, and it was a condition of the gift of these objects that I agreed to fulfil those requests.

The first of the objects is a double-cast bronze casket the size of a swan's egg, which sits heavy in the hand. It is a kist and what it contains is toxic. Its maker wrote his demons down on a sheet of paper: his hatreds, fears and losses, the pain he had inflicted on others and the pain others had inflicted on him – all that was worst in his mind. Then he burned the paper and sealed the ashes inside the casket. Then he double-cast the casket, giving it a second layer of bronze to increase the strength of the containment. That outer layer of bronze became pitted and encrusted in the process of its casting, such that it seemed to resemble either the surface of a planet or the weather above it. Then he drove four iron nails through the casket's centre, cutting off their ends and filing them flush. It is an exceptionally powerful object, which possesses a ritual intensity of creation. It could have been fashioned at any point in the past 2,500 years, but it was made only recently.

I was given the casket on the condition that I disposed of it in the deepest or most secure underland site that I reached – a place from which it could never return.

The second of the objects is an owl cut from a slice of whalebone. It is a talisman and what it connotes is magic. The minke whale from which the owl was taken had washed up dead on the shoreline of a Hebridean island. One of its rib bones was smoothed into cross-sections, each less than half an inch thick and six inches high. One of those cross-sections was then cut into the form of an owl with four bold strokes of a blade: two strokes for the eyes, and two for the wing lines. It is an exceptionally beautiful object, which possesses an Ice Age simplicity of making. It could have been fashioned at any point in the past 20,000 years, but it was made only recently.

I was given the owl on the condition that I carried it with me at all times in the underland, to help me see in the dark.

PART ONE

Seeing (Britain)

2
Burial
(*Mendips, Somerset*)

The bones of a child lie in darkness on a ledge of limestone. Sunlight has not seen this child for over 10,000 years. In that time, calcite has flowed like silver varnish from the rock around, chrysalizing the body.

A January day in 1797 and two young men are out rabbit-catching in the Mendip Hills of Somerset. They flush out a rabbit on the slope of a ravine. The rabbit runs and finds refuge in a jumble of boulders. The men are hungry; they want the rabbit. So they pull away some of the rocks – and are 'surprised with the appearance of a subterraneous passage'. They enter the passage, which leads them steeply into the limestone of the scarp and then opens into a 'large and lofty cavern, the roof and sides of which are most curiously fretted and embossed'.

Winter sun follows them down the passage and lights up the chamber. It is, they see, a charnel house. On the floor and ledges to their left are scattered bones and complete skeletons, 'lying promiscuously, almost converted into stone'. The relics shine with calcite, and dusting some of the bones is red ochre powder. A single large stalactite hangs from the chamber's ceiling which, when struck, rings like a bell, its peal echoing in the cave-space. The stalactite has reached down and begun to absorb one of the skeletons; embedded in it are a skull, a thigh bone and two teeth with the enamel still intact.

Also present in the cave are animal remains: the teeth of a brown bear, a barbed spear-point made from a red deer antler, and the bones of lynx, fox, wildcat and wolf. Votive objects have been interred here, too: sixteen periwinkle shells pierced so that they will hang spiral-outwards when worn against the body as a necklace; and a nest of seven pieces of fossil ammonite, the ends of their arcs rubbed smooth.

The human bodies, it will later be established, are more than ten millennia old, and among them are children and infants as well as adults. All show signs of chronic malnutrition. The adults stood little more than five feet tall. The children's molars were scarcely worn. Slowly, it becomes clear to those who study this mysterious place – now known as Aveline's Hole – that, far back in the Mesolithic, the cave was used as a cemetery over a period of around a century. Much of the world's water was then still locked up by glaciation. Sea levels were much lower. What we now call the Bristol Channel and much of the North Sea did not exist; one could walk north from the Mendips to Wales on dry land, or eastwards over Doggerland to France and the Netherlands.

The evidence from Aveline's suggests a shifting group of hunter-gatherers taking that area of the Mendips as their home range over two or three generations, and using the chamber as their mausoleum. These people – whose lives were short and unthinkably hard, who suffered from paucities of food and energy – made the effort and took the care to carry the bodies of their dead to this difficult hillside site, to place them within the chamber, to leave significant objects and the bones of creatures with them, and to open and then reseal the entrance with each new burial.

These wandering, hungry people wished for a secure location in which to entomb their dead – a place to which they could return over time. No comparable cemetery is known to have been established in Britain for another 4,000 years.

We are often more tender to the dead than to the living, though it is the living who need our tenderness most.

~

'Mendip is mining country,' says Sean. 'It's also caving country. But above all it's burial country. There are hundreds of Bronze Age funeral barrows spread across this landscape, some joined with monuments and henges into large-scale ritual complexes. In one of the barrows an antiquarian called Skinner found an amber bead with a bee trapped inside it, preserved right down to the hairs on its legs.'

Late afternoon, early autumn, unseasonable heat. Air shimmering in the sun, car doors scalding to touch. But it is cool as a pantry in Sean and Jane Borodale's house, set down in the shadows of a quiet side-arm of Nettlebridge Valley. Board games are piled in teetering stacks in the porch. Mint, thyme and rosemary flourish in pots by the porch. A large ammonite is embedded in the front doorstep, polished by decades of footfall. And in the garden, hanging from the outstretched wings of a towering wooden totem pole, are the flayed skins of two men.

'Those are our caving suits,' says Sean, waving towards the skins. 'Strictly speaking, they're chemical hazmat suits. I sourced them from eastern Europe. They're ideal for our needs. You'll see.'

Sean, Jane and their two boys have lived in this fairy-tale cottage for several years. The former owner held seances here, believing she could speak through the veil to the dead. To the west of the house a wrinkled field rises up the scarp before disappearing into ash woods on the ridge line. A stream gurgles off the scarp and past the house.

I have come to the Mendips to learn how to see in the dark. Sean knows the Mendips profoundly well, above ground and below. He is a bee-keeper, a caver, a walker and a remarkable poet. He has curling black hair and is very gentle. For several years he has been working on a long series of poems or voicings that emerge from – and in some cases are written within – the underland of the Mendips: their lead mines, iron workings and limestone quarries, their many burial sites, their Cold War bunkers and the countless miles of natural cave and tunnel that honeycomb their bedrock. Sean is compelled by the great descent stories of underworld mythology – Dante and Virgil, Persephone and Demeter, Eurydice, Orpheus and Aristaeus (the keeper of bees) – and by the associated visionary powers of darkness and blindness. The poems he writes about the underland feel to me both unearthed and unearthly. In them deep time is given utterance, earth is stirred, stone speaks. In them, too, the dead are quickened briefly back to life by the poet's attention.

The Mendips rise south of Bristol and west of Bath. From their southern edge on a clear day, Glastonbury Tor can be seen across the water-bearing flatlands of the Somerset Levels. From west to east they stretch almost thirty miles, tapering down towards the sea at the Bristol Channel. Their geology is elaborate, but predominantly they are a limestone range – and limestone land, wrote Arthur Conan Doyle, 'is hollow . . . land; could you strike it with some gigantic hammer, it would boom like a drum, or possibly cave in altogether and expose some huge subterranean sea.'

The first fact of limestone is its solubility in water. Rain absorbs carbon dioxide from the air, creating a mild carbonic acid – just sharp enough to etch and fret limestone, given time. This fretwork deepens into limestone's surface perforations of gryke and clint, and also its hidden labyrinths of rift and chamber. Streams shape stone with their energy. Thermal waters rise from within the earth, biting rock into form. Limestone landscapes are rich with clandestine places. They have the unexpected volumes of a lung's interior. Portals give access to their extensive underland: pots and sinkholes, swallets where streams vanish into their own beds. The great writer and cartographer of the west of Ireland, Tim Robinson, knows the deceptions of limestone better than almost anyone. After living on and mapping limestone for more than forty years, he concludes: 'I do not trust space an inch.'

'Let me show you the garden,' says Sean.

The cottage's land drops down to the valley's main stream. We stop at its bank. The water is so clear it can hardly be seen. Small trout fin in the current.

'It's a petrifying stream,' Sean says. 'There's so much calcium carbonate dissolved in it that any twigs or leaves snagged there soon pick up a white crust of stone.'

Green-black damselflies dance on the current. Horseflies cruise for blood.

'Look at this,' Sean says, pointing upwards. Where the lowest bough of an old alder meets its trunk, one end of a curved metal blade protrudes. The rest of the object is lost below the bark.

'It's a scythe. Someone hooked it up on here many decades ago and forgot about it. So the tree absorbed the blade, growing around it while the handle rotted away.'

In the vegetable garden, tucked into the lee of a blackthorn hedge, are two beehives the colour of red ochre. Sloped landing boards lead up to the dark hive mouths. Bees alight on the boards, crawl into the hives, whirr out again.

Everywhere I look there is evidence of burial and excavation. Badger setts, molehills, bee tunnels, the engulfed scythe, the hives, the entrances to mine adits. Even the house, set back into the dolomite slope, is part cave.

'I didn't understand the Mendips until I began to explore them from below,' Sean says. 'Almost everything here involves the underworld somehow: quarrying, mining, caving. Bronze Age lead mining. Coal mining by the Romans. Quarries for limestone grit, so big they have a spiral ramp cut to a narrow core, in order that the lorries can get up and down, like an industrial version of Dante's descent in *The Inferno*. And basalt quarries to supply hardcore for top-dressing roads.'

A dragonfly rustles past.

'Then there are the burial sites – Bronze Age bowl barrows mostly, but Neolithic long barrows too, and of course, at Aveline's, the Mesolithic chamber. Medieval and early modern graveyards, and then our own still-growing cemeteries. This has been a funerary landscape for over 10,000 years. It's a terrain into which we have long entrusted things, as well as from which we have long extracted things.'

~

'To be human means above all to bury,' declares Robert Pogue Harrison in his study of burial practices, *The Dominion of the Dead*, boldly drawing on Vico's suggestion that *humanitas* in Latin comes first and properly from *humando*, meaning 'burying, burial', itself from *humus*, meaning 'earth' or 'soil'.

We are, certainly, a burying species as well as a building species – and our predecessors were buriers too. In a cave system called Rising Star in the limestone of South Africa a team of palaeoarchaeologists led by six women has discovered fossilized bone fragments thought to belong to a previously unknown early human relative, a species now named as *Homo naledi*. The disposition of this dark matter in two deep-set chambers suggests, remarkably, that *Homo naledi* was already interring its dead underground some 300,000 years ago.

In burial, the human body becomes a component of the earth, returned as dust to dust – inhumed, restored to humility, rendered humble. Just as the living need places to inhabit, so it is often in the nature of our memory-making to wish to be able to address our dead at particular sites on the Earth's surface. The burial chamber, the gravestone, the hillside on which ashes have been scattered, the cairn: these are places to which the living can return and where loss might be laid to rest. The grief of those who have been unable to locate the bodies of their loved ones can be especially corrosive – acid and unhealing.

We give bodies and their residues to the earth in part as a means of safekeeping. Burial often aspires to preservation – of memory, of matter – for time behaves differently in the underland, where it might be slowed or stayed. Early in his profound meditation on inhumation and history, *Urne-Buriall* (1658), Thomas Browne describes the discovery – in the sandy soil of a field near Walsingham in the 1650s – of 'between fourty and fifty Urnes . . . not a yard deep, nor farre from each other'. Each of the urns contained up to two pounds of human bones and ash, as well as offerings: 'peeces of small boxes, or combes handsomely wrought, handles of small brasse instruments, brazen nippers, and in one some kind of *Opale*'. Browne refers to the dark interiors of these buried urns as 'conservatories' – that is, spaces of conservation, insulated from what he calls 'the piercing Atomes of ayre' that corrupt the upper world. He represents each urn as a bright chamber of memory, secured in the 'nether part of the Earth'.

Limestone, in particular, has long been a geology of burial – in part because it is so common globally, in part because its erosive tendencies create so many natural crypts into which bodies may be laid, and in part because limestone is itself, geologically speaking, a cemetery. Limestone is usually formed of the compressed bodies of marine organisms – crinoids and coccolithophores, ammonites, belemnites and foraminifera – that died in waters of ancient seas and then settled in their trillions on those seabeds. These creatures once built their skeletons and shells out of calcium carbonate, metabolizing the mineral content of the water in which they lived to create intricate architectures. In this way limestone can be seen as merely one phase in a dynamic earth cycle, whereby mineral becomes animal becomes rock; rock that will in time – in deep time – eventually supply the calcium carbonate out of which new organisms will build their bodies, thereby re-nourishing the same cycle into being again.

This dance of death and life that goes into limestone's creation is what makes it without doubt the liveliest, queerest rock I know – and the human burials it holds have sometimes echoed these vibrancies, and the multi-species makings that have brought limestone into being.

Around 27,000 years ago, on a limestone hillside overlooking what is now the Austrian Danube, two babies, dead at birth, were placed side by side in a freshly dug round hole. Their remains were wrapped in animal hide, and the space around them was packed with red ochre, into which were mixed yellow beads of ivory. A shelter was then constructed to protect them from the crushing embrace of the earth: a scapula from a woolly mammoth, propped up as a bone shroud on pieces of tusk.

Twelve thousand years ago in a limestone cave above the Hilazon River in what is now northern Israel, a grave was prepared for a woman in her forties. An oval hole was dug in the cave floor, and its sides were walled with limestone slabs. Her body was placed in the grave, curled against the northern side of the oval. Two stone martens, their brown and cream fur sleek in the low light, were draped over her: one across her upper body, one across her lower. The foreleg of a wild boar was laid on her shoulder. A human foot was placed between her feet. The blackened shells of eighty-six tortoises were scattered over her. The tail of an aurochs was put near the base of her spine. The wing of a golden eagle was opened over her. She had become a wondrous hybrid – a being of many beings. At last, a single large plate of limestone was pulled over the hole, closing this compound creature inside her chamber.

On a limestone outcrop near the Somerset village of Stoney Littleton, around 5,500 years ago, a chambered tomb was constructed. It remains present in the landscape: low-slung and turf-roofed on a slope of the hill, the beckoning mouth of its main entrance marked by a vast lintel stone and two flanking door jambs of single upright slabs. Set into the western jamb is the cast of an ammonite almost a foot in diameter.

And across ten millennia – since those first hunter-gatherer bodies were placed in the chamber discovered by the rabbit-catching boys – humans have buried their dead in the limestone uplands of the Mendips. There are some 400 Bronze Age round barrows in the Mendips, dating from around 2500 BC to around 750 BC. Most are clustered together, and most contained – until they were plundered or ploughed out – a single inhumation and the grave goods that were left with it. The bodies were typically placed in a stone-lined kist or collared urn under the dome of earth. The accompanying grave goods included pottery cups, barbed flint arrowheads, a bronze dagger, amber-headed pins, and beads of jet and shale. Their inclusion in the barrows speaks of a belief, widespread among cultures, that burial is a form of onwards journey to an afterlife where earthly items will be needed.

~

Sean and I walk back up to the cottage, step over the ammonite set into the door sill and enter the white-walled kitchen. It's a relief to be back in the cool of the house after the garden's heat. Jane smiles in welcome.

'You're here on a good day for the cottage,' she says. 'In summer, it's a dream. But in the other three seasons of the year, when the north wind blows straight down that valley, in one gable and out the other, it's impossible to keep warm. We lose the light so fast too. By early afternoon in full winter we're in deep shadow, cold shadow.'

That afternoon we sit, talk, drink tea. On the table is a blue-and-white china plate, Russian in its decorative style, showing a steam train emerging from a tunnel into winter fields. Two peasant figures walk by the trackside, carrying bundles of sticks on their backs, and the train trails a rooster-plume of steam that rises up into the blue dusk sky before bending back into the tunnel mouth.

Jane and Sean's two boys, Louis and Orlando, are playing Minecraft on a computer in a corner of the room. I go over to join them. They are mining hard, pickaxing down towards bedrock in search of precious minerals.

'We don't want redstone, we need obsidian,' says Louis.

'We want to fight the Ender Dragon!' says Orlando.

'We're building a portal to the Nether!' says Louis.

'Let's go caving,' says Sean.

~

Evening light now, thick as amber, pouring east across the land.

Over a stile, through a field thronged with yellow ragwort to where the grass sinks into a collapsed cone, sixty feet or so at its widest point. Horses in halos of flies.

The sinkhole's sloped sides are lush with rosebay willowherb. Its belly is scrubbed with elder. Two wood pigeons clatter away at our approach. In the lowest point of the dip is an entrance to the Mendip underland.

A small blockhouse protects a dark mouth in the limestone. Though I have been into cave systems before, I find swallowing is suddenly difficult, as if I have a pebble in my gullet. My scalp swarms with bees. Sean is calm, eager to get under.

The entry is awkward – a body-bending downwards wriggle before a drop into a pot that feels locked, a closed cylinder of space. Our pupils widen to well-shafts in the darkness, until we pop on our beams. Sean leads and is off, lies down, moves head first into a small gap in shadow at the pot's base. I watch his twitching legs slowly disappear, and when his feet have gone I drop to join him. Face forced into wet gravel, moving along by squirm, a sense of the rock as a hand pressing down first on the skull, then the back, then the whole of the body, a moment spent briefly in its grip – and then I am out and with Sean at the top of a twelve-foot notch where a waterfall has run for thousands of years, cutting this narrow channel to the rift below. We down-climb the notch, facing inwards, feet slipping on the wet rock, me going first then spotting Sean as he descends. The rift turns, turns again – and then opens dramatically out.

We are in an awesome space. We track our beams along its roof and walls, scoring out its dimensions. The portal through which we squeezed has become a gorge, hollowed by the work of water over immensities of time. The sides of the gorge are great curves of grey limestone, cross-struck with calcite streaks like lightning flashes.

We move on down. Car-sized blocks of stone have fallen from the roof into the torrent bed and must be clambered around. The slope steepens. The ceiling gleams with star points: stalactite blebs, catching and condensing our torchlight. And then suddenly from either side of the gorge fall two avalanches of stone, waves of boulders and rock fragments crashing down upon us – but somehow frozen in mid-sweep, cantilevered out over our heads. I see that the fragments are all glued together by calcite. Time is starting to play tricks. Movements that have been stilled for thousands of years seem as if they might recommence without warning. My nerves tingle as I pass between the hanging waves of stone. The actions of my body feel jerky, triggering.

Up on the surface, horses flick at flies, caterpillars seethe on ragwort, the sun lowers to dusk. People drive home from work, radios on, windows down.

Beneath all of this, Sean and I pass under two further stone arches. The gorge's floor is slicker now. An awareness grows in us of a big drop somewhere ahead. I feel pulled on like water, as if I might flow down that slope and over the unseen edge. The acoustics change; echoes grow. Warned, we stop just short of a brink. At our feet the gorge floor falls away in a cliff, the base of which we cannot see.

'This feels like the Nether to me, Sean,' I say.

'Let's take a few minutes here,' says Sean.

We sit on boulders, flick off our head-torches. Afterlives of light at first, ghost-patterns on the retina: ferns and leaves. Then the darkness settles and trues, so that when I hold my hand an inch from my eyes I know its presence only from the sound and heat of breath on palm. A heavy black curtain has fallen between Sean and me, then hardened into a wall of stone, such that we are soon in different underlands altogether.

We tend to imagine stone as inert matter, obdurate in its fixity. But here in the rift it feels instead like a liquid briefly paused in its flow. Seen in deep time, stone folds as strata, gouts as lava, floats as plates, shifts as shingle. Over aeons, rock absorbs, transforms, levitates from seabed to summit. Down here, too, the boundaries between life and not-life are less clear. I think of the discovery of the bones in Aveline's, shining with calcite, *lying promiscuously, almost converted into stone* . . . I slip out the whalebone owl, feel the Braille of its back, the arcs of its wings, thinking of how it had taken flight from a whale's beached ribs. We are part mineral beings too – our teeth are reefs, our bones are stones – and there is a geology of the body as well as of the land. It is mineralization – the ability to convert calcium into bone – that allows us to walk upright, to be vertebrate, to fashion the skulls that shield our brains.

Sean flicks his light back on. Glare and blink. There is the cliff again at our feet, water streaming down its face. It is possible that we will find our way to the base of the waterfall later in the journey, so we decide to fix a rope down it now, in case we need to ascend it from below. We find a boulder and loop the centre of the rope around its back, then Sean hammers a chockstone into place with the heel of his hand to prevent the rope riding up and over the boulder when weight comes onto it. I lap-coil the rest of the rope, tie off the two ends, and after two warm-up swings – one, two, *three!* – hurl the coils over the edge.

Hiss, thrum, shiver of snakes in the torchlight, whip-slap as falling rope cracks tight against stone.

'Now,' Sean says, 'we just need to find the way down and round.

There's a side passage somewhere up to our left, according to the maps I've seen, but it's a case of choosing the right one.'

We climb back up the belly of the gorge, away from the lip, moving upstream through the ghost-torrent, probing the left-hand side of the gorge with our torch-beams. There are three visible side passages. We try each in turn.

One spins us around in its twists before curving back at last to end in a wide window overlooking the waterfall, with an unclimbable drop below. The second is a rift entered by a squeeze that we have to repeat when the passage deads out. The third takes us far from the main chamber, and we have to count the turns in our minds, muttering them to ourselves (*first left, first right, second right*) so that the sequence can be reversed if we have to return – which we do.

There is one possibility left: a small entrance near the roof of the chamber, which can be reached only by the traverse of a cascade of damp flowstone, itself set high above the gorge bed. We clamber up to the cascade's edge, and consider the traverse. It is an intimidating crossing. We can rope up, but there is nothing to secure the belayer: one slip and we'll both go.

The cascade is a baroque structure. Flowstone is the name given to the calcite deposits that precipitate out of minerally saturated water as it runs over the slopes of limestone caves. You might imagine flowstone as a kind of white candle wax, gradually hardening as it runs, though it is built up over spans of time rather than by brief incandescence. Because of the gradual nature of its formation, flowstone sets into elaborate ruches and folds – elephant-skin gathers of texture, wrinkled stockings. Flowstone is very beautiful to look at and very hard to grip.

People don't often die caving, but it can be a hell of a job to get someone with a broken leg back up from deep in a rift. The fall from the cascade isn't necessarily a death fall, but it is definitely a double-leg-breaker. Twenty-five feet, perhaps. We know it's the right route, though, because Sean's head-torch has picked out a line of marks traversing near its high point, where earlier boots have cracked the calcite to the consistency of mint cake.

Little demons of worry bite at my stomach as we start out over the cascade. Steady steps, testing the take of each foot, like trying to walk across a slope of wet stone ropes, leaning down to touch the bosses with fingertips for balance, *slowly, slowly, slowly* . . . and then Sean is over and I am over and we are into the entrance near the roof of the chamber, laughing with relief – and a new region of the labyrinth is open to us.

We let gravity lead us through it, taking always the downwards path where the tunnel splits, until the echoes tell us that our passage is approaching broad space – and then there we are at the base of the waterfall, and there is the rope we threw down earlier.

But the rope is stuck. It has jammed behind the belay boulder and won't run through to us evenly, making easy movement up it impossible as we climb. All we can do is tie off to it, climb, release, then tie off again. It offers some protection from a fall; better than nothing. I lead. The rock is wet; the climb has a couple of tricky moments. I am glad we threw the rope down. Sean comes up after me and we rest together at the top of the waterfall, mustering energy for the return. I am cold now, chilled to the bone by the dark, the wet and the stone.

Up the gorge, up the notch, through the squeeze, the smell of green growing in the nose, up into the belly of the elder-filled dip of land, and up to the level of the fields, the horses, the swooping swallows, out of the Carboniferous and into the Anthropocene.

Sundown on the surface. Pupils shuttering to pinpricks. Colour is preposterous, gorgeous again. Blue is seen utterly as blue, green known fully as green. We are high on hue, high on the wild noise of the wind, high on the last of the sunlight that glosses the streamers of the veering swallows, high on the huge vault of the sky and the boiling clouds it holds.

We walk, still blinking, to the road in our orange hazmat suits. A family drives past in a shiny Land Rover, the children in the back seats swivelling their heads to look at these aliens who seem to have been dropped from high in the sky but who have in fact emerged from deep within the earth.

~

The most notorious story in British caving history involves a twenty-year-old Oxford philosophy student called Neil Moss. It is still, in my experience, a story that some people in the Peak District do not like to discuss, nearly sixty years on.

On the morning of Sunday, 22 March 1959, Moss set off as part of an eight-person exploratory trip into the further reaches of Peak Cavern, a system near Castleton in Derbyshire. The first half-mile or so of Peak Cavern is an open show-cave, into which tourists and locals have wandered since the early nineteenth century, not least to hear choral recitals sung from the 'Orchestra', a natural gallery of limestone set high in the 'Great Chamber'.

Half a mile into Peak Cavern, however, the terrain becomes far more serious. The roof of the cave drops to leave only a wet crawl-space known as the Mucky Ducks, which floods in heavy rain. After the Mucky Ducks comes a long, low rift called Pickering's Passage, leading to a right-angled bend guarded by an eyehole of stone just wide enough to admit a human. After the eyehole comes a thigh-deep lake and beyond that a small chamber, from the floor of which descends a shaft around two feet wide at its mouth. It was this fissure that the team had come to explore, hoping it might lead further into the maze of passages under the White Peak.

Moss, a tall and slim young man, was given the lead. An elektron caving ladder was dropped down the shaft, and Moss lowered himself into it. The shaft remained near vertical for around fifteen feet, then shallowed and twisted before making a sharp elbow-bend back to the vertical. With some difficulty, Moss negotiated the elbow-bend and descended the subsequent section – only to discover that the shaft then became choked with boulders. It had deaded out.

He could feel the boulders shifting beneath his feet, but there seemed no further possibility of descent. So he began to re-ascend. Just below the elbow-bend Moss lost his footing on the ladder, slipped down a little – and found himself wedged.

He could not bend his knees to regain purchase on the rungs of the ladder, which were anyway greasy with mud. His arms were pinned close to his body by the sides of the shaft, and his hands scrabbled vainly for grip on the slick limestone. The ladder seemed also to have shifted across the space of the shaft, perhaps dragged by the movement of the boulders at the shaft's base, further blocking upwards progress. The fissure had him fast – and with each movement he made, his entrapment tightened a fraction.

'I say,' he called up to his friends in the chamber, some forty feet above him, 'I'm stuck. I can't budge an inch.'

His friends presumed that the problem could be solved by dropping Moss a line and hauling him free. But they only had a light hand-line with them, rather than a belay rope. The line was lowered and Moss somehow managed to secure it around himself. But when hauling began, the line snapped. It was lowered again, refastened. It broke again. And then again. The ladder itself could not be hauled up for fear of further jamming Moss.

Moss's panic rose. Every twitch of his body caused him to slide very slightly deeper into the shaft. He was indeed stuck – and he was also suffocating. With each breath Moss slightly depleted the limited oxygen supply in the shaft, and slightly increased the carbon dioxide content. Because carbon dioxide is heavier than oxygen it began to fill the shaft from its base upwards. The air became more and more foul, first in the shaft and then in the chamber above it.

By this time the alarm had been raised above ground, and what was at the time one of the largest cave-rescue attempts in history began. Radio bulletins were sent out on the BBC, and teams from the RAF, the National Coal Board and the navy, as well as individual civilian cavers, scrambled to the scene. Neil's father, Eric Moss, rushed to Castleton but was unable to proceed far into the cave. He waited nearby, stricken with fear, unable to assist. The shaft in which Moss was stuck was around 1,000 feet from the entrance, and all rescue equipment and personnel had to be moved awkwardly through the obstacles just to reach the top of the pot. Heavy oxygen cylinders were wrestled through the Mucky Ducks and pushed by head and hand along the boulder passage. Two young men hauled a twelve-volt car battery to provide energy for light. Soda lime was carried in to absorb the build-up of carbon dioxide. Hundreds of yards of telephone line were threaded through the system to link the fissure to the outside world. Three volunteers who tried to descend the shaft with a stronger rope lost consciousness and had to be pulled out themselves. A fourth man managed to get to the rope around Moss's chest, but pulling on the rope only worsened his already-tortuous breathing. By this point Moss had mercifully lost consciousness, stifled by his own exhalations.

One of the people who heard the news about Moss's plight was an eighteen-year-old typist from Manchester called June Bailey. Bailey – who was an experienced caver and very slender of figure – rushed to Castleton to try and help. She made the difficult journey to the shaft and agreed to attempt a rescue. She was instructed to break Moss's collarbones or arms if necessary, in order to free his shoulders from the stone's grip and allow him to be pulled out. While an RAF doctor up to his waist in mud hand-pumped oxygen down the shaft, Bailey tried to reach Moss – before she, too, was driven back by the fetid air.

On the morning of Tuesday, 24 March, Moss was officially declared dead. When Eric heard the news, he requested that his son's body be left in the shaft rather than that others endanger themselves trying to retrieve it.

Eric wished for a burial of a kind, however. So he sought permission from the coroner to have Moss's body sealed in the fissure that had killed him. Cement powder from the local works was carried into the cave, mixed with water from the thigh-deep lake, and then poured down the pot, entombing Moss in perpetuity. This section of Peak Cavern is now known as Moss Chamber.

~

It is full dark by the time Sean and I get back to the cottage. We hose down and then hang up our hazmat suits in the cool air of the garden, one on each wing of the totem pole. I whistle a song from *Rubber Soul* by the Beatles while we work.

Sean tells me about how he once climbed a wooded slope in Burrington Combe, opposite Aveline's, and found an entrance to a chamber, the aperture big enough for him to put his head into but too tight to admit his body.

'I called out into it,' says Sean, 'and the chamber answered, singing a different note back to me.'

I am sleeping in the attic room. It runs the length of the house. Head-height beams of crooked elm brace the space, tunnelled by boring beetles into galleries I cannot see. Each gable end has a small oak-framed window set into it, through which cool night air moves. Books stand in tall piles on the floor because the whitewashed plaster walls are too sloped for shelves. Before sleep I read Harrison's *The Dominion of the Dead*. I copy out a few sentences from early in the book:

> For the first time in millennia, most of us don't know where we will be buried, assuming we will be buried at all. The likelihood that it will be among our progenitors becomes increasingly remote. From a historical or sociological point of view this is astounding. Uncertainty as to one's posthumous abode would have been unthinkable to the vast majority of people a few generations ago.

Tawny owl cries float into the room from the woods around. That night I dream of being slowly absorbed by calcite, a varnish creeping over me, setting me in my place.

I am woken by cries from the garden. Dawn light. Through the gable window I can hear Louis running in the garden. I look out. He is barefoot and in his pyjamas, standing at the chicken coop.

'Mum! How many eggs do we need for breakfast?!'

The newspaper that morning reports that geologists have discovered buried seas of water in the Earth's mantle. Four times as much water might be locked up there in a mineral called ringwoodite as is currently held in all the world's oceans, rivers, lakes and ice put together.

~

Over the days that follow, Sean and I move from place to place in the Mendips. Sean is teaching me undersight – how to see the underland's subtle entrances, its disguised extents. The heat persists, building but not breaking. The earth longs for rain but we do not, for rain will rush through the cave systems, making them too dangerous to enter.

Up on wooded ground, where the bracken grows above head height and an old pine plantation has run into what feels like wildwood, we follow deer paths to a small escarpment, at the base of which a cave mouth beckons us under the stone. Ferns mark the entrance, hooped with bramble. Ivy climbs the cliff. A red admiral basks where light falls, slowly opening and closing its wings. Scrambling under the escarpment, we enter an alarming space. A scree-slope tilts to a flat lower chamber. Big blocks of rock hang from the rift's fractured roof. We descend to the chamber, and crouch there.

This is clearly a strong place – and it has drawn humans to it for thousands of years. Ritual depositions have happened here: the bodies of humans and animals were thrown or placed into the rift, probably during the Neolithic. Bronze Age relics have also been found, and at some point in the sixteenth or seventeenth century the stone near the entrance was marked with red painted figures. They are thought to be protection marks – apotropaic inscriptions made to avert evil. Are they designed to stop evil from entering this underland space, I wonder down there in the rift, or to stop it getting out?

Another day, near the highest point of the Mendip plateau, Sean and I walk what is known as 'gruffy ground'. 'Gruffy' means 'rough', 'rugged', and gruffy ground is the relic landscape of lead-mining activities dating back more than 2,000 years. Small-scale Roman mining left behind hundreds of small heaps of tailings; in the eighteenth century these were reheated to melt out any residual lead ore. This double working of the landscape has left the ground humped with small hills of toxic slag, over which grass has thickly grown, shunned by grazing animals which sense its contamination.

We walk that lush and poisonous little valley to a viewpoint. The air is lightly hazed. Sean picks out the landmarks: the Bristol Channel, the rise of Dartmoor to the south-west, Hinkley Point nuclear power station squat on the coast, and below us the spreading flatlands of the Somerset Levels, where we know – through the startling precision of tree-ring dating – that in 3807 BC Neolithic people cut and split oak into planks, bound them together, supported them with cross-poles, and laid them as a trackway over marshland, joining high ground to high ground.

Kites turn above us, and above the kites turn buzzards. A telecoms mast relays signals through the air, through our bodies. Down in the Levels a fire is burning from within a stand of willows, and its smoke plume rises straight up in the still air. The sun beats on us. I close my eyes, see red and gold tendrils.

'It's way too hot above ground,' says Sean. 'Let's go somewhere cooler.'

So we do. It will be one of the most unnerving spaces I have ever entered.

~

Over field and down into bower of elder and old ash, moss plushing rock to soft gold-green. Follow the stream bed through gorse and bracken, setting fieldfares flaring to the west with chatter and crackle. Swallows skimming meadows on the fly, blowy warmth in a north-east wind. On and into the deep-set hollow, a last nod to the sun – to the light falling through leaves in nets, to the buzzard drifting over – and then we are down a hole in the stone-cold soil, worn to a swallet by the run of a stream, into the earth's gullet, into the black bite of a polished stone-vice set carelessly and wondrously with the spirals of ammonites and the bullets of belemnites, and down into trouble.

Sean leads, lowering into a six-foot shaft. I follow, drop into darkness, and find him on his knees. There is just space for the two of us hunched together here. Ahead of us is the shoulder-wide entrance to the ruckle.

'This is a space granted by collapse,' says Sean quietly, admiringly.

A ruckle is a group of boulders that have caved against one another, blocking a section of passage, but through the gaps of which a path might just be traced. Ruckles are delicate, unpredictable structures. Without disturbance, a ruckle might hold its position for tens of thousands of years. But an earth tremor might shake it into a new order in an instant. Or a human touch might cause a boulder to move, shifting the whole stack, trapping foot or hand – or just, terribly, locking one in.

Crouched in that bare space, my heart hammers warning in my ears. I reach out and place a hand on the black rock of the first boulder, and the cold volts into me like a current, surging up my arm, petrifying me.

The stone of the ruckle is beautiful, I think – a dark limestone, glittering in the torchlight like ice – and then I see that even the air in the spaces between the boulders seems somehow to shine, so that it is, really, impossible not to move on into the ruckle.

And there is a hint as to how to navigate the maze – for hanging from the first boulder is a line of white nylon cord. This is an 'Ariadne's thread', left by earlier cavers and named after the ball of wool that Theseus was given by Ariadne to unravel behind him, laying a track back to safety as he twisted down the dark passageways of the Minotaur's lair.

'After you,' whispers Sean to me, gesturing with a wave of his hands towards the cord, and giving as much of a bow as he can muster from that cramped position.

'No, please, really, after you,' I whisper in reply, bowing back.

Sean rolls his eyes and leads off, easing head first through an opening little more than twenty inches wide. His feet vanish. I follow.

On and in and down, sliding through each black mouth of each new turn in the ruckle, following the white thread, bending body to fit space, curling against the cold stone, trying to push as little as possible against any boulder, trying somehow to *evaporate* myself such that I become a gas that can flow through this place without touching a surface. But instead I am aware of my clumsy bag of bones and blood, of the need to lever myself onwards with elbow and knee, to push off with foot and pull on with fingers, and each touch of rock feels a risk, a tap that might spring the trap of the ruckle – until at last Sean eases through a gap and I hear him sigh in space, and I slide through to join him in a chamber big enough almost to stand in, and the roof above us is solid once more.

'Hell,' I say, breathing hard.

'Yes,' says Sean.

To our left is a passageway cinching to a shoulder-width black circle. Ahead of us, drawing my eye and tightening my throat, two leaning ten-foot slabs of black rock, more marble than limestone, run off into shadow, angling towards one another.

This is a bedding plane, formed when the rock was being laid down as sediment on a seabed. Strata movement has prised the sides of the plane apart millions of years later, water has worked to burnish an absence between them, and our onwards route is into this deep time space, this deep time vice.

We enter the bedding plane with trepidation, leaning back on the lower angle of stone, and sliding ourselves onwards into the darkness, the upper plane tilting out and over us. There is no danger of collapse here but the sense of confinement is severe. We *submit* ourselves to the bedding plane, until at last it tightens to a silted-up sump that is not the end of the passage for water, but is surely the final affordance for our stubborn, unshrinkable human bodies.

In that vanishing point, neither of us speaks. Language is crushed. We are anyway too busily engaged building structures within ourselves that might house our spirits, for the pressure here is immense, a weight of rock and time bearing down upon us from every direction with an intensity I have never experienced before, turning us fast to stone. It is a fascinating and terrible place, and not one that can be borne for long.

We return to the edge of the ruckle, knowing we must pass back through it – and there lies the end of our thread, our white clue. Without it we could hardly retrace our route through the boulder labyrinth. It would be like memorizing a fifty-word tongue-twister on the way down and then reciting it in reverse on the way back up.

I lie down to lead, I follow the thread, and each tiny room in the ruckle opens onto the next as it should, in turn, in order. I pass through the last of the gaps, and as I lift myself into the entry shaft

I feel the snap of the black stone's jaws at the empty air below my toes, and then I am out of the swallet and into the hollow, and warm air is rolling around me, and my bones grow again in the storm of light and ferns furl their green over and into me and moss thrives on my skin and leaves teem in my eyes, and Sean and I sit laughing, knowing for those few moments that to understand light you need first to have been buried in the deep-down dark.

We emerge from the hollow, step clear of the elder and the ash. The sunshine is so thick I want to lie belly up in it, floating as if in a salt-rich sea. After the bedding plane, our sight lines are huge. Silhouetted on the horizon above us are two round grassed domes.

Sean points up at them. 'Those are some of the Priddy Nine Barrows,' he says.

These are hay-making days in the Mendips, and there is the ripe reek of cut grass in the air. Where hay has been lifted and black-wrapped as bales, aftermath already shows as green shoots in the gold stubble. Sean and I walk together uphill towards the barrows from the cave of the ruckle, along a holloway with sides fifteen feet high from bed to hedge top.

A charm of goldfinches flitters away, the birds' high song glittering around us. I am moved by the generosity of colour and space in this ordinary land. Here in the Mendips I have seen how thin the border is between the upper and lower worlds, and how hard it can be to pass in either direction.

The holloway leads to a gap in a stone wall, then out onto a meadow over which a warm westerly blows. The burial barrows lie in a line along the slope. Sean and I cross the meadow, happy in each other's silence and glad of each other's company. We reach the first barrow and lie there in the long grass, backs to the hill's back, sun hot on skin.

Meadowsweet, knapweed, scabious. Everything is shiveringly strange. Flies on the grass blades are exotic as tigers – eyes of a thousand ruby hexagons, wings of the finest filigree. We lie so still that a grasshopper lands inches away, and I watch its drumsticks quiver as it drags leg over wing case, stridulations streaming out. I think of the makers of these barrows, choosing this high place as their site of burial. The construction of the kists, the casting of the collared urns, the burning of the bodies, the building of the barrows.

Eight of the nine barrows were excavated in a single week by the Reverend John Skinner and his men in 1815, the exhumation motivated by a combination of antiquarian interest and grave robbery. All were found to contain at least one cremation. One of the barrows held the wealthiest burial found anywhere in the Mendips: a woman who had been pregnant, missing her pelvis but buried with beads of amber and faience, a copper awl and an elaborate dress fastening. Twenty-four years after he plundered the Priddy Nine Barrows, Skinner would shoot himself in the face. It is thought that his friends succeeded in concealing his suicide, thus allowing his body to be buried in the consecrated ground of his Somerset parish at Camerton. *We are often more tender to the dead than to the living, though it is the living who need our tenderness most . . .*

Sean tells me a story. Modern archaeologists excavating a Bronze Age barrow in a Mendips wood find the remains of a woman placed in a funerary urn. The barrow had already been ruptured by deep ploughing when the cemetery was planted with trees early in the twentieth century, but the urn somehow survived. The archaeologists disinter the urn, and study the remains of the woman that it holds. Once their work is done, one evening while white moths flit in the shadows of trees, they rebury the woman's remains in a replica urn. As they do so one of them speaks a blessing at the graveside – a reburial ritual performed across the space of several thousand years, spoken out of respect and also, perhaps, out of apology.

Sean and I stand up in the warm wind and we follow the barrows, walking past each mound in turn until we have reached the end of the line, the last of the nine. From there we return to the first of the barrows, and we lie again on its slope, talking and not-talking. Beneath us is the earth and the kists it holds, and below that is the limestone and the rifts it holds.

We lie on the barrow's turf for so long that, when we leave, I look back and see that we have pressed imprints of our own bodies into the grass of that burial site, leaving outlines of what is to come.

3
Dark Matter

(*Boulby, Yorkshire*)

More than half a mile under the earth, in a laboratory set into a band of translucent silver rock salt left behind by the evaporation of an epicontinental northern sea some 250 million years earlier, a young physicist is trying to look into a void.

He sits watching a computer screen, close to a large silver cube. The cube's name is DRIFT and it is a breath-catcher. The young physicist is trying to catch the faint breath of a particle wind sent blowing across space from a constellation called Cygnus, the Swan, many light years distant from Earth.

The young physicist is searching for evidence of the shadowy presence at the heart of the universe: a presence so mysterious that it has thus far engulfed almost all of our attempts either to investigate or to represent it. The name we have given to this presence – which refuses to interact with light, which may not even exist – is 'dark matter'. And the only place the young physicist can conduct his enquiry is down here in the underland, shielded from the surface by 3,000 feet of halite, gypsum, dolomite, mudstone, siltstone, sandstone, clay and topsoil.

It is a paradox of his work that in order to watch the stars he must descend far from the sun. Sometimes in the darkness you can see more clearly.

~

In the early 1930s a Swiss astronomer called Fritz Zwicky was studying galaxy clusters through the telescopes of the California Institute of Technology when he noticed an anomaly of extraordinary implications. Clusters are groups of gravitationally bound galaxies, and Zwicky's work involved measuring the speeds of revolution of individual galaxies in their orbits around the core of the cluster, in order to weigh the cluster as a whole. What Zwicky observed was that the galaxies were revolving much faster than expected, especially towards the outer reaches of the cluster. At such speeds, individual galaxies should have broken their gravitational hold on one another, dispersing the cluster.

There was, Zwicky determined, only one possible explanation. There had to be another source of gravity, powerful enough to hold the cluster together given the speeds of revolution of the observable bodies. But what could supply such huge gravitational field strength, sufficient to tether whole galaxies – and why could he not see this 'missing mass'? Zwicky found no answers to his questions, but in asking them he began a hunt that continues today. His 'missing mass' is now known as 'dark matter' – and proving its existence and determining its qualities is one of the grail-quests of modern physics.

How, though, to hunt for darkness in darkness? How to seek a substance that has mass and therefore exerts gravity, but that does not emit light, reflect it or block it? Since Zwicky, the evidence for the existence of dark matter has been gathered largely by inference: the detection not of the matter itself, but of its presumed influence on luminous entities, observable objects. To perceive matter that casts no shadow, you must search not for its presence but for its consequence.

It is now known, for instance, that dark matter affects the rotation curves of spiral galaxies, causing all bodies within such a galaxy to revolve at comparable rates, regardless of their distance from the galaxy's gravitational centre. It is also known that dark matter bends light as it passes around a galaxy, causing what is referred to as 'gravitational lensing'. Mass curves space, Einstein showed in his General Theory of Relativity, and light follows those curves of space – as when it passes around a massive entity such as a galaxy. But just as Zwicky's galaxies rotated too fast, so light also bends too greatly to be due only to the visible components of a given galaxy. There must again, therefore, be more mass than that which can be seen. This imperceptible, space-curving, light-lensing massive presence that surrounds a visible galaxy is known to astrophysicists as a 'dark-matter halo'.

What these observations and others like them suggest is that only around 5 per cent of the universe's mass is made of the matter we can touch with our hands and witness with our eyes and instruments. This is the matter of stone, water, bone, metal and brain, the matter of which the ammoniacal storms of Jupiter and the rubble rings of Saturn are made. Astronomers call this 'baryonic matter', because the overwhelming share of its mass is due to protons and neutrons, known to physicists as 'baryons'. A little over 68 per cent of the universe's mass is presumed to be made of 'dark energy', an enigmatic force that seems to be accelerating the ongoing expansion of the cosmos. And the remaining missing 27 per cent of the universe's mass is thought to be made up of dark matter – the particles of which almost wholly refuse to interact with baryonic matter.

Dark matter is fundamental to everything in the universe; it anchors all structures together. Without dark matter, super-clusters, galaxies, planets, humans, fleas and bacilli would not exist. To prove and decipher the existence of dark matter, writes Kent Meyers, would be to approach 'the revelation of a new order, a new universe, in which even light will be known differently, and darkness as well'.

Dark-matter physicists work at the boundary of the measurable and the imaginable. They seek the traces that dark matter leaves in the perceptible world. Theirs is hard, philosophical work, requiring patience and something like faith: 'As if' – in the analogy of the poet and dark-matter physicist Rebecca Elson – 'all there were, were fireflies / And from them you could infer the meadow'.

Presently, the particle thought most likely to be the constituent of dark matter is known wryly as a WIMP – a weakly interacting massive particle. What we know of WIMPs suggests that they are heavy (up to more than a thousand times the weight of a proton), and that they were created in sufficiently vast quantities in the seconds after the birth of the universe to account for the missing mass.

WIMPs – like neutrinos, nicknamed 'ghost particles' – have scant regard for the world of baryonic matter. WIMPs traverse our livers, skulls and guts in their trillions each second. Neutrinos fly through the Earth's crust, mantle and solid iron-nickel core without touching a single atom as they go. To these subatomic particles, we are the ghosts and ours the shadow-world, made at most of a diaphanous webwork. The great challenge faced by physicists has been how to compel such elusive particles to interact with experiments; how to weave a net that might catch these quick fish. One of the solutions has been to go underground. Subterranean laboratories have been established around the world, dedicated to the detection of evidence that a WIMP or a neutrino has briefly interacted with baryonic matter. The experiments under way in these deep-sunk laboratories are all forms of ghost hunting, and they are located far underground because the surrounding rock shields the experiments from what physicists call 'noise'.

Noise is the trundle of everyday particles through the air, the din of the ordinary atomic world going about its business. Radioactivity is deafening noise. Cosmic-ray muons are noise. If you wish to listen for sounds so faint they may not exist at all, you can't have someone playing the drums in your ear. To hear the breath of the birth of the universe, you must come below ground to what are, experimentally speaking, among the quietest places in the universe.

Half a mile underground in an abandoned mine in Japan, set in a chamber of 250-million-year-old gneiss, a stainless-steel tank holds 50,000 tons of ultra-pure water. Watching the water are 13,000 photomultiplier tubes, forming a compound eye. The eye is looking for tiny flashes of blue light. These flashes are Cherenkov radiation, produced when an electron moves faster than the speed of light in water. Electrons reach such speeds when an atom is – occasionally – struck by a neutrino, the impact scattering the atom's electrons at velocities in excess of the speed of light. These scattered electrons are called 'annihilation products', and if those electrons are scattered in water then they briefly create a luminous blue cone around them as they move. The compound eye of the photomultiplier tubes therefore watches for trebly displaced evidence of the 'ghost particles': not the neutrino itself, or the atom it has struck, or the electrons it has dispersed, but the blue aura left by that ghost-struck atom – annihilation's afterglow. This buried chamber of gneiss is called an 'observatory', for although it is deep underground it is in fact scrying the stars: among its many other tasks is keeping watch for supernovae in the Milky Way.

Deep in a worked-out open-cut gold mine in South Dakota, super-cooled xenon is held in a six-foot-tall vacuum vessel, surrounded by 71,600 gallons of deionized water contained in a welded steel tank and watched by photomultiplier tubes for the displacement of a single photon and a single electron brought about by the strike of a WIMP. Xenon, a noble gas, has large atoms. When xenon is very cold it is very dense; those large atoms huddle together, thus presenting a greater cross-section to incoming particles and optimizing the chances of WIMP strikes. In a landscape where the earth was once raked and gouged in search of a highly valued rare metal, the search is now ongoing for a substance that is plentiful beyond imagination and of no worth at all.

And near the little village of Boulby on the Yorkshire coast, in a salt cavern far below the headworks of a potash and rock-salt mine that commenced operation in 1973, a dark-matter detection experiment is presently under way that is known by the acronym DRIFT: directional recoil identification from tracks.

~

Neil Rowley unrolls his map of the underland on his desk, and places four chunks of rock on its corners to keep it flat, naming each as he puts it down: sylvite, halite, polyhalite, boracite. He smooths the map out with his hands, working from the centre to the edges. Neil is a mine-safety specialist. He has worked in coal; now he works in potash. He likes W. H. Auden, he likes maps and he loves mining.

Neil's map records the roadways and refuge chambers of the Boulby mine. At first glance, it looks to me like the wings of a dragonfly, intricately veined and structured. Slowly my eyes key into its codes.

The north-east coastline of England is present as a faint grey line running across the map from north-west to south-east: a surface irrelevance, shown chiefly for purposes of orientation. At Boulby itself, two circles signify the two shafts that plunge into the bedrock, giving access to the tunnel network. From that centre point the tunnels fan out to the north-east and the south-west, forming the wings of the dragonfly. To the south-west, they spread under moor and dale, deep into North Yorkshire. To the north-east they spread under the North Sea, running out beyond the shipping lane and into open ocean.

This network of tunnels and roadways is collectively known as 'drift'. There are more than 600 miles of existing drift burrowed into the soft bands of halite (salt) and sylvite (potash) that stretch below sea and land, out to the mining faces where – every hour of every year – men and machines claw tons of potash from the seams, duct the potash onto hoppers and start the journey of this buried residue of a Permian sea up to the world's crop fields, where it will be spread as a fertilizer in both of the Earth's annual two springs, returning vital potassium to the growing cycle.

As the land below the Mendips holds a water-made labyrinth, so the land below Boulby holds a human-made maze. I have come from rift to drift.

On Neil's map, red lines signify drift cut through salt, black lines signify drift through potash. Yellow squares mark refuge chambers, dug into the side walls of the tunnels and armoured against heat by polyfoam outer walls. In the case of collapse or fire at depth, these are the fall-back sites.

At the tips of the wings – far under the sea and far under the moors respectively – thin green threads lick out. These are the lateral boreholes being drilled by mine geologists to test the lie and integrity of the deposits ahead of the workface. The information they return will determine the future directions of the mining, the future spread of the wings.

'You need to understand that the tunnel network is on a tilt,' says

Neil, drawing his finger across the map, from one end of the dragonfly's wing to the other. 'The drift tilts because the deposit tilts. The tunnels follow the potash, and the potash strata are inclined.'

Inland the potash deposit runs deeper, reaching a maximum depth of around 4,500 feet at the outermost limit under the moors. Seawards, it rises to a minimum depth of around 2,600 feet at the outermost point beyond the shipping channel. A temperature gradient follows the depth gradient. At 2,600 feet the air temperature is 35°C. At 4,500 feet it's 45°C. In both places the geothermal heat is so intense, and the moisture content of the air so low, that sweat evaporates before it can even be seen. Dehydration is rapid. For the miners it is like labouring in the Sahara at noon, in darkness.

'The men all carry cool-boxes with four litres of chilled water per shift,' Neil says. 'They have rehydration timetables throughout their shifts. Got to keep drinking. Much safer.

'Come on – let's see if we can catch a lift down there, find some dark matter, then we'll make the long drive out to the mining face under the sea.'

Ear defenders on. Respirator hooked at the belt. Numbered bronze triangle in pocket as proof of entrance: *Don't lose it now, you won't be allowed out* . . . Yellow cage door clangs shut, cage starts its drop, steady but still stranding the stomach. Roar of the fan-house fading away, cage speeding up. Halfway down a shudder and blast as the other cage crosses on its way up, squeezing air between cages with a *crash-whoosh* like two trains passing in opposite directions. Slow, slow, slow, bump, stop, cage door clangs open – and voices are yelling, 'Ears off, lights on! Ears off, lights on!'

Rock dust swirls in the air, thick enough to taste, salty on the tongue.

Black mouths of drift lead away under the ocean, into the Permian.

An airlock in a wall opens into a laboratory.

~

The young physicist sits at his computer, watching for signals from Cygnus. His name is Christopher Toth, and his white lab coat is too big for him. Christopher speaks with calm clarity. His manner is modest, gracefully gentle, and I wonder if this comes in some way from spending your days thinking through time so deep it stretches to the birth of the universe.

Along the walls of the laboratory, at intervals of every fifteen feet or so, black-and-yellow warning tape marks the outlines of what look like potential doorways, rising only to thigh level. Above each taped outline, a long-handled axe with a splitter blade is held in two hooks.

Salt has very low gamma radiation. Salt is a good insulator. Salt is radio-pure. Salt is an excellent substance in which to encase yourself if you want to study weakly interacting massive particles. But salt is also highly plastic. Salt *flows* over time. It creeps around. It sags. If you cut a chamber out of a seam of halite with 3,000 feet of bedrock above it, that chamber will slowly distort. The ceiling will dip, the sides will bulge. Gravity wants that space back. So the scientists working in the Boulby laboratory know they are operating in a temporary zone, with limited years of safe life. Deep time must be studied fast.

'Those are your emergency exits in case of a sudden slump in the halite,' says Christopher, mimicking the hand gestures of a flight attendant explaining the safety protocols, and pointing to the doorways marked with warning tape, 'here, here . . . and here. If the lab begins to collapse, you grab an axe, hack your way through the lab wall, then hack your way out through the salt to safety.'

He pauses, smiles. 'Well, that's the theory, at least.'

Several different kinds of underground experiment are presently taking place in the laboratory. One assays rock samples in order to research techniques for the long-term burial of radioactive waste. Another investigates a technology known as 'muon tomography', which makes use of highly penetrating charged particles (muons) produced by cosmic rays from space. Because of their ability to pass deep into rock, muons allow sunken structures such as the interiors of volcanoes and the hollow hearts of pyramids to be perceived. Muons offer a way of seeing through stone. These are all remarkable experiments. But the jewel in the experimental crown of the Boulby laboratory is DRIFT.

Christopher walks me towards a large object located at one side of the laboratory. 'This is my underground crystal ball,' he says – flourishing his hands like a magician revealing a trick – 'also known as the Time Projection Chamber.'

From the outside, the magnificently named Time Projection Chamber is disappointing to look at. Black bin liners are taped scruffily around a large metal-clad box.

'I see that bin bags make up the vital outer layer of your crystal ball,' I say.

'You mock,' replies Christopher, 'but duct tape and bin bags have proved crucial to more scientific breakthroughs than you'd imagine.'

He explains the experiment to me. 'We know dark matter is massy. Massively massy. So its particles, even though they're invisible to us, have mass, and if they have mass then they must at least occasionally collide with particles we can see. These collisions send nuclei scattering. Our first goal with DRIFT is to detect these collisions, and follow the nuclei as they scatter.'

He pauses. I wait. Trillions of neutrinos pass through our bodies and on through the Earth's bedrock, its mantle, its liquid innards, its solid core.

'Imagine watching a game of billiards in which the red balls are visible but the white isn't. Suddenly you see the red ball – an electron – move across the baize. By plotting the red ball's path, you might be able to backtrack, as it were, the path of the invisible white ball – the WIMP – that struck it. And from this, you might be able to learn more about the direction, mass and qualities of that white ball. We're looking to do this enough times, and with enough precision, to provide the signature of a dark-matter halo.'

At the core of the DRIFT device is a steel vacuum vessel a cubic metre in volume, criss-crossed by a meshwork of ultra-thin, highly charged wires spaced a millimetre apart. If a WIMP collides with the nucleus of an atom of ordinary matter inside the chamber, it causes an ionization track, which the meshwork of wires both intensifies and records. The track can therefore be reconstructed in three dimensions, providing information about the type and origin of the colliding particle. The wires are held within a low-pressure gas, the low-pressure gas within a conductive chamber, the conductive chamber within a steel neutron shield – and the whole unit is held in a band of halite left by the evaporation of an ancient sea.

I will learn over the years ahead that many such Chinese-box structures, with their multiple containment protocols, characterize storage procedures in the underland, from the falcon-headed Canopic stone jars of ancient Egyptian burial practice – into which were placed the vital organs of the dead, and which were themselves encased in a painted wooden chest, itself encased in a tomb, itself encased in a pyramid – to the concentric sheathing of spent uranium pellets from nuclear reactors; the pellets placed within rods of zirconium, the rods encased in a copper cylinder, the copper cylinder encased in an iron cylinder, the iron cylinder encased in bentonite clay rings, and the rings encased in the bedrock of a deep geological storage facility, sunk thousands of feet into gneiss, or granite, or salt.

Christopher leads me to his desk. The screensaver image on his computer is of the turquoise waters of Lake Louise in the Canadian Rockies. He shows me a diagram representing data returns from the Time Projection Chamber. It has lines of different bright colours, across which a fine black streak runs at an angle.

'This diagonal line is the path of an alpha particle,' Christopher says, following it with his little finger. 'He's a bludgeoning, portly gentleman who comes barrelling through our experiment, making a lot of noise as he goes. He's not of interest to us, except insofar as identifying his signal helps us know what we're *not* looking for.

'What we're trying to hear, instead, are the quiet whispers behind his boisterousness. Not even whispers, in fact; more like the faintest of breaths – of breathing. Down here in the salt is about the only place you could hear such breath. That breath is the sound of a weakly interacting massive particle passing through – and it leaves a fine trace. What we think is a WIMP-collision looks more like two small blips, one on each of two channels.'

With his fingernail he picks out two dots: one on a yellow line, one on pink. He pauses. His screensaver changes to an over-saturated image of a white-sand beach with palm trees, lapped by a lapis-lazuli sea. The WIMP wind from Cygnus blows through our bodies.

'This data is very beautiful once you get used to it,' he says. I nod in agreement.

'Right now,' Christopher says, 'you are looking into the absolute smallness of the universe with pinpoint accuracy, peering down at the most minute of scales. Those coloured lines are our magnifying lens.'

Then he says – as if the phrase has just entered his head without warning, scoring a trace as it passes through – 'Everything causes a scintillation.' He pauses.

'Why are you searching for dark matter?' I ask.

'To further our knowledge,' Christopher replies without hesitation, 'and to give life meaning. If we're not exploring, we're not doing anything. We're just waiting.'

He pauses again. I wait. The screensaver on his computer changes to Yosemite in the autumn, with early snow on the top of El Cap. Christopher does not speak.

'Is the search for dark matter an act of faith?' I ask him.

He waits for me to elaborate – he has heard the question before, wants more before he answers. His screensaver changes to the desert dunes at Sossusvlei in Namibia.

I think of Rievaulx Abbey west of Boulby, where in a fertile river valley Cistercian monks founded and built a space in which to hold Mass. Out of ironstone they made an airy structure of soaring buttresses and vaulted ceilings. Their abbey was one among a network of such sites spread around the world, in which prayers were offered to a presence disinclined to disclose itself to the usual beseechings.

On the hillsides above the abbey geological forms known as 'sliprifts' slowly open and close in the rock, emitting warm air from deep within the earth, such that on cold days the hillside itself appears to be breathing – as if the land itself were alive. Thousands of years before the Cistercians arrived in those valleys, Neolithic and Bronze

Age peoples entered the darkness of the slip-rifts to carry out rituals that may have been sacrificial and were surely devotional, interring body parts amid the stones of the rifts; another kind of annihilation product.

I remember the Wind Cave system in the Black Hills of South Dakota, sacred to the Lakota Sioux people and close to the American dark-matter detection laboratory set deep in the worked-out gold mine. From the opening to Wind Cave, which extends for more than 130 miles below ground, air rushes or is drawn with such force that it can strip hats from heads. In the Lakota creation stories it is from Wind Cave that humans first emerge into the upper world, where they are astonished by colour and space.

'My sense,' I say to Christopher, 'is that the search for dark matter has produced an elaborate, delicate edifice of presuppositions, and a network of worship sites, also known as laboratories, all dedicated to the search for an invisible universal entity which refuses to reveal itself. It seems to resemble what we call religion rather more than what we call science.'

'I grew up as a very serious Christian,' Christopher says. 'Then I lost my faith almost entirely when I found physics. Now that faith has returned, but in a much-changed form. It's true that we dark-matter researchers have less proof than other scientists in terms of what we seek to discover and what we believe we know. As to God? Well, if there were a divinity then it would be utterly separate from both scientific enquiry and human longing.'

He pauses again. It is not that this thinking is hard for him – he has moved down these paths before – but that he is picking each word with care.

'No divinity in which I would wish to believe would declare itself by means of what we would recognize as evidence.' He gestures at the data read-out. 'If there is a god, we should not be able to find it. If I detected proof of a deity, I would distrust that deity on the grounds that a god should be smarter than that.'

'Does it change the way the world feels?' I ask him. 'Knowing that 100 trillion neutrinos pass through your body every second, that countless such particles perforate our brains and hearts? Does it change the way you feel about matter – about what matters? Are you surprised we don't fall through each surface of our world at every step, push through it with every touch?'

Christopher nods. He thinks. His screensaver changes to the limestone towers at Guilin, seen near dusk such that they are backlit in ways that are considered widely appealing on Instagram and other large-scale image-sharing platforms.

'At the weekends,' Christopher says, 'when I'm out for a walk with my wife, along the cliff tops near here, on a sunny day, I know our bodies are wide-meshed nets, and that the cliffs we're walking on are nets too, and sometimes it seems, yes, as miraculous as if in our everyday world we suddenly found ourselves walking on water, or air. And I wonder what it must be like, sometimes, *not* to know that.'

He pauses, and it is clear that he is thinking now beyond the confines of the salt cavern, beyond even the known limits of the universe.

'But mostly, and in several ways, I'm amazed I'm able to hold the hand of the person I love.'

~

Neil wanted to drive the Paris–Dakar Rally back in the day. Neil is steering a stripped-down doorless Ford Transit van in a subterranean desert maze more than 600 miles in extent, Neil is a matter of weeks from retirement, and Neil doesn't give a shit.

We take the ramps fast enough to lift up as we come over them. We leave the tunnels behind us clouded with dust. Instead of slowing down for the corners, Neil just leans on the horn. *Paaaaarp!* He's a man passionate about mine safety; he's also a man passionate about fun. I like him a lot.

I hang off the roof handle with my left hand, lean forwards and brace myself with my right hand against the dashboard. I clench my jaws to stop my teeth clattering.

'Between the main shaft, where the lab sits, and the production districts, there's barely anyone except at shift change,' says Neil. 'If they're coming our way, we should see their lights from a long way off.'

The roadways are cut from the halite, with ramps leading up into the potash seams. The sides of the roadways glimmer a little in the light, like ice. We're trucking through pure salt. The tunnels are of standard dimensions – 3.8 metres high, 8 metres wide – and their ceilings are regularly reinforced with bolts the length of a man, to slow the slump.

'The potash is more fissile,' says Neil. 'Cracks more easily. You don't want to run roadways through it unless you have to. Halite tends to sag rather than shatter. Much safer.'

Thump! Paaaaarp!

'These main roadways have two years or so in them before they start to scrunch up. We prop them up with wood stacks. Wood's better than steel: it squishes rather than snapping. Much safer. Still, sometimes we lose a district before it's mined out. So it goes.'

Neil has a disconcerting habit of turning to me as he talks, keeping one hand on the top of the wheel but no eyes on the road. Sometimes he rotates the steering wheel with his palm, as though he is buffing a car's panel work in small arcs. Wax on, wax off. 'It's not like a coal mine, where you're always worried about combustion of the coal dust in the air,' he says. 'Here the salt dust acts like a dry-powder fire extinguisher. Much safer.

'The last death down here was in the 2000s, caused by a low-velocity explosion at the production face: 500 tons of rock came down in a recently mined roadway, pushed the machine back, and the machine crushed a man to death. No one's died down here this decade.'

A few months later a popular miner called John Anderson will be killed in a gas blowout.

We ramp up into a potash seam. Neil brakes the van to a halt in a swirl of dust, jumps out, cracks a fat flake of potash off the tunnel wall and hands it to me. It is pink as meat and flecked with silver mica. It is surprisingly light, almost buoyant in the hand.

'Lick it,' says Neil. It fizzes on my tongue. It tastes of metal and blood. I want to eat it all.

A stream of water runs down a wall of the tunnel from a crack in the ceiling. Neil points upwards. 'We've just crossed the coastline! We're under the sea now!

'Halite and sylvite are both soluble in water,' says Neil. 'This poses problems when you mine below the ocean. We have to pump the mine continuously to keep it workable: 1,000 gallons a minute, giving us an electricity bill of about £3 million per year. The Russians and the Canadians have both lost potash mines to flooding in the past.

'We had a big flood not so long back: 3,500 gallons a minute, running for eight weeks. We thought we'd lost the mine for a while.

Then it slowed as it self-sealed; don't quite know why. Nothing to say it won't start up again.'

'How reassuring.'

We get back into the Transit. 'How's this for a job, eh?' Neil asks no one in particular. 'I get paid to do this!' He slams the pedal to the metal, we lurch back in our seats and hammer on down the drift.

Neil's navigational powers impress me. He has no map, there are no signs, but he shows no hesitation at any of the dozens of junctions we meet.

'If you were to die,' I say, 'just hypothetically speaking, how would I get out of here?'

'If in doubt, follow the wheel tracks,' he yells. 'And if I cop it, just keep the wind in your face and you'll find the way out!' He points up again. 'We're out beyond the shipping lane now. Imagine those captains in charge of their boats, with never a clue we're careering about below them!'

It takes us another twenty minutes to reach the production face. Neil parks at the side of a tunnel, behind two other transits, straightening up the wheels as neatly as if he is on a suburban street.

Dust smogging the air; the tunnels ahead forking out; flickering lights and shadow-movement. The walls of the tunnels are inscribed with gouged patterns: spirals, cross-hatchings. They look like the cuts of a creature trying to claw its way out of a trap, or the ritual petroglyphs of a tribe.

'Production District 887 – the limits of the seam,' says Neil. 'The test-probes suggest the seam exhausts itself more or less here. Once this district is mined out, there'll be no more north-westwards progress; we'll look to the eastern and south-eastern edges of the undersea drift.'

Two teams of men are sitting at tables, drinking and eating. In the blackness I can see only the glowing strips of their hi-vis jackets. It's a scene from *Tron*. The men look up, nod a greeting, get back to their food. There are dozens of penises scrawled in biro and marker pen on the white wipe-clean PVC of the tabletop.

Left down one tunnel, right down another. Noise increasing; dust increasing. Halogen light-beams slice through choking air. Screaming noise of metal on mineral.

A huge red-black machine, low-slung and sharp-toothed as a Komodo dragon, is feeding at a face of rock. The dragon is controlled via a thick black rubberized cable, as if on a dog leash. From the lizard's arsehole extrudes a long thin stream of potash ore on a conveyor belt, ducted back towards a hopper to begin its journey to the world's fields.

The lizard-machine feeds at the face, the conveyor belt continues to trundle ore towards the hopper, and I am struck by a sense of the creatureliness of the mining operation: the avid clawing at the rock, the tunnel network that has been created. I remember cross-sections I have seen of the interiors of termite mounds and ants' nests, rabbit warrens and mole runs. Neil's map of the mine, with its hundreds of miles of intersecting drift, is just the plan of another animal's burrow-complex, bored out in search of resources.

What curious partners they have become in the darkness, the mine and the laboratory, oddly echoic of each other's operations. The geologists sending their probes out into the rock ahead, hoping to detect and pursue the most remunerative seams. The physicists watching for the arrival of knowledge, pure knowledge, the *sylvite* of knowledge, hard to reach, worth nothing, hoping to detect the missing portion of the universe: dark matter, a yield that cannot be sold.

Neil leans close again, cups his hands to shout into my ears above the noise of extraction. 'Those face-mining machines? They cost £3.2 million each. The engines are modified, obviously, to prevent sparking. We bring them down in sections in the lift shaft, assemble them in build-up bays, and then drive them out to the production face, towing a generator behind them. It takes them three days to trundle the seven or so miles out here to where they begin work.'

The strains of the work are intense, the lifespan of the machines short. 'When one of them reaches the end of its useful days,' says Neil, 'it's not cost-effective to bring it back up. It'd take the place of ore in the upshaft, and that's too expensive. So instead the machine gets driven into a worked-out tunnel of rock salt, and abandoned there. The halite will flow around it as the tunnel naturally closes up.'

It is an astonishing image: the translucent halite melting around this cybernetic dragon – the fossilization of this machine-relic in its burial shroud of salt.

I remember the pit ponies about which Emile Zola had written, brought down as foals into France's great nineteenth-century coal mines. The foals would not see daylight again. They grew in the mines, were fed there, were worked to death there, and their stunted bodies were left in side tunnels, awaiting the burial of collapse.

In the halite strata that underlie the New Mexico desert, an underground facility known as the Waste Isolation Pilot Plant has been excavated, designed for the long-term disposal of transuranic radioactive waste arising from the research and production of nuclear weapons. More than 2,000 feet below the desert surface, a burial site has been created for thousands of silver steel drums packed with nuclear waste. Because the waste remains radioactive for thousands of years, it generates heat. This heat will increase the plasticity of the halite – and so once each chamber is replete, the warmed halite should creep around the barrels, securing them for the deep time future.

I am briefly filled with a longing to step into a side tunnel myself, lie down and let the halite slowly seal me in for five years or 10,000 – to wait out the Anthropocene in that translucent cocoon.

~

In 1999, at a conference in Mexico City on the Holocene – the epoch of Earth history that we at present officially inhabit, beginning around 11,700 years ago – the Nobel Prize-winning atmospheric chemist Paul Crutzen was struck by the inaccuracy of the Holocene designation. 'I suddenly thought this was wrong,' he later recalled. 'The world has changed too much. So I said, "No, we are in the Anthropocene." I just made the word up on the spur of the moment. But it seems to have stuck.'

The following year, Crutzen and Eugene Stoermer – an American diatom specialist who had been using the term informally since the 1980s – jointly published an article proposing that the Anthropocene should be considered a new Earth epoch, on the grounds that 'mankind [*sic*] will remain a major geological force for many millennia, maybe millions of years to come'. As the Pleistocene was defined by the action of ice, and the Holocene by a period of relative climatic stability allowing the flourishing of life, so the Anthropocene is seen to be defined by the action of *anthropos*: human beings, shaping the Earth at a global scale.

The scientific community took the Crutzen-Stoermer proposal seriously enough to submit it to the rigours of the stratigraphers. In 2009 the Anthropocene Working Group of the Subcommission on Quaternary Stratigraphy was created. It was charged with delivering two recommendations: firstly, whether the Anthropocene should be formalized as an epoch and if so, secondly, when its 'stratigraphically optimal' temporal limit should be located, i.e. when it could be said to have begun. Among the baselines considered by the group have been the first use of fire by hominins around 1.8 million years ago, the start of agriculture around 8,000 years ago, the Industrial Revolution, and the so-called 'Great Acceleration' of the mid twentieth century, when the nuclear age dawned, massive increases occurred in terms of resource extraction, population growth, carbon emissions, species invasions and extinctions, and when the production and discard of metals, concrete and plastics boomed.

What signatures our species will leave in the strata! We remove whole mountain tops to plunder the coal they contain. The oceans dance with hundreds of thousands of tons of plastic waste, slowly settling into sea-floor sediments. Weaponry tests have dispersed artificial radionuclides globally. The burning of rainforests for monoculture production sends out smog-palls that settle into the soils of nations. A nitrogen spike, indicated in ice-cores and sediments, will be one of the key chemical insignias of the Anthropocene, caused by the mass global use of synthetic nitrogen-rich fertilizers and by fossil-fuel burning. Biodiversity levels are crashing worldwide as we hasten into the sixth great extinction event, while the soaring number of a small number of livestock species ensures the geological posterity in the fossil record of sheep, cows and pigs. We have become titanic world-makers, our legacy legible for epochs to come.

Among the relics of the Anthropocene, therefore, will be the fallout of our atomic age, the crushed foundations of our cities, the spines of millions of intensively farmed ungulates, and the faint outlines of some of the billions of plastic bottles we produce each year – the strata that contain them precisely dateable with reference to the product-design archives of multinationals. Philip Larkin famously proposed that what will survive of us is love. Wrong. What will survive of us is plastic, swine bones and lead-207, the stable isotope at the end of the uranium-235 decay chain.

There are many reasons to be suspicious of the idea of the Anthropocene. It generalizes the blame for what is a situation of vastly uneven making and suffering. The rhetorical 'we' of Anthropocene discourse smooths over severe inequalities, and universalizes the site-specific consequences of environmental damage. The designation of this epoch as the 'age of man' also seems like our crowning act of self-mythologization – and as such only to embed the technocratic narcissism that has produced the current crisis.

But the Anthropocene, for all its faults, also issues a powerful shock and challenge to our self-perception as a species. It exposes both the limits of our control over the long-term processes of the planet, and the magnitude of the consequences of our activities. It lays bare some of the cross-weaves of vulnerability and culpability that exist between us and other beings now, as well as between humans and more-than-humans still to come. Perhaps above all the Anthropocene compels us to think forwards in deep time, and to weigh what we will leave behind, as the landscapes we are making now will sink into strata, becoming underlands. What is the history of things to come? What will be our future fossils? As we have amplified our ability to shape the world, so we become more responsible for the long afterlives of that shaping. The Anthropocene asks of us the question memorably posed by the immunologist Jonas Salk: 'Are we being good ancestors?'

But to think ahead in deep time runs against the mind's grain. Try it yourself, now. Imagine forwards a year. Now ten. Now a century.

Imagination falters, details thin out. Try a thousand years. Mist descends. Beyond a hundred years even generating a basic scenario for individual life or society becomes difficult, let alone extending compassion across much greater reaches of time towards the unborn inhabitants of worlds-to-be. As a species, we have proved to be good historians but poor futurologists. While we have devised abbreviations for marking out deep time in the past – BP for 'before present'; MYA for 'million years ago' – we have no equivalent abbreviations for marking out deep time in the future. No one speaks of AP for 'after present', or MYA for 'million years ahead'.

The Anthropocene requires us to undertake a retrospective reading of the current moment, however – a 'palaeontology of the present' in which we ourselves have become sediments, strata and ghosts. It asks that we imagine a single figure: a hypothetical post-human geologist who – millions of years into the future, long after the extinction of our species – will examine the underland for what it reveals of the epoch of *anthropos*. This imaginary figure – our archivist, our analyst, our judge – is the contemporary version of the 'last man' presence that haunted nineteenth-century extinction narratives, or of Thomas Macaulay's 'New Zealander', sitting by the banks of the Thames in a London that has been overwhelmed by nature, ruminating on ruination.

Down at the chaos of the mine's production face, I think of the puzzles we are creating for our future geologist. I wonder how, millions of years on, she will interpret the fossil presence of the lizard-like mining machines of Boulby, manufactured in the Anthropocene and embedded in the strata of a 250-million-year-old seabed. How will she distinguish them as machines rather than as organisms? And what of the drift itself – the faint impress that this 600-mile maze will leave in the layers of halite and sylvite?

Geologists and palaeobiologists speak of 'trace fossils'. A trace fossil is the sign left in the rock record by the impress of life rather than life itself. A dinosaur footprint is a trace fossil. The enigmatic doughnut-shaped flints called 'paramoudra' are thought to be the trace fossils of a burrowing worm-like creature that lived vertically in the seabed during the Cretaceous, its breathing organs just above the level of the silt. Boreholes, funnels, pipes, slithers and tracks are all trace fossils – stone memories where the mark-maker has disappeared but the mark remains. A trace fossil is a bracing of space by a vanished body, in which absence serves as sign.

We all carry trace fossils within us – the marks that the dead and the missed leave behind. Handwriting on an envelope; the wear on a wooden step left by footfall; the memory of a familiar gesture by someone gone, repeated so often it has worn its own groove in both air and mind: these are trace fossils too. Sometimes, in fact, all that is left behind by loss is trace – and sometimes empty volume can be easier to hold in the heart than presence itself.

~

The return from the production face is a madcap rally-drive. Neil works the van even harder. Dust in the mouth, hitting the ramps at speed – *whump* – stomach in the mouth, then slam down onto the halite floor. We approach a corner. Neil hammers the horn. *Paaaaarp!* He hammers it again. Silence. Hammer. Silence.

'I must have shaken a circuit loose,' Neil says.

'That's been evident for a while now,' I say.

'Not to worry. We're coming back out. We've got right of way, at least in theory. I'll slow down a bit.'

He doesn't slow down at all.

'Watch out for oncoming headlights on the side walls! If I knock myself out, take over the wheel and head south-west!'

We pass two wrecked Transits in side tunnels, their bonnets crushed from unknown impacts, waiting to be absorbed by the halite. On we pummel through miles of tunnel, back at last to the yellow cage of the upshaft.

Soft whoosh and air-squeeze halfway up as the down-cage passes us. Jolt and slow as we near the surface. The shuffle of men readying themselves for the exit, thinking of shower, home, family, food, drink. Rattle of the door opening. Block-squares of light through steel gate lock-hatches. Smell of the sea, smell of the sun. Into the airlock, counted out one by one. Miners first. Respirator back on the peg. *Check*. Push the bronze triangle in at the window desk. *Check*. Clear.

Out through the door and into burning white day, blue billowing sky, sun glinting off windscreen and chain-link, tarmac and grass blade, dark matter nowhere and everywhere around me – and surfacing into this blinding light seems like stepping into ignorance.

~

Later I drive west over the moors for hours, winding home. The ling is in bloom and pollen glitters in the air. Marks of mining are everywhere I look, left by thousands of years of human boring into this northern landscape in search of materials: slate, lead, iron, copper, ironstone, silver, coal, fluorspar. Marks of burial too, left by thousands of years of humans interring their dead in the same terrain: medieval church cemeteries, burial mounds from the Neolithic, the Bronze Age, the Iron Age.

Near dusk I am in the ridge-and-fold limestone valleys of the

North Pennines. The easterly breeze of the morning has grown in force to a gale. At Rookhope I park and walk the mile or so up onto the moor above the village.

The wind at that height is chilling, though the late sun is still strong. Cottontails of bog-grass thrum in the wind, glowing like gas mantles. Four kestrels, strung in a ragged low line above the moor to my west, hold their positions with grace against the wind. I gorge on the glut of light, the fetch of space. Reaching a jumble of boulders I stand on the highest stone, face east and lean a little into the wind, feeling the push of its hand on my chest – holding me in part-flight, kestrelling me.

Time feels differently reckoned after the mine: further deepened, further folded. My sense of nature feels differently reckoned too: further disturbed, further entangled. Somewhere to my east, men are at work a mile below the moors, half a mile under the sea, cutting tunnels through the salt-ghost of an ocean to harvest its energy for crops as yet ungrown. A Time Projection Chamber is waiting for signals from Cygnus, the Swan, that might tell something of the birth of the universe, 13.8 billion years earlier. A labyrinth of drift is slowly closing up, lizard-machines and Ford Transits are being sealed into their tombs of salt – and through it all is passing a particle wind of WIMPs and neutrinos, to which this world is as mere mist and silk.

'At night, according to their accustomed watches, the stars traverse a path beneath the earth,' Bede had written in *The Reckoning of Time*, 1,300 years earlier, as he calculated the six ages of the Earth, and the seventh age to come. I think of the miners who worked the underland of these Pennine valleys through the nineteenth century, following seams holding metallic ores of silver, magnesium, lead and zinc. Where galena coated the sides of a rift, it could gleam as brightly as a mirror. The same veins held wondrous blossoms of fluorspar, crystals of which shone blue in ultraviolet light. Occasionally the miners hacked their ways into geodes the size of rooms, walled and roofed in crystals and metals. The flames of their lamps glittered off quartz, aragonite, dolomite, fluorspar, iron pyrites and galena – as if they had broken into a buried star-chamber down there in the crust.

A full moon has begun to rise. The sky is darkening to red and black, the moor is sinking to browns and silvers, and the valley is suddenly off-planet.

The first star shows, then others glimmer into view. I step off the boulders, begin the walk down off the ridge, when a skylark shoots up a yard or so from me, shocking my heart, and I put a hand down in the hollow from which it has flown in time to catch the trace warmth of its body before the cold steals it away. The lark rises up into the sky, its cascading song clear and present in the moment.

~

A long night drive on high moors and down over coastal plains, headlights sweeping heather on the corners, coning the sky on the uphills, and home at last after midnight to a house at the foot of a mountain. The sky is salted with stars.

I pad into the room where my youngest son, Will, is sleeping. The moonlight pouring through the thin curtain casts my shadow across the floor.

I stand over Will and he is lying so still that panic sluices coldly through me, my heart thumps in my ears and I reach my hand towards his mouth to feel for his breath, to search for proof of life in the darkness.

Nothing, no breath, *no breath* – and then there it is, on the exhale, drifting faint and warm on my skin, and I rest the back of my fingers for a few seconds on his cheek, feel the mass of his body.

Still there, my love?

Breathe.

Breathe again.

My heart slowing back down. Starlight silvering the fine down on the edge of his skin. Everything causing a scintillation.

4
The Understorey
(Epping Forest, London)

Occasionally – once or twice in a lifetime if you are lucky – you encounter an idea so powerful in its implications that it unsettles the ground you walk on.

The first time I heard anyone speak of the 'wood wide web', more than a decade ago now, I was trying not to cry. A beloved friend was dying too young and too quickly. I had gone to see him for what I took to be the last time. He was tired by pain and drugs. We sat together, talked. My friend was a woodsman. Trees grew through his life and thought. His grandfather's surname was Wood, he lived in a timber-framed house that he had built himself, and he had planted thousands of trees by hand over the years. 'I have sap in my veins,' he wrote once.

That day I read aloud a poem that was important to us both, 'Birches' by Robert Frost, in which climbing the snow-white trunks of birches becomes both a readying for death and a declaration of life. Then he told me about new research he had recently read concerning the inter-relations of trees: how, when one of their number was sickening or under stress, they could share nutrients by means of an underground system that conjoined their roots beneath the soil, thereby sometimes nursing the sick tree back to health. It was a measure of my friend's generosity of spirit that – so close to death himself – he could speak unjealously of this phenomenon of healing.

He did not have the strength then to tell me the details of how this below-ground sharing operated – how tree might invisibly reach out to tree within the soil. But I could not forget the image of that mysterious buried network, joining single trees into forest communities. It was planted in my mind, and there took root. Over the years I would encounter other mentions of the same extraordinary idea, and gradually these isolated fragments began to connect together into something like understanding.

~

In the early 1990s a young Canadian forest ecologist called Suzanne Simard, studying the understorey of logged temperate forests in north-west British Columbia, observed a curious correlation. When paper birch saplings were weeded out from clear-cut and reseeded plantations, their disappearance coincided with first the deterioration and then premature deaths of the planted Douglas fir saplings among which they grew.

Foresters had long assumed that such weeding was necessary to prevent the young birches (the 'weeds') depriving the young firs (the 'crop') of valuable soil resources. But Simard began to wonder whether this simple model of competition was correct. It seemed to her plausible that the paper birches were somehow helping rather than hindering the firs: when they were removed, the health of the firs suffered. If this interspecies aid-giving did exist between trees, though, what was its nature – and how could individual trees extend help to one another across the spaces of the forest?

Simard decided to investigate the puzzle. Her first task was to establish some kind of structural basis for possible connections between the trees. Using microscopic and genetic tools, she and her colleagues peeled back the forest floor and peered below the understorey, into the 'black box' of the soil – a notoriously challenging realm of study for biologists. What they saw down there were the pale, super-fine threads known as 'hyphae' that fungi send out through the soil. These hyphae interconnected to create a network of astonishing complexity and extent. Every cubic metre of forest soil that Simard examined held dozens of miles of hyphae.

For centuries, fungi had generally been considered harmful to plants: parasites that caused disease and dysfunction. As Simard began her research, however, it was increasingly thought that certain kinds of common fungi might exist in subtle mutualism with plants. The hyphae of these so-called 'mycorrhizal' fungi were understood not only to infiltrate the soil, but also to weave into the tips of plant roots at a cellular level – thereby creating an interface through which molecular transmission might occur. By means of this weaving, too, the roots of individual plants or trees were joined to one another by a magnificently intricate subterranean system.

Simard's enquiries confirmed that beneath her forest floor there did indeed exist what she called an 'underground social network', a 'bustling community of mycorrhizal fungal species' that linked sapling to sapling. She also discovered that the hyphae made connections between species: joining not only paper birch to paper birch and Douglas fir to Douglas fir, but also fir to birch and far beyond – forming a non-hierarchical network between numerous kinds of plants.

Simard had established a structure of connection between the saplings. But the hyphae provided only the means of mutualism. Its existence did not explain why the fir saplings faltered when the birch saplings were weeded out, or details as to what – if anything – might be transmitted via this collaborative system. So Simard and her team devised an experiment that could let them track possible biochemical movements along this invisible buried lattice. They decided to inject fir trees with radioactive carbon isotopes. Using mass spectrometers and scintillation counters, they were then able to track the flow of carbon isotopes from tree to tree.

What this tracking revealed was astonishing. The carbon isotopes did not stay confined to the individual trees into which they were injected. Instead, they moved down the trees' vascular systems to their root tips, where they passed into the fungal hyphae that wove with those tips. Once in the hyphae they travelled along the network to the root tips of another tree, where they entered the vascular system of that new tree. Along the way, the fungi drew off and metabolized some of the photosynthesized resources that were moving along their hyphae; this was their benefit from the mutualism.

Here was proof that trees could move resources around between one another using the mycorrhizal network. The isotope tracking also demonstrated the unexpected intricacy of the interrelations. In a research plot thirty metres square, every single tree was connected to the fungal system, and some trees – the oldest – were connected to as many as forty-seven others. The results also solved the puzzle of the fir–birch mutualism: the Douglas firs were receiving more photosynthetic carbon from paper birches than they were transmitting. When paper birches were weeded out, the nutrient intake of the fir saplings was thus – counter-intuitively – reduced rather than increased, and so the firs weakened and died.

The fungi and the trees had 'forged their duality into a oneness, thereby making a forest', wrote Simard in a bold summary of her findings. Instead of seeing trees as individual agents competing for resources, she proposed the forest as a 'co-operative system', in which trees 'talk' to one another, producing a collaborative intelligence she described as 'forest wisdom'. Some older trees even 'nurture' smaller trees that they recognize as their 'kin', acting as 'mothers'. Seen in the light of Simard's research, the whole vision of a forest ecology shimmered and shifted – from a fierce free market to something more like a community with a socialist system of resource redistribution.

Simard's first major paper on the subject was published in *Nature* in 1997, and it was from there that the subterranean network of tree–fungus mutualism gained its durable nickname of 'the wood wide web'. Her *Nature* paper was a groundbreaking publication, the implications of which were so significant that an entire research field subsequently formed to pursue them. Since then the scientific study of below-ground ecology has boomed. New technologies of detection and mapping have illuminated fresh details of this 'social network' of trees and plants. 'The wood wide web has been mapped, traced, monitored and coaxed,' as Simard puts it, 'to reveal the beautiful structures and finely adapted languages of the forest network.'

And among this new generation of the forest's linguists and mappers is a young plant scientist called Merlin Sheldrake. Truly, that is his name.

~

Merlin and I stand side by side in a beech coppice – the biggest I have ever seen, let alone entered. The stool is ten yards from one end to another, the tree perhaps 400 or 500 years old.

'I'd guess this hasn't been coppiced for at least half a century,' I say to Merlin.

Coppice shoots have grown, unlopped, into upright trunks,

raying up around the edge of the coppice's base and leaving a space in the centre easily big enough to hold us both. We stay there for a while, enjoying being inside this ancient tree, looking out at Epping Forest from between the grey-barked bars of our cage.

Two of the beech's lower limbs have melted into one another, their bark conjoining into a single continuous skin, their vascular systems growing and uniting. Living wood, left long enough, behaves as a slow-moving fluid. Like glacial ice – like the halite I had seen in Boulby, like the calcite I had seen in the Mendips, like stained glass in medieval churches which, over centuries, gradually thickens out at the base of each pane – living wood *flows*, given time.

'I've heard this called "pleaching",' I say to Merlin, patting the fused branches. 'The artist David Nash planted a circle of ash trees in a clearing in North Wales, then bent and wove the trees so that they grew not just next to one another but *into* one another, a dancing "Ash Dome", made of a meld of boughs and limbs.'

'Actually,' says Merlin, 'plant scientists have a technical term for this. We call it "snogging", or to give it its full name, "tree snogging".' He smiles. 'Well, not quite. The technical term is actually "inosculation", from the Latin *osculare*, meaning "to kiss". Inosculation means "to en-kiss". It can happen across trees and between species too.'

Though I know the word 'inosculation', I had not known its etymology; what seemed a chilly specialist term gains a passionate warmth, and feels true to this arboreal 'en-kissing', which makes it hard to say where one being ends and another begins. I think of Ovid's version of the 'Baucis and Philemon' myth, in which an elderly couple are transformed into an intertwining oak and linden, each supporting the other in terms of both structure and sustenance, drawing strength for each other from the ground through their roots – and tenderly sharing that strength through their en-kissing.

'This kind of merging happens below ground too,' says Merlin, 'but probably more intensely between the roots of the trees than between branches, because space is more limited below ground and the criss-crossing will be denser. And it happens *vastly* more profusely in the fungal networks, often between quite different species.' He follows the pleaching of the two branches with a finger.

'From being two hyphal tubes, two fungi are suddenly one, and things can start flowing between them, including genetic material and nuclei. This is why it's so hard to deal with species concepts in fungi, or even the question of what an organism is – because while fungi do the sex thing, they also have this *wildly* promiscuous horizontal transfer of genetic material that is unpredictable in a still ill-understood way.'

Merlin Sheldrake, as the oldest joke in mycology goes, is a fun guy to be around. During the days in which he conjures open the underland of Epping Forest for me, I ask more questions than I have of anyone for what feels like years. What he tells and shows me in that modest peri-urban forest reshapes my sense of the world in ways I am still processing.

The night of Merlin's birth was that of the Great Storm, 15 October 1987, when hurricane-force winds, gusting to strengths of 120 mph, capsized carriers, drove ferries ashore, and felled some 15 million trees – ripping up the forest floor across southern England and northern France and tilting it skywards in the form of root plates. The first full day of Merlin's life was Black Friday, when the Dow Jones suffered a record fall, wiping trillions off global wealth and triggering a crash in financial markets worldwide.

No, the omens of Merlin Sheldrake's arrival into the world were not auspicious. In Greek myth he would surely have been fated to be a force for destruction and ruin. But he was given a magical name and he grew into a magical person. He is tall, slim, and very upright in his bearing. He has tight curls of dark hair, intense eyes with full circles of white visible around each iris, and a wide, warm grin. He is also a formidable scientist, with a doctorate in Plant Science from Cambridge. There is something faintly antiquarian to him – a disinterest in disciplinary boundaries, a boundless curiosity – and something of the heroic-age plant hunter too. He puts me in mind of a cross between Sir Thomas Browne and Frank Kingdon Ward, collector of *Meconopsis betonicifolia*, the legendary blue poppy of the Himalayas.

It is typical of Merlin that he became fascinated from a young age not with the charismatic megafauna of the world, but instead with the undersung, underseen inhabitants of the biota: lichens, mosses and fungi. He studied them as an amateur teenage scientist, counting lichen species on gravestones and granite boulders, and trying to comprehend the subterranean architecture of fungal life – above-ground mushrooms as fruiting bodies that stand as mere fleeting allusions to immense underland structures.

'My childhood superheroes weren't Marvel characters,' Merlin once said to me, 'they were lichens and fungi. Fungi and lichen annihilate our categories of gender. They reshape our ideas of community and cooperation. They screw up our hereditary model of evolutionary descent. They utterly *liquidate* our notions of time. Lichens can crumble rocks into dust with terrifying acids. Fungi can exude massively powerful enzymes *outside* their bodies that dissolve soil. They're the biggest organisms in the world and among the oldest. They're world-makers and world-breakers. What's more superhero than *that*?'

~

Merlin and I set off on foot into Epping Forest one morning from a high clearing, heading roughly north, keeping the sun to the right of our line.

Epping extends to the north-east of London, and it is very far from a wildwood. It was first designated as a royal hunting forest in the twelfth century by Henry II, with penalties for poaching that included imprisonment and mutilation. Presently it is managed by the City of London Corporation, and has more than fifty bye-laws governing behaviour within its bounds – though the punishments are now fiscal rather than corporal. It is fully contained within the M25, the orbital motorway that encircles outer London. Minor roads traverse it, and it is never more than two and a half miles wide. Despite its small extent, Epping is easy to get lost in – a forest of forking paths to which, for a thousand years, the people of London and its surrounds have gone for shelter, sex, escape and a relic greenwood magic.

Growl of roads. Whirr of a low-flying bumblebee, stirring the leaf litter with its downdraught. Buzzard overhead, turning, mewing. Old coppice trees left uncut, hydra-headed pollards. A fallen log, thick with moss; small orange fungi sprouting from wet breaks in its grain. Where trees thin out and light falls, hundreds of green beech seedlings are pushing up through the litter, none more than an inch high. Five fallow deer appear between hollies ahead of us, the dapple of leaf-light flicking off the dapple of their flanks as they move through the understorey.

In the language of forestry and forest ecology, the 'understorey' is the name given to the life that exists between the forest floor and the tree canopy: the fungi, mosses, lichens, bushes and saplings that thrive and compete in this mid-zone. Metaphorically, though, the 'understorey' is also the sum of the entangled, ever-growing narratives, histories, ideas and words that interweave to give a wood or forest its diverse life in culture.

'What interests me most,' says Merlin, 'is the understorey's understory.' He points around at the beech, the hornbeam, the chestnut. 'All of these trees and bushes,' he says, 'are connected with one another below ground in ways we not only cannot see, but ways we have scarcely begun to understand.'

While studying Natural Sciences at Cambridge, Merlin read Simard's groundbreaking research into the wood wide web. He also read E. I. Newman's classic 1988 paper, 'Mycorrhizal Links between Plants: Their Functioning and Ecological Significance'. There Newman argued against the assumption that 'plants are physiologically separate from each other', proposing instead the existence of a 'mycelial network' that might link plants together. 'If this phenomenon is widespread,' wrote Newman, 'it could have profound implications for the functioning of ecosystems.'

Those 'implications' were indeed profound, and they fascinated Merlin. He already loved the alien realm of fungi. He knew that fungi could turn rocks to rubble, could move with swiftness both overground and underground, could reproduce horizontally, and digest food outside their bodies by means of metabolically ingenious excreted acids. He knew that their toxins could kill us, and their psychoactive chemicals could induce hallucinogenic states. The work of Simard and Newman, however, revealed to him that fungi could also allow plants to communicate with one another.

Merlin was taught as an undergraduate by Oliver Rackham, the legendary botanist whose research transformed our understanding of both the cultural and botanical history of the English landscape. Working with Rackham, Merlin found himself most intellectually attracted to places where orthodox evolutionary theory felt thinnest – and for him the thinnest places were where mutualisms were at work. Mutualism is a subset of symbiosis in which there exists between organisms a prolonged relationship that is interdependent and reciprocally beneficial.

'What fascinates me about mutualisms,' says Merlin, 'is that one would predict from basic evolutionary theory that they would be massively unstable, and collapse quickly into parasitism. But it turns out that there are very ancient mutualisms, which have remained stable for puzzlingly long times: between the yucca plant and yucca moths, for example, or of course between the bacteria that illuminate the bioluminescent headlamp of the bobtail squid, and the squid itself.'

'Of course,' I reply. 'The ancient glowing-bobtail-squid-and-bacteria mutualism.'

'The ultimate mutualism, though,' says Merlin, 'is between plants and mycorrhizal fungi.'

~

The term 'mycorrhiza' is made from the Greek words for 'fungus' and 'root'. It is itself a collaboration or entanglement; and as such a reminder of how language has its own sunken system of roots and hyphae, through which meaning is shared and traded.

The relationship between mycorrhizal fungi and the plants they connect is ancient – around 450 million years old – and largely one of mutualism. In the case of the tree–fungi mutualism, the fungi siphon off carbon that has been produced in the form of glucose by the trees during photosynthesis, by means of chlorophyll that the fungi do not possess. In turn, the trees obtain nutrients such as phosphorus and nitrogen that the fungi have acquired from the soil through which they grow, by means of enzymes that the trees lack.

The possibilities of the wood wide web far exceed this basic exchange of goods between plant and fungi, though. For the fungal network also allows plants to distribute resources between one another. Sugars, nitrogen and phosphorus can be shared between trees in a forest: a dying tree might divest its resources into the network to the benefit of the community, for example, or a struggling tree might be supported with extra resources by its neighbours.

Even more remarkably, the network also allows plants to send immune-signalling compounds to one another. A plant under attack from aphids can indicate to a nearby plant via the network that it should up-regulate its defensive response before the aphids reach it. It has been known for some time that plants communicate above ground in comparable ways, by means of diffusible hormones. But such airborne warnings are imprecise in their destinations. When the compounds travel by fungal networks, both the source and recipient can be specified. Our growing comprehension of the forest network asks profound questions: about where species begin and end, about whether a forest might best be imagined as a super-organism, and about what 'trading', 'sharing' or even 'friendship' might mean between plants and, indeed, between humans.

The anthropologist Anna Tsing likens the below-ground of a forest to 'a busy social space', where the interaction of millions of organisms 'forms a cross-species world underground'. 'Next time you walk through a forest,' she writes memorably in an essay called 'Arts of Inclusion, or How to Love a Mushroom', 'look down. A city lies under your feet.'

~

Merlin and I have been walking the forest for two hours or so when we reach one of Epping's great pollard beech groves. Pollarding – the pruning of the upper branches of a tree to promote dense growth – keeps trees alive for longer, indeed can enter them into an almost indefinite fairy-tale time of longevity. Here in the grove, long waving trunks yearn up to the sun. Through their leaves falls a green sub-sea light. It feels as if we are swimming through a kelp forest.

We stop and lie down for a while on the woodland floor, on our backs, not speaking, watching the trees' gentle movements in the breeze, and the light lacing and lancing from fifty feet or more above us. Where the pollards spread out to form the canopies, I realize I can trace patterns of space running along the edges of each tree's canopy: the beautiful phenomenon known as 'crown shyness', whereby individual forest trees respect each other's space, leaving slender running gaps between the end of one tree's outermost leaves and the start of another's.

Lying there among the trees, despite a learned wariness towards anthropomorphism, I find it hard not to imagine these arboreal relations in terms of tenderness, generosity and even love: the respectful distance of their shy crowns, the kissing branches that have pleached with one another, the unseen connections forged by root and hyphae between seemingly distant trees. I remember something Louis de Bernières has written about a relationship that endured into old age: 'we had roots that grew towards each other underground, and when all the pretty blossom had fallen from our branches we found that we were one tree and not two.' As someone lucky to live in a long love, I recognize that gradual growing-towards and subterranean intertwining; the things that do not need to be said between us, the unspoken communication which can sometimes tilt troublingly towards silence, and the sharing of both happiness and pain. I think of good love as something that roots, not rots, over time, and of the hyphae that are weaving through the ground below me, reaching out through the soil in search of mergings. Theirs, too, seems to me then a version of love's work.

Merlin gets up, walks towards the centre of the grove as if looking for something, then bends down and brushes away leaf litter and beech mast, to clear a patch of soil the size of a saucer. I get up and follow him. He pinches some of the earth and rubs it between his fingers. It smears rather than crumbling: a rich, dark humus, made of composted leaves.

'This is our problem when it comes to studying the fungal network,' he says. 'Soil is fantastically impenetrable to experiments, and the fungal hyphae are on the whole too thin to see with the naked eye. That's the main reason it's taken us so long to work out the wood wide web's existence, and to discern what it's doing.'

Rivers of sap flow in the trees around us. If we were right now to lay a stethoscope to the bark of a birch or beech, we would hear the sap bubbling and crackling as it moves through the trunk.

'You can put rhizotrons into the ground to look at root growth,' Merlin says, 'but those don't really give you the fungi because they're too fine. You can do below-ground laser scanning but, again, that's too crude for the fungal networks.'

I am reminded once more of how resistant the underland remains to our usual forms of seeing; how it still hides so much from us, even in our age of hyper-visibility and ultra-scrutiny. Just a few inches of soil is enough to keep startling secrets, hold astonishing cargo: an eighth of the world's total biomass comprises bacteria that live below ground, and a further quarter is of fungal origin.

'We know the network is there,' Merlin says, 'but it's so effortful to track it. So we have to look for clues to the labyrinth – find clever means of following its paths.'

I kneel beside him. I can see dozens of insects just in this small area, the names of most of which I do not know: gleaming spiders and red-bronze beetles battling over the leaves, a woodlouse curled up into a sphere, a green threadworm writhing through the humus.

'It's *roiling* with life,' I say to Merlin.

'That's just the visible life. Hyphae will be growing into the decomposing matter of this half-rotting leaf,' says Merlin, 'into those rotting logs and those rotting twigs, and then you'll have the mycorrhizal fungi whose hyphae grow into hot spots – all of them frothing and tangling and fusing, making a network that's connecting holly to holly but also to this beech, and to a seedling of something else over there, layering and layering and *layering* – until, well, it blows your computational brain!'

As Merlin speaks I feel a quick, eerie sense of the world shifting irreversibly around me. Ground shivering beneath feet, knees, skin. *If only your mind were a slightly greener thing, we'd drown you in meaning* . . . I glance down, try to trance the soil into transparency such that I can see its hidden infrastructure: millions of fungal skeins suspended between tapering tree roots, their prolific liaisons creating a gossamer web at least as intricate as the cables and fibres that hang beneath our cities. What's the haunting phrase I've heard used to describe the realm of fungi? *The kingdom of the grey.* It speaks of fungi's utter otherness – the challenges they issue to our usual models of time, space and species.

'You look at the network,' says Merlin, 'and then it starts to look back at you.'

~

ST. JOHN THE BAPTIST PARISH LIBRARY
2920 NEW HIGHWAY 51
LAPLACE, LOUISIANA 70068

In the underland of the hardwood forests of Oregon's Blue Mountains there exists a honey fungus, *Armillaria solidipes*, that is two and a half miles in extent at its widest point, and covers a total lateral area of almost four square miles. The blue whale is to this honey fungus as an ant is to us. It is a deeply mysterious organism: the largest in the world that we know of, and one of the oldest. The best guess that US Forest Service scientists have been able to offer for the honey fungus's age is between 1,900 and 8,650 years old. The fungus expresses itself above ground as mushrooms with white-flecked stems rising to tawny, gill-frilled cups. Below ground, where its true extent lies, *Armillaria solidipes* moves as rhizomorphs resembling black bootlaces, out of which reach the hyphal fingers of its mycelium, spreading in search both of new hosts which they might kill, and the mycelia of other parts of the colony with which they might fuse.

All taxonomies crumble, but fungi leave many of our fundamental categories in ruin. Fungi thwart our usual senses of what is whole and singular, of what defines an organism, and of what descent or inheritance means. They do strange things to time, because it is not easy to say where a fungus ends or begins, when it is born or when it dies. To fungi, our world of light and air is their underland, into which they tentatively ascend here and there, now and then.

Fungi were among the first organisms to return to the blast zone around the impact point in Hiroshima, the point from which the mushroom cloud had risen. After Hiroshima, too, images of the mushroom cloud began to appear ubiquitously in media and culture – the fruiting bodies of a new global anxiety. Scientists working in Chernobyl after the disaster there were surprised to discover fine threads of melanized fungi lacing the distressed concrete of the reactor itself, where radiation levels were over 500 times higher than in the normal environment. They were even more surprised to work out that the fungi were actively thriving due to the high levels of ionizing radiation: that they benefited from this usually lethal gale, increasing their biomass by processing it in some way. Ecologists in the US seeking to understand how American trees will respond to the stress of climate change have begun to focus on the presence of soil fungi as a key indicator of future forest resilience. Recent studies suggest that well-developed fungal networks will enable forests to adapt faster at larger scales to the changing conditions of the Anthropocene.

'Learning to see mosses is more like listening than looking,' writes the ethnobotanist Robin Wall Kimmerer; 'mosses . . . issue an invitation to dwell for a time right at the limits of ordinary perception.' Learning to see fungi seems even harder – requiring senses and technologies that we have yet to develop. Even to try and think with or as fungi is valuable, though, drawing us as it does towards lifeways that are instructively beyond our ken.

Certainly, orthodox 'Western' understandings of nature feel inadequate to the kinds of world-making that fungi perform. As our historical narratives of progress have come to be questioned, so the notion of history itself has become remodelled. History no longer feels figurable as a forwards-flighting arrow or a self-intersecting spiral; better, perhaps, seen as a network branching and conjoining in many directions. Nature, too, seems increasingly better understood in fungal terms: not as a single gleaming snow-peak or tumbling river in which we might find redemption, nor as a diorama that we deplore or adore from a distance – but rather as an assemblage of entanglements of which we are messily part. We are coming to understand our bodies as habitats for hundreds of species of which

Homo sapiens is only one, our guts as jungles of bacterial flora, our skins as blooming fantastically with fungi.

Yes, we are beginning to encounter ourselves – not always comfortably or pleasantly – as multi-species beings already partaking in timescales that are fabulously more complex than the onwards-driving version of history many of us still imagine ourselves to inhabit. The work of the radical biologist Lynn Margulis and others has shown humans to be not solitary beings, but what Margulis memorably calls 'holobionts' – collaborative compound organisms, ecological units 'consisting of trillions of bacteria, viruses and fungi that coordinate the task of living together and sharing a common life', in the philosopher Glenn Albrecht's phrase.

Little of this thinking is new, however, when viewed from the perspective of animist traditions of indigenous peoples. The fungal forest that science had revealed to Merlin and that Merlin was revealing to me – a forest of arborescent connections and profuse intercommunication – seemed merely to provide a materialist evidence-base for what the cultures of forest-dwelling peoples have known for thousands of years. Again and again within such societies, the jungle or woodland is figured as aware, conjoined and conversational. 'To dwellers in a wood almost every species of tree has its voice as well as its feature,' wrote Thomas Hardy in *Under the Greenwood Tree*. The anthropologist Richard Nelson describes how the Koyukon people of the forested interior of what we now call Alaska 'live in a world that watches, in a forest of eyes. A person moving through nature – however wild, remote . . . is never truly alone. The surroundings are aware, sensate, personified. They feel.' In such a vibrant environment, loneliness is placed in solitary confinement.

There in the grove with Merlin, I recall Kimmerer, Hardy and

Nelson, and feel a sudden, angry impatience with modern science for presenting as revelation what indigenous societies take to be self-evident. I remember Ursula Le Guin's angrily political novel, set on a forest planet in which woodland beings known as the Athsheans are able to transmit messages remotely between one another, signalling through the medium of trees. On Athshe – until the arrival of colonists committed to the planet's exploitation – the realm of mind is integrated into the community of the trees, and 'the word for world is forest'.

~

Four hours into our walk and Epping is playing the usual tricks of forests: disorientation, echoes, a refusal to repeat itself. Often I think we are reversing a path we have already walked, only to find ourselves led into a new area of heath, an unfamiliar grove or thicket. We kick up invisible spores spread by last autumn's fungi, breathing them into our lungs. We wander so far north that we run out of forest, rebound off the M25, hop a barbed-wire fence, and come to rest in a field that looks as if it is privately owned. We aren't lost, exactly, but we want to know where the forest widens again.

So I use my phone to summon the satellite network, and pull up a hybrid map of the forest. Sixty-three distinct chemical elements including rare earth metals and minerals mined mostly in China interact within the casing of my device. A blue lanthium dot pulses our location. I pinch and splay the screen to get the right scale. The map shows that the forest flares green to the south-west, so that is where we head – crossing a busy road and then pushing deeper into the trees until we can hardly hear car noise.

In a dry part of the forest, on rising ground, with old pines, beeches and an understorey of holly, we stop to eat and drink, sitting among snaking pine roots. I tell Merlin about Boulby Mine, about the dark-matter lab, the halite tunnels, the men at the face, the geologists sending their probes forwards, falling back from the face, questing in the dark.

'It's so similar to the way fungi work,' Merlin says, 'always prospecting for the most resource-rich or beneficial region, pushing on where they sense benefit. They fan out and if they find a decent seam in one place they die back from the poor areas and concentrate their efforts elsewhere.' He takes my notebook and pen, and draws a diagram of the classic hyphal structure: a branching fan in which it is hard to speak of a main or originary stem, only of shoots and offshoots.

In the second year of his doctorate, Merlin went to conduct fieldwork in the Central American jungle – on Barro Colorado Island, located in the man-made Gatun Lake of the Panama Canal.

'I was so ready to leave the lab for the jungle,' he says. 'In a molecular biology lab you're in near-total control of these little worlds; you're the giant puppet master making your subject dance to your tune. In the field, though, you're *inside* your subject matter and the power relation is totally different.'

On the island Merlin joined a community of field biologists, all of them dancing to the tune of the jungle. He operated under the watchful eye of a grizzled evolutionary biologist called Egbert Giles Leigh Jr, who lived on the base and received new arrivals in his book-lined study, where he played Beethoven on his gramophone and drank whisky-no-ice-no-mixer. This benevolent Kurtz was the island's archive and overseer.

Some of the science undertaken on the island was methodologically high risk. There was a young American scientist researching what Merlin called 'the Drunken Monkey Hypothesis'. Her plan was to collect monkey urine after the monkeys had feasted on fermenting fruit, and to assess the urine for intoxication levels. The problem was that the monkeys tended to urinate from high up in the trees. So she developed a wide-mouthed funnel with which to catch the falling liquid.

'Just to be clear,' I ask, 'she was getting drunken monkeys to pee down a funnel from the canopy?'

'Absolutely – and that was very laborious work. She also seemed, you might say, an unlikely candidate for this particular kind of research.'

Then there was someone nicknamed 'the Bumblebee Guy', who trapped bumblebees and stuck adhesive radio-trackers to their abdomens in order to be able to map their feeding and pollinating patterns of movement.

'But the adhesive didn't stick so well, because the bees were hairy and the air was humid,' says Merlin, 'so he then had to capture his bees and shave tiny patches of their abdomens, in order to stick the trackers on more reliably.'

There was also 'the Lightning Guy', who studied the effects of lightning strikes on below-ground ecologies, and tried to induce site-specific strikes by firing crossbow bolts trailing copper wire at storm clouds.

'Sounds like a carnival out there,' I say.

'Basically, what you quickly discovered,' Merlin says, 'was that if your experiment wasn't good enough, the jungle would screw it up.'

During his second season on the island, Merlin became interested in a type of plants called 'mycoheterotrophs' – 'mycohets' for short. Mycohets are plants that lack chlorophyll and thus are unable to photosynthesize. As such they are entirely reliant on the fungal network for their provision of carbon. Some are white, some tinged lilac or violet.

'These little ghosts plug into the fungal network,' Merlin explains, 'and somehow derive everything from it without paying anything back, at least in the usual coin. They don't play by the normal rules of symbiosis – but we can't prove they're parasites. You could imagine them as the hackers of the wood wide web.'

Merlin focused on a genus of mycohets called the *Voyria*, a group of gentians known as 'ghost plants', the flowers of which studded the jungle floor on Barro Colorado Island like pale purple stars. Working with local villagers, he carried out a painstaking census of the soil in a series of plots, sampling and sequencing the DNA of hundreds of root samples taken both from green plants and the *Voyria*. The census allowed him to determine what species of fungi were connecting with which plants – and thereby to make an unprecedentedly detailed map of the jungle's social network.

'I only stumbled across the importance of *Voyria* by accident,' he says, 'wandering around one day looking for something else, when I realized that they had almost vanished from a plot from which we'd increased the phosphorus input. That's how my breakthrough began. Science is full of this stuff: full of happenstance and stumbles and getting knackered and crazy in the field or the lab. It's so weird to me how science always presents its knowledge as *clean*.'

A green woodpecker yaffles in the distance.

'I have this plan,' Merlin says, 'that for each formal scientific paper I ever publish I will also write its dark twin, its underground mirror-piece – the true story of how the data for that cool, tidy hypothesis-evidence-proof paper *actually* got acquired. I want to write about the happenstance and the shaved bumblebees and the pissing monkeys and the drunken conversations and the fuck-ups that *actually* bring science into being. This is the frothy, mad network that underlies and interconnects all scientific knowledge – but about which we so rarely say anything.'

~

Late in the day we come to a lake in the woods, where a mudbank slopes into shallow water.

Fish sup in the shadows. Moorhens bicker. The lake-bed belches gas bubbles. Merlin and I sit facing the setting sun, enjoying its warmth.

Two dog-walkers approach, looking hopeful. 'Do you know where the visitors' centre is? We're lost.'

'No, we're lost too,' I say happily.

We trade best guesses, share what information we have, and they wander off.

Sitting quietly in the sun, by the lake, I think about the ways we seek to make meaning of the wood wide web. Both of the two main models of interpretation that Merlin has told me about – the 'socialist' and the 'free-market' models – smuggle a very human politics into a more-than-human science. According to the 'free-market' model, the connected forest is to be understood as a competitive system, in which all entities act out of self-interest within a cost-benefit framework, regulating one another by means of 'sanction and reward' systems. According to the 'socialist' model, by contrast, trees act as carers to one another, sharing resources through the fungal network, with the well-off supporting the needy.

I ask Merlin about this question – about how the politics of representation press especially hard on mycorrhizal studies. It seems to me that what is at stake is not only the relations of nature but also the nature of relations.

'You're exactly right. In my field, discourse choice forcefully shapes research directions. "Sanction and reward", for instance, is a central technical concept in mycorrhizal studies, not just an ornament of speech. The metaphor *drives* the scholarship. I read research papers with titles like "Unequal Goods Shared under Common Terms of Trade".'

'That sounds as if it was commissioned by the Ayn Rand Think Tank,' I say.

'Indeed. Awful. Politically, I'm obviously inclined to dislike the language of biological free-marketry far more than the socialist version,' says Merlin. 'Why should we expect fungi and plants to behave as humans started to behave economically in the eighteenth century, with the emergence of the limited liability corporation? I find it *so* bizarre. It's one reason I love the *Voyria*. They demand immediately that you go beyond cost-benefit analysis when thinking about plant life.

'But I'm also sceptical of the socialist dream of fungi as sharing and caring, a rose-tinted vision that sees trees as nurses, every tree a carer to every other, with "mother trees" recognizing and talking with their kin, and "injured trees" selflessly passing on their legacies to neighbours before they die.

'I'm tired of both of these stories,' Merlin says as we leave the lake. 'The forest is always more complicated than we can ever dream of. Trees make meaning as well as oxygen. To me, walking through a wood is like taking a tiny part in a mystery play run across multiple timescales.'

'Maybe, then, what we need to understand the forest's underland,' I say, 'is a new language altogether – one that doesn't automatically convert it to our own use values. Our present grammar militates against animacy; our metaphors by habit and reflex subordinate and anthropomorphize the more-than-human world. Perhaps we need an entirely new language system to talk about fungi . . . We need to speak in spores.'

'Yes,' says Merlin with an urgency that surprises me, smacking his fist into the palm of his hand. 'That's *exactly* what we need to be doing – and that's *your* job,' he says. 'That's the job of writers and artists and poets and all the rest of you.'

~

Potawatomi, a Native American language of the Great Plains region, includes the word *puhpowee*, which might be translated as 'the force which causes mushrooms to push up from the earth overnight'. In 'all its technical vocabulary', Robin Wall Kimmerer notes, 'Western science has no such term, no words to hold this mystery.'

Kimmerer herself is a member of the Citizen Potawatomi Nation. A speaker of what she calls 'fluent botany', she is careful to distinguish this from what she refers to as 'the language of plants' – that is to say, the language that plants speak, as opposed to the language that is used to speak of plants. Kimmerer does not disdain the precisions of botanical lexis, which 'polishes the gift of seeing', but she finds it also to be of necessity a lexis of objectification and distancing, with something missing beneath its finely faceted surface. That missing something is predominantly the acknowledgement of life in the more-than-human world, an indifference which is grained into language not just at the level of individual words, but at the deeper-down levels of grammar and syntax.

In Potawatomi, by contrast, almost all words declare the animacy or inanimacy of that to which they refer. The language is predisposed to recognize life in otherness, and also to extend the reach of that category of 'life' far beyond its familiar limits in Western thought. In Potawatomi, not only humans, animals and trees are alive, but so too are mountains, boulders, winds and fire. Stories, songs and rhythms are all also animate, they *are*, they *be*. Potawatomi is a language abundant with verbs: 70 per cent of its words are verbs, compared to 30 per cent in English. *Wiikwegamaa*, for instance, means 'to be a bay'. 'A bay is a noun only if water is *dead*,' writes Kimmerer:

> trapped between its shores and contained by the word. But the verb . . . releases the water from bondage and lets it live. 'To be a bay' holds the wonder that, for this moment, the living water has decided to shelter itself between these shores, conversing with cedar roots and a flock of baby mergansers.

Like Kimmerer, I wish for a language that recognizes and advances the animacy of the world, 'the life that pulses through pines and nuthatches and mushrooms . . . well[ing] up all around us'. Like Kimmerer, I relish those aspects of discourse that extend being and sentience respectfully and flexibly beyond the usual bearers of such qualities. Like Kimmerer I believe that we need, now, a 'grammar of animacy'. A modern predisposition to regard animacy as anomaly runs through what the poet Jeremy Prynne once called 'mammal language', by which he means the language that is used by humans, encoding intent, agency and muscular power deep in its grammar.

The real underland of language is not the roots of single words, but rather the soil of grammar and syntax, where habits of speech and therefore also habits of thought settle and interact over long periods of time. Grammar and syntax exert powerful influence on the proceedings of language and its users. They shape the ways we relate to each other and to the living world. Words are world-makers – and language is one of the great geological forces of the Anthropocene.

Projects have recently been started around the world to gain even the most basic of vocabularies for the experiences of life and death in the Anthropocene. These stuttering attempts to speak what it is we are doing have generated ugly new terms for an ugly epoch: 'geotraumatics', 'planetary dysphoria', 'apex-guilt'. Such words feel like futile forms of nominalism, a hopelessly hyperactive pointing and naming. They stick in the throat in two ways: they are difficult to utter and hard to swallow.

Only one of these recent coinages resonates with me: 'species loneliness', for the intense solitude that we are fashioning for ourselves as we strip the Earth of the other life with which we share it. If there is human meaning to be made of the wood wide web, it is surely that what might save us as we move forwards into the precarious, unsettled centuries ahead is collaboration: mutualism, symbiosis, the inclusive human work of collective decision-making extended to more-than-human communities.

You look at the network, and then it starts to look back at you . . .

Writing of mycorrhizal fungi, Albrecht proposes that we rechristen the Anthropocene, naming it instead the Symbiocene – an epoch characterized in terms of social organization 'by human intelligence that replicates the symbiotic and mutually reinforcing life-reproducing forms and processes found in living systems . . . such as the wood wide web.'

The word for world is forest.

~

That evening, in a deep part of the woods, far from a road, near an Iron Age earthwork and an old pollard beech grove, on a slope of high ground nicknamed 'Friendship Rise', Merlin and I settle down for the night. We scrape a shallow fire-pit, haul dead birch trunks around it for seats, and set a small fire going with a tinder of leaves and kindling of twigs, in contravention of Epping Forest bye-laws, with murmured apologies to the Corporation of London.

Merlin opens his rucksack and brings out a small decoction bottle containing a dark moss-green liquid. He shakes it.

'Coca extract. Home-made. The perfect pick-us-up after a day among the leaves.'

He reaches again into his bag, and brings out another bottle.

'Home-made honey mead,' he says.

He reaches in again, and brings out a third.

'Home-made cider,' he says.

The brown glass of the bottle carries a single white label, on which is written the word 'Gravity'.

'I pressed this from some windfall apples that had fallen from Newton's apple tree in Cambridge. It's quite hard to get to, that tree. It's in Trinity College. Security is reasonably tight. The scrumping had to be done under cover of night. I wish I could have brought us a bottle of the first ever batch we made. That was from apples scrumped from Darwin's orchard at Down House. You can probably guess the label name for that batch.'

'Evolution.'

'Bingo.'

People begin to emerge from the shadows of the trees in ones and pairs: friends of mine and friends of Merlin, friends of our friends, invited by social network, by text, by phone, zeroing in on our location using GPS. One brings a harmonica, two bring guitars, and

Merlin's brother brings two sets of bones and a small set of hand drums.

Moths dance around the flames. Satellites blip above us. The red landing lights of planes, visible through the crown shyness, cut paths between the leaves. I have a strong sense of the forest looming around and over and below us.

I drink Merlin's coca decoction, feel my mind rapidly sharpen. The fire works its magic of storytelling and conviviality. People talk, reestablish existing connections, make new ones, bring into being a temporary community in that fire-braced forest space. I show the whalebone owl and the bronze casket, explain about their giving and the obligations they bring with them. Merlin and I tell some of our day's understories. Merlin speaks, like Tsing, of the soil as a city, a city beneath our feet in which countless species and kinds of matter are busy interacting.

A young man whose nickname is 'the Hand Owl' plays bluegrass on his cupped hands alone, hooting and whooping. Folk songs are sung – 'Nine-Pound Hammer', 'Seven Drunken Nights', 'Brown Trout Blues' – with people passing chorus lines and verses from one to another and back again. Merlin plays the bones, clacking a beat for each new song. The night chills us and the fire warms us.

Drums, songs, stories. The trees shifting, speaking, busy making meaning that I cannot hear. Fungi writhing in the birch logs, in the soil.

I sit with my back to a birch log, feet to the fire, next to Tara. Tara is tall, gentle of speech, Greek. She is a singer. She grew up on a small island in the Mediterranean. She learned song and voice from a Russian émigré who had been washed up on the island by the tides of history. She tells me about the consequences for the island of the refugee crisis: the networks of support that were put in place for the refugees, but also the resistance of the islanders who saw the crisis as a threat to their ways of life.

'There reaches a point where you see other humans drowning, or washed ashore with nothing,' says Tara, 'where there is no possibility other than to help with all your heart. It isn't kindness, exactly – because there is less element of choice than people think, and for this reason it is less noble.'

Later, Tara sings a sad song from her island, and my heart breaks a little. The flames die to purring embers.

I am too tired to see the fire to its end, so I wander away into the forest to find somewhere to sleep. Looking back I can see only orange glow, shadows cast against the surrounding trunks – then the firelight dwindles until it is lost in the forest's dark.

I find myself in a grove of pollard beeches atop a prehistoric earthwork. Under one of them, children have built a den of sticks and boughs, resting them against a low branch to form a crooked timber tent that is long enough for me to sleep in. It is an invitation I cannot refuse, so I creep inside the den and lie down, looking up through its slats at branches, stars, satellites. I feel myself suddenly – strongly – surrounded by beings whose ways of relating to one another are dimly but powerfully perceptible, as if seen through thick gauze. The sensation is at once comforting and lonely-making.

Owl hoot. Dog bark. Back in the clearing the fire dims, song falls silent. The canopy of the pollards spreads above me, whispering in the night breeze. *There's something you need to hear . . .* Seeking sleep, my mind follows leaf to branch, branch to trunk, trunk to root and from there down along the hyphae that web the earth below.

SECOND CHAMBER

Down in the labyrinth under the old ash tree with the riven trunk, a new passage is chosen and followed.

This water-worn rift bends into the earth, each fresh curve emerging from its predecessor as pleat is shaken from pleat in the unfurling of a cloth. As the rift deepens, its sides lean closer to one another and its roof drops until, just where it seems the rift will narrow to impassability, it opens abruptly into a big new chamber.

Sound echoes off this chamber's walls and light flickers across them, and where the light falls it shows the stone to be alive with more scenes from the underland. The scenes here are those of hiding, sheltering and finding – dispersed across time and space, but bound again by strange echoes.

An artist is at work a thousand years ago on a painting that will become part of a menologion for an emperor. The painting shows a mountain rising from a desert landscape. The sky above glints with gold leaf. The bedrock is blue-grey. From the mountain's slopes rise two cypress trees and an evergreen oak. The mountain's side has been cut away by the artist so that what it contains can be seen. In the shadows of its interior seven men are asleep. The rock encases and shelters them. They wear loose robes of grey, red, blue, tawny-brown, purple. They lie close together. Some are barefoot, some shod. There is a brotherliness to their pose – a tenderness to the way one of the sleepers rests his hand on the brow of another. These are the Seven Sleepers of Ephesus, known in Arabic as *aṣḥāb al kahf*, the 'people of the cave', and theirs is a story of waiting in darkness, within rock, until it is safe to emerge. Their story recurs in the Christian and Islamic traditions; it is there in the Qur'an, in the Roman Martyrology. The young men, fleeing religious persecution in the city of Ephesus, enter a cave mouth that leads them deep into the mountain. In that den of night, exhausted by flight, they lie down and sleep. They will sleep for 300 years – and when they emerge all danger will be gone.

Cold sleet on old slate. Grey air, grey stone. Hawthorn bushes crabbed in the low ground. A single holly bush, its berries reddening. Winter in a slate quarry hollowed into a mountain, almost 2,000 feet above the sea. The work here is brutal, killing. The quarrymen have long died of explosions and falls, the slate-cutters of lung disease. The workers walk here each weekend from their homes, following paths marked by lines of white stones. They sleep together in barracks through which the wind blows, two men to a bed, curled up for warmth. For this privilege they must pay the quarry owners. Sometimes at night, they sing chapel hymns together. So it has been for almost 200 years here – an asymmetry of power and suffering. Something odd is happening in this quarry now, though. Men from a ministry have come, and they have paid for five of the caverns cut inside the mountain to be converted into bunkered treasure rooms. Small brick houses have been constructed inside the caverns, their interiors air-conditioned and temperature-controlled. And up the old quarry road have laboured lorries containing hundreds of large, thin packages. The packages are paintings: Claude's *Landscape with David at the Cave of Adullam*, Piombo's *Raising of Lazarus*, Van Dyck's eleven-foot-high version of Charles I mounted on a horse, work by Gainsborough, Hogarth, Constable, Turner and Monet. All of these paintings have been taken under armed guard from the National Gallery in London and transported to this cored-out Welsh mountain, to be placed in these brick chambers under 300 solid feet of 400-million-year-old slate – safe, surely, here of all places, from the bombs of the Luftwaffe.

Nuclear fears crackle the world's air. The Cuban Missile Crisis is only weeks past. Flashbulbs pop and crowds cheer as a man enters a limestone rift near Nidderdale in Yorkshire (Nidder – a variant of 'nether'; to 'nidder' or 'nether' is to 'keep down', to 'press under'). The rift opens into a complex cave system, its full extent unexplored. The man wishes to study the effects of chronic darkness and the absence of visible time upon body and mind. He wishes also to show the people of Britain that 'if we need to go into caves in a nuclear war, all we need do is wrap up warm and take down a lot of food'. There, in the underland, he believes it will be possible to wait out the radioactivity above ground until it is safe to emerge. He pitches a tent by a stalactite. The man intends at first to spend a hundred days in the underland, but without the cycle of day and night he loses circadian knowledge, and his body instead falls into a rhythm only of need – sleep when necessary, in short bursts. He emerges after 105 days below ground, to find a world unburned by nuclear fire.

Inside a tent made from shrapnel-torn sheets of white plastic a shaft has been dug into sandy soil. It drops fifty feet vertically, at which point a tunnel just high enough for a man to stand in extends laterally for 900 feet, to where a similar shaft rises to the surface, its mouth also concealed by a tent. The two shafts are separated by a national border. The illegal tunnel is a means of evading a punitive blockade on the movement of goods across this boundary. There are hundreds of similar tunnels, riddling the underland below the frontier, and along them are smuggled supplies of food, clothes, hardware, people, livestock and weaponry. When war flares here, as it does often, air-strikes target the tunnels with fighter planes dropping one-ton bombs in an attempt to destroy what is buried. But the tunnels are relatively cheap to make, swift to repair, profitable to run – and lifelines to the community blockaded behind the border. So they must be dug, though diggers lose their lives each year through cave-ins and bomb strikes.

A summer's day in Connemara in the west of Ireland. A woman wades into the water of a bay, walking the slick stones with the confidence of custom. She is an artist and among her subjects are the submerged dark depths of the human mind, and those points at which mythical and physical landscapes powerfully converge. She has always been at ease in water and she has taken to swimming in the sea each day – sometimes straight out offshore for half a mile, sometimes into a sea cave to the north of the bay. She has also begun holding her breath and diving to the bed of the bay, carrying sardines with her, with which bait she has learned to tempt conger eels out of their lairs in the rocks. These powerful creatures, some as long as she is, come snaking out of their holes to take the sardines she offers. Some even allow her to stroke them. It has become important to her art to encounter these uncanny beings in their own realm: a confrontation with what lies beneath, a befriending of fear. She remembers the words of Wittgenstein, who came to live on this same coastline in order to undertake some of his most intense philosophizing: 'I can only think clearly in the dark, and here I have found one of the last pools of darkness in Europe . . .'

A door is set into a wedge of concrete which angles back into a mountainside, high on an Arctic island. The roof of the portal radiates an otherworldly green – an installation of prisms is reflecting the aurora borealis that are shimmering in the polar night sky. Prophecies of the world's end in fire have receded; now the eschatology is one of ongoing breakdown rather than apocalypse. The end times are here, present, all around now, no longer deferrable. The heavy doorway opens into a tube of corrugated metal sloping deep into the mountain, far above sea level. This is a doomsday vault, made to survive for as close to eternity as can be reached on Earth. The frosted vaults of this eschaton, quarried from the limestone of the island, hold not people – but seeds. Life in fabulous abundance is here, chilled to dormancy: 90 million seeds; 860,000 crop varieties, 120,000 different strains of rice alone. Squash, alfalfa, sorghum, pigeon peas, foxtail millet, some of the earliest strains of Levantine wheat and durum, more than 10,000 years old. The outside of this mountain has no trees – just scant coverings of lichen, moss and little more. Inside, frost-flowers bloom on the vault walls. The seeds bide their time.

On the Anatolian plateau, where ash spewed by volcanoes 30 million years ago has hardened into a rolling terrain of cones and dips, a man is rebuilding his house. He decides to knock down a wall where it stands flush with the bedrock tuff – and behind the wall he discovers a chamber. Off the chamber runs a passage – and the passage leads down to a subterranean city. The city has eighteen different levels arranged over 300 vertical feet, and it offers shelter for up to 20,000 people. There are chambers for storing food, water, wine and oil. There are sleeping rooms, communal rooms, cooking rooms and tombs. Stone doors can be rolled across key openings to isolate areas in case of attack. Air moves by means of dozens of vertical ventilation shafts, and thousands of lateral tubes dispersing air between the individual chambers of the city – and through its centre runs an underground river.

The man feels he has stepped into a fable. The name that will be given to this discovered city is Derinkuyu, meaning 'Deep Well'. Its excavation is thought to have begun in the fourth century BC, and for more than a thousand years it has provided a place in which persecuted minorities might hide until trouble has passed. From a far chamber of the city, a five-mile-long passage connects this city to another such sunken city, even larger in extent. The man has stumbled into an invisible city – no, a network of invisible cities. *There may be more than a hundred such settlements as yet undiscovered, sleeping forgotten beneath the surface of the landscape.*

PART TWO

Hiding (Europe)

5
Invisible Cities

(*Paris*)

The map runs to sixteen laminated foolscap pages, or about ten square feet in area when I tile the pages together. I have been given it on the condition that I do not pass it on. It is not like any map I have ever seen and I have seen some strange maps in my time. The plan of the above-ground city is traced carefully but in pale silver-grey ink, such that if you trick your eyes to read only for the grey, you can discern the outline of this upper city as a spectre-architecture: the faint footprints of apartment blocks and embassies, parks and ornamental gardens, boulevards and streets, the churches, the railway lines and the train stations, all hovering there, intricate and immaterial.

The map's real content – the topography it inks in black and blue and orange and red – is the invisible city, the realm out of which over centuries the upper city has been hewn and drawn, block by block. This invisible city follows different laws of planning to its surface counterpart. Its tunnelled streets often kink and wriggle, or run to dead ends. Some of them curl back on themselves like whips. At junctions, three or four tunnel-streets might splay out. There are slender highways running almost the length of the tiled map, from south-west to north-east. There are inexplicably broken grids of streets, or hubs where the spokes of different tunnels meet. Coming off some of the tunnels are chambers, irregular in their outlines and with dozens of small connecting rooms.

The invisible city exists across multiple levels of depth, each connected to the other by staircases and wells. These sites of juncture between the levels are marked on the map with orange rings (for wells with rung ladders), blue rings (for wells with sheer sides) and segmented dark blue circles (for staircases). Deeper-down layers and systems are shaded in darker inks. I learn to let my eyes go lazy, so that one level swims above another and I can perceive the different strata of the under-city.

The map's place names traverse a range of cultural registers, from the classical to the surreal to the military-industrial. The Room of Cubes. The Passage of the Claustrophile. The Boutique of Psychosis. Crossroads of the Dead. The Clinic of the Aliens. The Chamber of Phantoms. The Medusa. The Glazery. The Maze of Montsouris. The Bermudas. The Shelter of the Little Leaves. The Monastery of the Bears. Bunker under the Mountain. The Cabinet of Mineralogy. The School of Mines. The Chamber of Oysters. Ossa Arida. Stairways of the Ossuary. Room Z.

Affordance is specified on the map in handwritten cursive words: 'Low', 'Quite low', 'Very low', 'Tight', 'Flooded', 'Impracticable', 'Impassable'. More detail is occasionally given: 'Humid and unstable region (sometimes flooded)'; 'Beautiful gallery, vaulted and corbelled'. '*Chatières*' – cat-flaps – mark a point of lateral transition between tunnel and tunnel, or between tunnel and chamber. Other captions gloss contact-sites between the upper city and the invisible city – 'Hole to the sky' – or between levels: 'Tiny hole in the ground debouching into a dangerous lower level'. Scattered around the map are little inked skulls-and-crossbones, and laconic warnings of danger: 'Cave-in'; 'Open well: dangerous'; 'Collapsing ceiling'.

Here and there, boxed-out cartouches offer stories of individual sites. A blue compass rose with an orange northwards arrow is laid over an empty section of each page, and each page is given a district name. The typeface is a fine, seriffed font I do not recognize. The overall aesthetic is coolly contemporary, the cartography itself an elegant compression of a hard-to-map region. I admire the work of its anonymous makers. On the cover page of the map, a link is given to an 'Encyclopedia of the Underground World'. Authorship is attributed only to a collective called 'Nexus' – 'the connection or connections between the parts of a system or a group of entities'.

~

What can I tell you about my time in the invisible city? That it is the longest I have ever gone without seeing sunlight. That one night, or perhaps it is one day, we listen to 'Dig for Fire' by the Pixies, laying a phone against the wall of a tunnel so the limestone booms the track back to us and lifts our spirits and makes me smile. That the evening we emerge, the Draconid meteor showers come, showing as silver scratches in the sky.

And that on the day we first go down into the invisible city, castle-clouds mass over the lowlands to the north of our entrance point. Flat fields, square-steepled church towers, lines of poplars, red-tiled farms. Lowlands, flatlands. My last sight of the sun is a westerly blaze under rain clouds, part hidden by a huge cone of earth of obscure function. To the east the cloud base is low and level. Streamers of rain fall grey on a distant village and the sun sets behind that earthwork.

Later, we push at dusk through a door in a wall marked '*Interdit d'entrer*', slip through a hole in a chain-link fence, scramble down a cutting side to a railway line, and crunch along the tracks towards the brick arch of a tunnel. The cutting banks are tangled with acacia trees and wild clematis. Apartment blocks rise above the cutting on both sides, so tall they seem to lean over the space. Once in the railway tunnel we keep between the tracks, because what little light there is glints on the metal and shows us the way, as the floor lights promise to do in aeroplanes that fill with smoke.

Sounds come from up ahead, and a young woman in a white dress with long blonde hair and a porcelain face emerges from the shadows, walking down the tracks towards us. She doesn't blink or pause and so we step to left and right of the tracks and let her walk on. She passes through us in silence without breaking her stride and ghosts off towards where, in the far distance, I can just see the tunnel arch of low light through which we have come, fringed brightly with green.

We crunch on. Ahead in the darkness is a flock of fireflies: soft orange lights bobbing in the black air. The fireflies neither advance nor retreat, and their light causes the brickwork of the tunnel to flit and shine. We draw closer and bodies gradually attach themselves to the lights, and we see that they are not fireflies but devils, for the lights are the twinned, bared flames of carbide lamps mounted on the foreheads of people milling around one side of the tunnel.

When we are fifty yards from the people with their devil horns of light, I see a woman sit down on the tunnel floor, turn sideways, raise her arms above her head and join her palms like a diver readying to jump – then disappear feet first into the invisible city.

~

Between 1927 and 1940 – the year in which he sought to flee France into the safety of Spain, only to commit suicide in a hotel room in the Pyrenean border village of Portbou – Walter Benjamin compiled one of the most extraordinary city-texts ever written. The *Passagen-Werk*, as it is known in German – *The Arcades Project* in English – is a fragmentary, unfinished meditation on the topography, history and humanity of Paris, running to more than a thousand pages at the time of Benjamin's death. Its form may be compared to a constellation or galaxy, the individual stars of which he drew together over more than a decade, collecting notes, quotations, aphorisms, stories and reflections in dozens of dossiers that he called *Konvolute* – 'convolutes' in English, meaning 'coils', 'twists', 'enfoldments' – each of which was identified by a letter.

Rather than writing a linear history of Paris, Benjamin sought to create a kaleidoscope, the crystals of which might fall into fresh patterns with each new reader, even each new reading. His book – if it can even be called a book, given the extent of its incompletion – was a gigantic, futile, magical attempt at historical comprehension, which understood the city's past in part to be a collective dream and the city's structures to possess a metaphysical aura as well as a material presence.

Throughout *The Arcades Project*, scenes from Paris's past flicker back into being. 'It is more arduous to honour the memory of anonymous beings than that of the renowned,' Benjamin remarked in the preliminary notes for his essay 'Theses on the Philosophy of History'; the 'construction of history is consecrated to the memory of the nameless'. An early experiment in what has become known as 'history from below', Benjamin's Paris commemorates these 'anonymous beings'; it is peopled by quarrymen, prostitutes, convicts, soldiers and shopkeepers, as well as aristocrats, politicians and artists. He made his book from scraps, gleaning it into existence as an archive of the stories of the city's unsung masses rather than those of its rulers.

Benjamin himself would be buried in a common and unmarked grave near Portbou, his cause of death given as morphine overdose, the date of his death given as 25 September 1940. The day before his suicide he had walked over the mountains from France, stopping every ten minutes on the ascent to rest his already-strained heart. His companions on the climb had to help him reach the final ridge of the crossing – but from there the party could look down into Spain and the shimmering Mediterranean, which appeared to them like a blue mirage. The next day, however, Benjamin was told that he would not be permitted onwards passage through Spain, and would instead be handed over to local French officials the following day. He knew that this meant his subsequent surrender to Nazi authorities and then, as a Jew, near-certain death. That night he killed himself with morphine tablets he had brought for such a necessity from Marseilles.

Benjamin is commemorated at Portbou by a simple, powerful monument which itself takes the form of a series of passages. The first of these is a descent into the underland. A long rusted steel tunnel slopes into the coastal bedrock from a small square at the entrance to the town cemetery. The visitor steps into the shadowed mouth of the tunnel as if entering Hades or Avernus. At the end of the staircase, however, is found not darkness but light: a sheet of glass seals the tunnel, preventing onwards progress, but giving a view out to a glittering sea channel where the currents form a whirlpool, its spiral remade with each fresh tide.

The work that Benjamin left unfinished at the time of his suicide is itself continually new-making. To enter *The Arcades Project* by one of its thousands of access points is to enter a labyrinth of passages that do not seem ever to repeat their routes. Like the city it describes, it offers a multitude of courses through its levels. It deals not in plots but in patterns, echoes, memory-ghosts and tangled subtexts. Reading it, you come to feel bodiless and boneless – able to traverse time by means of the book's subtle *chatières*, its secret passages.

It is clear that Benjamin's imagination was strongly drawn to enclosed and underground spaces: the warren of the covered 'arcades' themselves, as well as the caverns, crypts, wells and cells that existed beneath Paris. Taken together, these sunken spaces comprise what Benjamin called a 'subterranean city', shadow twin to the 'upper world', and dream-zone to its conscious mind. 'Our waking existence is a land which, at certain hidden points, leads down into the underworld,' he wrote, memorably:

> the realm from which dreams arise. All day long, suspecting nothing, we pass by these inconspicuous places, but no sooner has sleep come than we are eagerly groping our way back to lose ourselves in the dark corridors.

Benjamin's obsessive tracing out of this hidden terrain was to him an endeavour of historiography as well as geography that, if completed, might offer a 'key' to the 'underworld' of the European past. He took as his precursor and partial inspiration in this project the Greek peripatetic Pausanias, who spent years on foot mapping the porous points of the Greek landscape – springs, fissures, gorges – and characterizing them as a system of portals where upper and lower realms interlocked. Benjamin was fascinated by the existence of such portal points in the city. He wrote of the need to 'make some sign to the world one is leaving' when a threshold to the underworld was traversed, of the 'hatchway[s] leading from the surface to the depths', and of the *penates* that 'guard the threshold' and 'protect and mark the transitions'.

The most subterranean of the convolutes in *The Arcades Project* is Convolute C, which contains Benjamin's work on both the catacombs and the quarry voids of Paris. It is in Convolute C that Benjamin proposes his vision of Paris's invisible city, filled with 'lightning-scored, whistle-resounding darkness'. 'Paris is built,' he wrote there, in a passage I have not forgotten since first reading it in my early twenties:

> over a system of caverns . . . this great technological system of tunnels and thoroughfares interconnects with the ancient vaults, the limestone quarries, the grottoes and the catacombs which, since the early Middle Ages, have time and again been entered and traversed.

~

Down in the railway tunnel, we reach the firefly-devils. They are standing around, smoking and talking, and all are wearing carbide lamps: carbide canisters belted at their waists with pipes leading up to the burners strapped to their heads. From the burners hiss two horns of bare orange flame, low in temperature but high in luminosity. They nod diabolic greetings to us, murmuring in French and English.

Down at track level, where one side of the tunnel begins to rise, is a ragged hole in the ground, just wide enough to admit a person. A few yards to its right I can see the outline of what was once a similar hole, now plugged with fresh-looking concrete.

I have come to the catacombs with two friends – let us call them Lina and Jay. Jay is a caver keen to extend his explorations into city systems. He is droll, unflappable and strong. Lina is the leader of our group and she has been here many times, sometimes staying down for up to a week at a stretch. She is passionate about the catacombs, especially about preserving and documenting their swiftly changing features through photography and record-keeping. She is a curious combination of tentative above ground and bold below. She wears scarlet lipstick, brightly coloured berets, and she ties her curly brown hair back to keep it out of trouble in the tunnels. Coming into the catacombs seems to offer her a new personality. The invisible city is a place where she can go to be herself, or other than herself. Here Lina is calm, cool and knowledgeable. I feel fortunate to be with her.

'The cataflics came down and filled that one up,' says Lina, pointing to the plugged hole at track level. 'So we brought a jackhammer and a generator down and opened up this new one. It's probably the safest way in right now, but we'll plan to exit by a manhole, whenever we come out.'

She gestures back up the tunnel. 'Take a last look back at the light, because you won't be seeing the sun again until next week. Let's go.'

Lina eases herself feet first into the ragged hole, raises her arms above her head and disappears. Jay does the same. I think of Benjamin's practice of marking the passage into the under-city, of making 'some sign to the world one is leaving', and I look once to the distant arch of light, then lower into the labyrinth.

~

Much of the Île-de-France sits on Lutetian limestone, which accumulated chiefly during the Eocene, when the region was for around 5 million years an area of calm bays and lagoons of seawater. Marine life thrived and died in abundance there, settling on the seabed as silt that was eventually compressed into stone. Lutetian limestone is an excellent building material: ranging from warm grey to caramel yellow in tone, durable and cuttable to a clean edge.

All cities are additions to a landscape that require subtraction from elsewhere. Much of Paris was built from its own underland, hewn block by block from the bedrock and hauled up for dressing and placing. Underground stone-quarrying began in earnest towards the end of the twelfth century, and Parisian limestone grew in demand not just locally but across France. Lutetian limestone built parts of Notre-Dame and the Louvre; shipped on Seine barges into the river network, it became a major regional export.

The residue of over 600 years of quarrying is that beneath the south of the upper city exists its negative image: a network of more than 200 miles of galleries, rooms and chambers, organized into three main regions that together spread beneath nine arrondissements. This network is the *vides de carrières* – the 'quarry voids', the catacombs.

Quarrying techniques changed surprisingly little over time. Shafts were driven sixty feet or so down to the limestone layers, then tunnels were cut laterally from there, following the strata. Where larger rooms were excavated, pillars of stone were left unquarried to support the ceilings. The standard tunnel was cut to six feet high and three wide: enough to accommodate a man pushing a barrow filled with stone. Dynasties of quarrymen came and went, passing down skills from father to son, extending the maze over centuries. Fatalities were relatively rare as the stone was not prone to collapse – but daily exposure to mineral dust, and the brutal grind of the heavy lifting, led to ruined lungs and bodies.

For centuries, quarrying was ill-regulated and largely unmapped. Then in the mid eighteenth century, the extensive undermining began to have consequences for the upper city, causing subsidence sinkholes known as *fontis* that were reputed to be of diabolic origin. The quarry voids had begun to migrate to the surface; the under-city had begun to consume its twin. In 1774 a *fonti* engulfed, in a matter of seconds, pavements, houses, horses, carts and people. The site of the sinkhole was, of all places, the Rue d'Enfer – the Street of Hell. Several minor cave-ins followed, and panic spread in the city at the unknown extent of the invisible danger.

Louis XVI responded shortly after his accession by creating an inspection unit for the 'Quarries Below Paris and Surrounding Plains', headed by a general inspector called Charles-Axel Guillaumot, and tasked with regulating the quarries for the purposes of public safety. It was Guillaumot who initiated the first mapping of the void network, with a view to consolidating existing spaces and regulating further quarrying activities. A subterranean town-planning system was established whereby chambers and tunnels were named in relation to the streets above them, thus creating a mirror-city with the ground serving as the line of symmetry. 'Paris has another Paris under herself,' wrote Victor Hugo in *Les Misérables*, 'which has its streets, its intersections, its squares, its dead ends, its arteries and its circulation.'

It was also Guillaumot who, in the mid 1780s, oversaw the idea of using the quarry voids for purposes of storage. And what urgently needed storing was Paris's dead. The city's earliest significant burial grounds, established during the Roman era, were located on the southern outskirts of the city as it then stood. But as Paris spread it took to burying the majority of its bodies in cemeteries within its bounds, notably in the main Cemetery of Saints-Innocents near the central marketplace of Les Halles. The result over centuries was a growing glut of the dead. Saints-Innocents became the resting place for millions of bodies. In an attempt to maximize the available space, ancient remains were exhumed from the earth and their bones were sorted and packed into galleries known as *charniers*, built within the curtilage of the cemetery. The main area of the cemetery was also built up with soil carted in from elsewhere, forming a dome of earth up to six feet above the previous ground level. But this, too, soon had a surfeit of rotting bodies.

The Parisian dead were pressing hard upon the Parisian living. In 1780 a basement wall in a property adjoining Saints-Innocents gave way under the weight of the mass grave behind it, and bones and earth spilled into the domestic space. A radical solution was clearly needed – and it was at last seen that the quarry tunnels could be that solution, offering as they did a sepulchre of great volume.

So started one of the most remarkable episodes of Paris's history. In 1786 the process began of evacuating the city's cemeteries, crypts and tombs of their dead, and transferring the remains of more than 6 million corpses to the quarry region known as the Tombe-Issoire, soon to become Les Catacombes, on what was then the Montrouge Plain. A grim, ritualized production line was established for this task, involving diggers, cleaners, stackers, drivers, porters and overseers. Every night for years, horse-drawn funerary wagons containing the bones of the disinterred dead, covered with heavy black cloths, preceded by torchbearers and followed by priests who chanted the Mass of the Dead, clopped through the streets from the cemeteries to the Tombe-Issoire, where they disposed of their contents. Down in the tunnels, workers sorted the remains of the dead, filing them by bones into space-efficient ricks and stacks. Minor forms of folk art emerged in the disposition of these bones: serried ranks of femurs, their gleaming lines separated by rows of skulls, all turned eye sockets outwards.

A century later the photographer Felix Nadar would pioneer low-light photography techniques down in these ossuaries. One of his best-known photographs shows a worker pulling a bone wagon. It is an unsettling image. The wagon's wheels are wooden, its sides made of rough-hewn planks in which the grain stands clear. The man's face is scarcely visible, bleached out by the flash, and he wears a wide-brimmed leather hat and a loose white smock-shirt which has been – like his trousers – sewn together from patches. Underfoot he tramples ribs and tibias, and from the bone heap in the wagon white skulls stare over his shoulder at the tunnel space ahead. Later, Nadar would step into the basket of a hot-air balloon and photograph Paris from above, becoming also a pioneer of high-altitude photography – the first person ever to make images of a city from a moving vessel above it as well as in shadows deep beneath it.

The deposition of bones into the catacombs continued over the course of the nineteenth century, but quarrying dwindled away as the best limestone deposits became worked out. From the 1820s the quarry voids were put to a new use as mushroom fields: damp and dark, they provided the perfect growing spaces for fungi, which sprouted from rows of horse manure. Adaptable quarrymen made a career move into mushroom farming, and a subterranean Horticultural Society of Paris was founded, its first president being a former general inspector of the mines. By 1940 there were some 2,000 mushroom farmers working underneath Paris. During the Second World War the French Resistance retreated into sections of the tunnels in the months following occupation. So did civilians during air raids – and so, too, did Vichy and Wehrmacht officers, who constructed bombproof bunkers in the maze under the sixth arrondissement.

After the war, the cult of the catacombs began to grow. Increasing numbers of people were drawn down into them for purposes of concealment, crime or pleasure. These users of the network became known as 'cataphiles' – 'lovers of the below'. In 1955 access to the catacomb network was made illegal, outwith a small area of show ossuaries that were kept open for purposes of tourism. Attempts to police the space were formalized: specialist police – quickly nicknamed 'cataflics' and 'catacops' – were trained in the network's geography. Barrier walls were built across main subterranean routes, and the entrances to the network (tunnels, gates, manholes) were welded and locked shut. But the cataphiles kept coming. For the labyrinth offered a space where Paris's subcultures could go to grow. It became – and still is – what the anarchist-theorist Hakim Bey calls a 'Temporary Autonomous Zone': a place where people might slip into different identities, assume new ways of being and relating, become fluid and wild in ways that are constrained on the surface.

The arrival of the Internet further boosted cataphilia. Chatrooms and websites enabled cataphiles to share and curate information about the network. Cataphiles went by subterranean pseudonyms online – 'Styx', 'Charon' – and mildly fetishized the pseudo-covert nature of their activities. An unofficial cataphile uniform declared itself: thigh-high waders, small waterproof backpack, hoodie and head-torch. Serious cataphiles carried manhole-cover keys at their belts. There was a street of cafes and pizza establishments where it was – is – usual to see dozens of people in dark-green waders, waddling down the street or sitting at cafe tables, like a convention of trout fishermen far from any river. A commune culture emerged, with its own honour codes. The rules were few, and clear. Respect the past of the catacombs. Take out what you take in. Resources are to be shared, even with strangers. No selling and no buying:

barter-exchange or gift are the only acceptable modes of transaction. Help is to be given wherever necessary. Create with care – and do not destroy.

Some of the cataphiles went down to party. Others, though, became fascinated by the layered histories of the space. An unofficial 'university' of the catacombs was established, dedicated to the restoration, preservation and mapping of the network, and to the formal archiving of its stories. Once, a pop-up cinema was established in one of the chambers, and themed films were shown over several weeks – Vertov's *Man with a Movie Camera*, Lynch's *Eraserhead* – until it was shut down by the cataflics. New rooms continue to be dug by cataphiles and new nameplates added to tunnels. Work groups are established to add fresh layers to the catacomb palimpsest: large graffiti murals, new carvings, a sword buried in a stone, or mosaic works involving thousands of tiles.

Among the most resonant emblems of the contemporary catacombs is a sculpture known as *Le Passe-Muraille* ('The Passer-through-Walls') after a short story of the same name by Marcel Aymé, about a man who discovers he can pass through solid surfaces – only to become trapped when his powers desert him just as he is stepping out of a wall. The sculpture shows the man at this moment of simultaneous liberation and entrapment – his face, torso and one leg pulled clear from the masonry, his back and hands still immured. He is caught between realms, unsure whether to proceed into air or retreat into stone.

~

I come feet first through the ragged hole and drop into a straight-running tunnel, its ceiling sturdily arched. The limestone walls writhe with graffiti: antifa slogans, zombie skulls with popping eyes, tags, names.

'The further in we get, the better the wall art gets,' says Lina. 'In the Salle de la Plage you'll see Hokusai's *Wave*. Let's move. We've got miles to cover, and it's good not to linger near the entrance. Plus – there's the Bangra to negotiate first, which will slow us down.'

'The Bangra?'

'You'll see. We need to find somewhere to sleep tonight, in the next few hours. We've a long day moving north tomorrow, which may well bring some obstacles.'

I like the sound of sleep. I am exhausted by nerves and travel. My stomach flips a little at the mention of those obstacles. In the mountains I am used to having full foresight, to laying plans and assessing hazards myself. Down here, I am in Lina's hands – and the extent of my clairvoyance runs as far as the next turn in the tunnels.

Lina leads off, Jay follows and I bring up the rear. Lina moves fast, settling to a quick march down the dry tunnels. 'You have to move quickly if you want to cover any ground, get far in,' she calls over her shoulder. Soon the floor of the tunnel begins to muddy up, then dips into black water.

'Welcome to the Bangra,' calls Lina over her shoulder. 'It acts as a kind of airlock, or water-lock. Stops most people who get this far from coming any further.'

She wades into the murky water. We follow. It deepens rapidly to waist height. Our head-torch beams bob on the water.

'Feel at the edges of the tunnel with your feet,' says Lina, 'there are ridges there you can walk on.' She's right, and this lifts me further out of the water but moves my head closer to the ceiling. I have to crook my neck as I edge on through the water, which presses cold on my legs.

We slosh past flooded junctions, with tunnels cutting perpendicular to ours. I glance left and right; they disappear into darkness. I am beginning to comprehend something of the extent of the system.

The water level lessens, then shallows to nothing, and we are on firm ground again. Lina increases our pace. She doesn't pause at junctions: takes turns without hesitation. The unerring nature of her direction-finding reminds me of Neil's driving in the undersea maze at Boulby, hammering on with no doubt as to the true course.

We have been moving for a couple of hours when Lina stops, checks a mark on the wall and turns into a narrow side tunnel.

'Down here,' she says. 'This is where we'll sleep. It's called the Salle des Huîtres – the Oyster Room. The quarrymen used to shuck oysters down here; easy food for them, kept well in the pocket, all-natural packaging.'

Twenty yards down the tunnel is a roughly square hole cut into the right-hand tunnel wall, about four feet off the ground and about a foot and a half across.

'Welcome to your first *chatière*,' says Lina. '*Chatière* means cat-flap, as well as something less polite than that. There's a technique to getting through them. I'll show you.'

She posts her pack through first. Then she leans into the *chatière* as far as she can go with the top half of her body, feels backwards with her feet until she reaches the far wall of the tunnel, then walks her feet blindly up the wall, bracing her body until she is horizontal: head and shoulders in the *chatière*, feet against the far wall. Then she bends her knees, braces, and kicks off the wall, like a swimmer doing a turn in a pool, driving herself into the *chatière* and pulling herself onwards and through. I watch her feet disappear, impressed.

'After you,' I say to Jay, bowing. He mimics Lina's technique perfectly.

Of my own entry, let me say only that it is far less elegant and far more painful.

I pull through and find myself in a low-ceilinged room, five feet high at its highest, with chisel marks visible on the stone. The main chamber has a stone table thick with white candle wax. In its centre stands a plastic bong, bubblegum pink and shaped like a foot-long penis. Oyster shells have been arranged around it. The floor is covered in small spill-heaps of grey powder: the spent waste from carbide lamps. Leading off the chamber is an open doorway to a neighbouring room, off which another room leads. We explore the rooms: a dozen or so, roughly organized around a supporting central trunk of stone.

'People will probably come to use the party space later in the night,' says Lina. 'If we want any sleep we should get as far from it as we can.'

So we set up camp in a distant room. Its ceilings are low, three or four feet high at the most. We move about it on hands and knees. The air swirls with rock dust, which I can taste on my tongue and feel on my eyes. The upper city seems very distant.

On a clean-cut wall of stone close to the entrance to our room are lines of cursive handwriting in black ink or paint. They record the names of quarrymen, the dates of completion of rooms and tunnels, and numbers of metres of stone cut on different days. Years are written by the different lines, starting in the late 1700s and running to the early 1800s. There is a pride to the making of this archive – and care has been taken to preserve it, too.

'Respect for the ways this place was made is key down here,' says Lina. 'The community largely polices itself. If you disrespect the space and its history, word gets around, and life becomes difficult for you.'

In a niche in the chamber's main wall squat three big, jowled monkeys carved from stone blocks. Their eyes are holes. They regard us, impassive and sightless. A spider crawls out of the right socket of the central monkey, the boss monkey.

Other walls of the room are skilfully decorated with modern graffiti, including animals and human faces. Lina lights six tea-candles, placing one in each of the eye sockets of the monkeys in the niche, and their flames set the graffiti cave art flickering. The russet and black swirls now possess their own movement in the candlelight, seeming to shift within the stone. I can see how the graffiti artists worked the textures and shapes of the rock into features of their images, much as the prehistoric cave artists of Lascaux did: a curve of stone swelling the belly of a creature, an embedded shell serving as the eye or nose of a face.

I crawl to the back of the room and find that it extends into a low cave-like space, a couple of feet high and wide enough for a body. I settle down for the night there, oddly comforted by the sense of enclosure. *I found a Hollow place in the Rock like a Coffin . . . exactly my own Length – there I lay & slept – It was quite soft* . . . I take the bone owl and the bronze casket-egg out of my pack, and place them near my feet. I already know this isn't the place to leave the casket, but I am glad to have the owl with me. Sixty solid feet of stone extend above me. I think of passing through the openness of northern France that morning – of the sunset behind the unexplained earthwork.

We talk for a while in the candlelight, struck into closeness by the oddity of our dormitory. Then silence falls as tiredness does, with stealth and force. I drift into Escher-dreams of stairways that lead back on themselves, tunnels folding like Möbius strips, shifting rooms, and monkey gods with flames for eyes.

~

We think of cities as lateral but of course they are also vertical. Cities extend upwards into the air by means of buildings, elevators and controlled airspace, and they extend downwards by means of tunnels, escalators, basements, graveyards, wells, buried cabling and mine workings. Just as a mountain does not end at its summit or its foothills, but extends instead into the weather it creates in the air above it, and the orogeny of the rocks that have raised it, so a city does not cease either at its foundations or the spires of its tallest buildings.

Yes, each city has its invisible city, as Italo Calvino suggests in his great story of that name. Calvino's story is itself cleverly nested within tellings of tellings, stories of stories, so that the text possesses multiple versions of itself. In what is to me the most memorable section, the narrator describes the impossible city of Eusapia, in which inhabitants of the living city have accompanied 'an identical copy of their city, underground', a 'Eusapia of the dead' which can be accessed only by a confraternity of hooded brothers – though over time the symmetry between upper and lower cities becomes so acute that 'in the twin cities there is no longer any way of knowing who is alive and who is dead.'

Long before Calvino wrote his story, in one area of the Parisian catacombs, a quarryman and former soldier named Beauséjour Décure dedicated his spare time to carving intricate scale models of the Minorcan town of Port-Mahon into the living limestone. His work is eerily precise, his imagined edifices grand despite their size. He carved the town's front walls and main gate, its portal set within five receding frames of stone; he carved one of its great pillared buildings – neoclassical in style, with echoes of Pharaonic Egypt – raised off the rock, but with descending vaulted walkways that zigzag down into the stone, hinting at further depths buried out of sight within it. Seeking to bring more people to view his carvings, Décure set out to open an access stairway – but while carrying out his excavation he was killed by a cave-in.

Cities have long been vertical. When Christopher Wren excavated the foundations of Old St Paul's after the Great Fire he found a row of Anglo-Saxon graves lined with chalk-stones, beneath which were pre-Saxon coffins holding ivory and wooden shroud pins. At a still greater depth were Roman potsherds and cremation urns, red as sealing wax and embellished with greyhounds and stags, and beneath those were the periwinkles and other seashells that spoke of the ocean that had once covered the area. Below the San Lorenzo Maggiore cathedral in Naples, writes geographer Wayne Chambliss, 'there is a layer of urban stratigraphy containing an earlier, wholly intact iteration of the city. Streets, apartment complexes, storefronts, all filled in centuries ago and built over, have been unearthed below ground.'

The extent of our cities' verticality is growing rapidly. With the rise in the number and size of the planet's cities since the mid twentieth century, and with the development of new technologies, the heights and depths of our cities have stretched to astonishing magnitudes: Pierre Bélanger estimates that the 'infrastructure that supports urban life' now spans from '10,000 metres below the sea to 35,000 km above the surface of the earth'. Stephen Graham also documents this elongation of city space into the air above and the earth below:

> Complex subterranean spaces below major cities . . . are themselves three-dimensional labyrinths which stack and intertwine infrastructures and built spaces as deeply as many cities rise into the sky . . . Major cities are thus increasingly organised as multi-level volumes both above and below ground.

Such a densely stacked modern cityscape leads, inevitably, to a new geography of inequality that requires reading in vertical terms. Broadly speaking, wealth levitates and poverty sinks. Privilege prefers to distance itself from the mess of the street by means of altitude – those fiftieth-floor infinity pools, those rooftop penthouse suites – and tends only to delve below ground when that delving offers security or privacy (the deep-sunk document stores maintained by US security firms such as Blackwater, for instance, or the oligarch basements dug down into the London clay at Mayfair and other high-end, low-rise residential districts of London).

Poverty, by contrast, pulls people down, pools low. This is the verticalization of wealth and power foreseen by H. G. Wells in his 1895 novel *The Time Machine*, with the mining subterranean Morlocks and the delicate above-ground Eloi. Today, in the network of storm drains beneath Las Vegas, a shaken-out sub-population has taken up residence, including people struggling with addictions as well as homelessness. When rain does fall in that glittering desert city, and flash floods fill the storm drains, the livelihoods and sometimes the lives of these people are washed away. In India's cities, sewers and septic tanks are often cleared by thousands of daily-wage workers, lowered on ropes to scoop out by hand and bucket the human waste, trash and congealed fat that accumulate in such places. When a manhole cover is first shifted to give access to a sewer, workers are glad to see flies and cockroaches swarm from the opening: it means that toxic gases have not gathered to a lethal degree. The life expectancy of these men is around ten years less than the national average. Hundreds die each decade by asphyxiation or drowning, their deaths typically unrecorded and unrecompensed.

Poverty and powerlessness have characterized the history of Paris's tunnels, too. Walter Benjamin worked hard in *The Arcades*

Project to retrieve the obscured histories of those spaces. He documents how, for instance, after the June Insurrection of 1848, those who had been taken prisoner were moved around the city by means of the quarry tunnels and the catacombs, shifted along the network from fort to fort in order to keep them secure and invisible. The catacomb labyrinth had become what we would now call a 'dark site' – an extra-jurisdictional space where special rendition of political prisoners could occur, out of public sight and mind.

I find Benjamin's historiography to be compassionate in its preservation of the details of the underground experiences of these prisoners, and others like them. 'The cold in these underground corridors [was] so intense that many prisoners had to run continually or move their arms about to keep from freezing,' he notes of the insurrectionists, 'and no one dared to lie down on the cold stones.' He preserves, too, moments of solidarity and companionship: how 'the prisoners gave all the passages names of Paris streets, and whenever they met one another, they exchanged addresses'; and how, in the eighteenth century, prisoners waiting in incarceration cells to be chained as oar-slaves in the galleys that worked the Seine would sing to one another, communicating by melody in the blackness.

~

We sleep late the next morning, and eat chocolate for breakfast while the monkey gods watch us through charred eyes.

'Time to move,' says Lina. 'We've got a rendezvous this evening way north of here, with some friends of mine, in the Salle du Drapeau. There'll be good things if we can get there – but that's going to depend on the stability of the ceilings, and whether there have been any collapses since I was last along that route. And before then, there are places I want to reach.'

We push back through the *chatière* – feet first this time, bending around and down to find footing in the passage – and then it's off at Lina's quick-march pace, charging the dry tunnels, wading the wet ones, picking carefully past well-shafts, pushing north by north-west. Again I am amazed at Lina's ability to navigate without consulting the maps we carry. She seems to have internalized this three-dimensional maze, or developed a below-ground mental GPS.

Late in the morning we drop down a set of stone staircases, moving between levels of the labyrinth, to a point named on the maps as Hell Well.

'Here's Hell,' says Lina. 'This is not an easy place in any sense.'

She points to a low tunnel leading off the main passage; an access adit perhaps two feet high. 'Through there,' says Lina. 'You go first, Rob. You'll need to lie on your back to make it.'

I lean back, reach under and ahead of me, find the adit's edge with my fingers, pull through, look up and stop dead.

I am in a vertical shaft and above me is a suspended wall of clay and earth, perhaps ten feet high, into which hundreds of human bones are embedded: skulls, ribs and limbs. In the belly of the well below are hundreds more fallen bones. It is a point where a burial ground has begun to disgorge its contents down through a breach in the tunnel network. The rough limestone from which the shaft has been hewn is also visibly thick with bodies – whelks and spiral shells that are fossilized uncrushed in the stone's sediment – and I have a sudden sense of both cities, upper and lower, as a single necropolis. *The city of the dead antedates the city of the living . . . is the forerunner, almost the core, of every living city . . .*

Lina and Jay pull themselves one by one into Hell Well. Afterwards, we speak little as we continue to traverse the passageways. In that region of the catacombs bones are profuse. There is no order to death here, no names or memorialization, just containment. Occasionally we pass under a vertical circular shaft leading up through bedrock to a manhole cover in the street. Some have ladder rungs. I pause under one and can see distant glimpses of light, hear faint clanks as the cover is moved by the footfall of pedestrians going about their upper-world business.

Once, in a long and boneless tunnel, I see flames flickering far ahead of us. Then the flames abruptly disappear. Lina sees them too, but when we reach the point of their vanishing there is no side tunnel into which they could have turned. 'The lamps of other cataphiles,' says Lina uncertainly. 'Though I don't see where they could have gone.' Then she smiles. 'Or perhaps the ghost of Philibert Aspairt, lost down here in 1793 and not discovered until eleven years later. Dead, obviously. Arguably the world's first urban explorer, and probably the worst.'

~

For some years before coming to the catacombs, I had been finding my way into the subculture of urban exploration: this was how I had come to know Lina. Urban exploration might best be defined as adventurous trespass in the built environment. Among the requirements for participation are claustrophilia, lack of vertigo, a taste for decay, a fascination with infrastructure, a readiness to climb fences and lift manhole covers, and a familiarity with the varying laws of access across different jurisdictions. Among the sites favoured by urban explorers are skyscrapers, disused factories and hospitals, former military installations, bunkers, bridges and storm-drain networks. A serious explorer needs to be content sitting on the counterweight of a crane 400 feet above the street, or skanking along a sewer twenty yards under the asphalt. Urban explorers shun the *Sturm und Drang* of mountains. Their thrills are niche, their epiphanies mucky. Rumours circulate about entry points which might give access to unseen spaces. Secrets are jealously guarded, closely shared.

The subculture has its subcultures. Just as certain climbers prefer granite to gritstone, and certain cavers prefer wet systems to dry ones, so explorers have their specialisms: there are bunkerologists, skywalkers, builderers, track-runners, 'drainers'. Most explorers start out in ruins, though: these tend to be the easiest sites to access, and the cheap aesthetic pay-offs – the pathos of abandonment, the material residue of inscrutable histories – are rapid, usually claimed in the coin of photographs. Ruinistas dig 'derp' (explorer's slang for 'derelict and ruined places'). Detroit was the world mecca for derp, until it became a city-sized version of Don DeLillo's 'most photographed barn in America', cloaked in a haze of voyeuristic ruin-porn imagery (HD stills of dusty ballrooms and atria, with artfully scattered detritus in the foreground, images that erase a hundred aspects of that city's hope and despair).

Urban exploration is international in its geography, with groups, crews and chapters around the world. There are a surprising number of female explorers, and the class base is mixed, drawing often on a disaffected and less legally obedient demographic. In Brisbane an explorer known as Dsankt boats into the underland like some latter-day Charon, boarding small skiffs in the rivers on the city's outskirts, following tides up the intake valves and into the abstract zones of the sub-city. In Canada, an explorer has penetrated the network of surge pipes that service the Ontario Generating Station at Niagara Falls: huge-gauge tunnels of riveted iron, with water-filled penstocks dropping vertically down from their floors. In the white sandstone under Minneapolis, digging teams work in shifts to open routes into new caves. In New York City, explorers ride the buses with faces pressed to the window glass, scouting out trunk conduits and side pipes by means of their street-level outlets, scribbling maps on notepaper or tablets as they go. In Madrid, drainers track the disappearance of streams and creeks as they reach the fringes of the city and get culverted down and under.

At the avant-garde of urban exploration are the infiltrators, the 'real' explorers, who tend to be more stimulated by systems and networks than by single sites, and who cherish the challenge involved in accessing super-secure locations. Like extreme climbers, infiltrators experience what Al Alvarez called 'feeding the rat', in his classic essay on climbing and fear. They are obsessives: they develop tunnel vision. They run tracks in the brief gaps between trains, they take dinghies down storm drains, they lift-surf – and occasionally they die. At its more political fringes, urban exploration mandates itself as a radical act of disobedience and liberation: a protest against state constraints on freedom within the city. Just as psychogeographers in the original Parisian Situationist vision of Guy Debord sought to discover astonishment on the terrain of the familiar by breaking out of the grooves of behaviour defined by capital, so politicized urban explorers present their trespasses as activism that 'recod[es] people's normalised relationships to city space'.

There are aspects of urban exploration that leave me deeply uneasy, and cannot be fended off by indemnifying gestures of self-awareness on the part of its practitioners. I dislike its air of hipster entitlement, its inattention towards those people whose working lives involve the construction, operation and maintenance – rather than the exploration – of these hidden structures of the city. I am sceptical of the dandified nature of its photographic culture, which seems chiefly to refocus the problems of Caspar David Friedrich's iconic 1818 painting, *Wanderer above a Sea of Fog*. And I feel uneasy at the opportunities urban exploration holds for insensitivity to those people who have no choice but to exist in contexts of dereliction and ruin.

Other aspects of the subculture have come to compel me, though, and so I began – cautiously – to spend increasing amounts of time with those who self-identified as explorers. I was especially struck by the manic systematicity of much explorer practice – its devotion to making visible 'the black box of infrastructure' and the 'dark fibre' of modern information exchange. I liked urban exploration's awareness of the porosity of the city's fabric, the proliferation of portals, rifts and drifts it perceived, and also its sense of sub-cities – like natural underlands – as spaces existing in long-term, slow-motion flux. And I was fascinated by urban exploration's pre-modern precedents, and the ways they intersected with histories of poverty and hope within cities: the Victorian mudlarkers and sewage pickers, for instance, who wandered the tunnels and outfalls of London's sewer system, holding their lamps aloft in the stench, sieving gold teeth and pearl earrings out of the ordure.

The poet and naturalist Edward Thomas seems to be far distant from urban exploration, but one day I discovered a passage from a 1911 essay in which Thomas imagined an abandoned London where he was free to adventure through the city's infrastructure, both above and below ground. 'London deserted would become a much pleasanter place,' wrote Thomas with a misanthropic flourish. 'I like to think what mysteries the shafts, the tubes, the tunnels and the vaults would make – and what a place to explore!'

~

Early that afternoon, Lina leads us to a vault the name and location of which I am forbidden to disclose.

We wall-kick one by one through a high *chatière* and find ourselves crouching in a desert zone. The floor is a rolling dunescape of stone-sand, compacted together and hardened over the centuries.

In places the dunes rise close to the ceiling. In places they meet it. Elsewhere they dip to leave crawl spaces a couple of feet in height, just large enough to admit a human body. From where we crouch there are seven or eight possible routes leading away from us, and each one in turn splits and spreads away. It is a threatening maze, and it reminds me of the boulder ruckle under the Mendips. Here, though, there is no Ariadne's thread to be followed.

Lina is brisk, all business. 'We leave our bags here. It's impossible to move with them on where we're going. Follow me.'

She slithers off on her belly along the dunes, heading for one of the right-hand crawl spaces. Jay and I slither after her, using feet and hands to move ourselves on, lizarding up and over a pass in the dunes where there is only just room to ease my skull between floor and ceiling. I try to move fast enough to keep Lina's boots in view.

Another rise in the dunes brings us still closer to the roof, and I feel my skull scrape on rock as I ease through, head turned sideways for clearance, face pressed against the stone-sand. Lina pauses for thought at only one junction, and then on we snake for ten minutes, until a dune slopes away to a black rabbit-hole, down which we each go head first.

I pop out of the rabbit-hole into a *Wunderkammer.*

We are in a chamber that is cuboid in form, each of its vertices perhaps twelve feet long. Its walls are of clean-cut yellow stone, its floor is curiously swept and it contains nothing at all except for a slender staircase of stone, which advances out of the far wall like the approach to a ziggurat. Each step of the staircase is marked on its side in black handwriting. Placed in the dead centre of each step of the staircase is a sample of stone, crystal or metal, each a different colour: white sandstone, yellow sandstone, quartz, limestone.

Lina looks rightly proud to have found the chamber and to be able to show it to us. 'This we call the ——,' she says. 'There are other examples of similar rooms spread out through the labyrinth, but this is the best, and also the least known.'

It is a Cabinet of Mineralogy – a teaching room from the period when the catacombs were part of the real estate of Paris's School of Mines. The room has been more or less undisturbed since it was closed at some point in the early 1900s. There is an austerity to the room's structure, and an elaborate care to the ritual placings of the samples – each on its own swept step.

We sit for a while in the Cabinet of Mineralogy. We eat, drink, rest, talk. Lina tells stories of her explorations in the vertical city. She describes climbing one of the chimneys of the Battersea Power Station, and then leaving the power station by means of an underground tunnel system, out of which she popped up in the middle of the Chelsea Flower Show, grimy and wide-eyed among the aspidistras.

Lina's great wish as an explorer is to enter the Odessa catacombs. Odessa, like Paris, is a city built on limestone, and it contains the world's most extensive sub-urban quarries. Some 1,500 miles of tunnel make up Odessa's invisible city, sinking to a depth of 160 feet over three levels. I have seen maps of the Odessa labyrinth. More even than the Parisian network, it has the improvised appearance of an organism or organisms: the branching structure of coral, perhaps. When the Germans were closing in on Odessa during the Second World War, the Soviets left behind Ukrainian rebel groups hidden under the city in the catacombs. Some of these stay-behinds remained below ground for over a year, suffering from malnutrition, malaria and vitamin deficiencies, occasionally surfacing to seek information or make attacks. Cat-and-mouse games were played between occupiers and the rebel groups; the Germans gassed and bombed the tunnel systems in an attempt to kill the Ukrainians. After the war in Odessa, the underworld moved into this underland, and smugglers and criminals enlarged the network for their own purposes.

'The Odessa tunnels make ours here in Paris look like a side-show,' says Lina. 'But it's dangerous there. Especially for a woman. There are bad stories about what can happen there, about what *has* happened. Definitely murders, probably at least one death through simply getting lost.'

Jay tells a story of taking a party of three novices caving in Aggy, a Welsh cave system, access to which begins with a notorious long rift so narrow that it is almost impossible for a human who has entered it either to turn around or to be passed. That day, said Jay, one of the novices got stuck in the rift, and began to panic. Her name was Luna, and she was a professional dominatrix who plied her trade from a basement beneath Baker Street.

'I'd assumed, given her day job, that she was at ease both with confinement and underground space,' says Jay. 'Not so, as it turned out. It took three hours to rescue her. I couldn't get past her, so I had to exit the system another way, then come back along the rift so

I could speak to her face to face, calm her down, help her work out how to free herself and keep moving forwards. Then I had to reverse down the rift myself, moving backwards, keeping her talking the whole time. I distracted her by asking for details of her dungeon's price list. An eye-opening variety of services were on offer down there.'

'Enough of this,' snaps Lina. 'We have an appointment to keep in the Salle du Drapeau.'

~

Among the people with whom urban exploration brought me into contact was a Californian called Bradley Garrett. Bradley saw cities more vertically and porously than anyone else I knew. To his eyes, the city was full of portals – service hatches, padlocked doorways, manhole covers – that lay unseen in plain sight. The usual constraints on urban motion – enforced by physical barriers, legal proscription or internalized notions of property rights – didn't tend to restrict Bradley. To him the city's accessible space extended far down into the earth (sewers, bunkers, tunnels) and far up into the air (skyscrapers, cranes), with the street level only serving as a median and somewhat tedious altitude.

We first met each other early one afternoon at London Bridge. Bradley had thick-rimmed black glasses, a goatee and moustache, and chin-length dark brown hair that he banded back into a ponytail. His speech mixed West Coast dude-isms with the gnarly syntax of cultural theory. 'London Bridge is hollow, like all big bridges,' he said, tapping his foot on a utility hatch set into the pavement about two-thirds of the way along the bridge. 'There's a control room at the north end; if you get into that, you can cross the Thames *inside* the bridge. It's neat. Come, I'll show you.'

At the north end we hopped a low iron gate and descended a staircase that led down the bridge's flank. Set into the flank was a steel security door sporting a chunky yellow padlock. The door looked as if it could withstand a lightsaber attack, and wore various notices explicitly forbidding entrance. Bradley pulled a ring of keys out of his pocket, murmured to himself as he sorted between them, chose one, leaned close and the hasp of the lock clicked open. He ushered me inside and shut the door with a soft clang behind us.

'That's some bunch of keys you have there,' I said. Bradley flicked on a head-torch. We were in a control room of some kind. Zinc venting, ducts and multicoloured wiring lashed with cable ties leading out of the room along a crawl space. Two wall-mounted dashboards with analogue switches and twist dials.

'So – if you follow this ducting south out of here down the crawl space, then you're fully inside the bridge,' Bradley said. 'Keep going all the way over the river, and you reach a much bigger control room at the south end. Hit the exit bar on the emergency door there from the inside, and you can let in who you want. When we made a film about exploring a few years ago, called *Crack the Surface*, that's where we held the premiere. We had eighty-six people, a generator, a screen, a projector and a whole lot of beer. It was a great party!' We slipped back out and Bradley locked up. Two passing men in suits gave us puzzled looks but didn't break stride.

Bradley's disobedience towards the usual rules began early. He grew up in a rough neighbourhood in Los Angeles, and was stabbed in the belly as a teenager. 'It grew me up, that stabbing,' he said. 'Got me out of trouble, oddly. Made me long to get out of those streets, to somewhere more open.' In 2001, aged nineteen, he co-founded a skateboard shop in the city of Riverside. He sold out to his partner two years later, and used the money to study maritime archaeology in Australia. Then – in search of some seriously empty space – he moved back to northern California and began working for the US Bureau of Land Management, specializing in the archaeological heritage of Native American groups. Then he moved to Mexico, spending three summers as an archaeologist excavating a post-classical-era village, and camping on the edge of a *cenote* – one of the flooded sinkholes that plunge into Mexico's natural limestone underland.

'It was such a place to live, Rob,' Bradley said as we walked through London. 'Bats came pouring out of the *cenote* in their hundreds at dusk each night, and pouring back just before dawn. The leathery noise of their wings kept time for me. That *cenote* was understood by the local indigenous people to be an access point to the Mayan underworld, to Xibalba. In Mayan, Xibalba means "place of fear". The whole limestone underworld of Mexico is a massively devotional terrain. Down there, where the water levels have risen, you sometimes swim past sunken altars, entrances to religious chambers that have been cut out of the stone.'

He described Xibalba to me as it was represented in K'iche' Maya myth. Even within the wider context of torturous underworld myth-zones, Xibalba sounded like a brutal realm. It was heavily staffed by demons with names like 'Flying Scab' and 'Stabbing Demon'. Just to reach Xibalba, you had to cross a river filled with scorpions, a river filled with blood and a river filled with pus. If you were lucky enough to make it that far, you were then tested in the six deadly Houses of Trial, including 'Bat House', filled with flesh-eating bats, 'Razor House', filled with unpredictably moving blades, and 'Jaguar House'.

'You can probably guess what that one was filled with,' said Bradley.

After Mexico, Bradley moved to London, where he wandered over disciplinary boundaries into cultural geography. Studying for a doctorate, he became fascinated by urban exploration, and decided to immerse himself ethnographically in its subculture. His research method was nothing if not committed. He spent four years embedded with a group of London-based explorers only ever identified by aliases (Patch, Winch, Marc Explo), with whom he learned the ropes, and ticked off the London ascent and descent classics including Battersea Power Station, Millennium Mills and the buried river of the Fleet.

Two years later, Bradley's group merged its efforts with another exploring team to form the London Consolidation Crew, which soon became known for the audacity and ambition of its exploits. The intensity of their activity increased and the rats inside them grew, fed regularly on adrenaline. In that time Bradley took part in more than 300 trespass events in eight countries. In America, he climbed a Chicago skyscraper in a storm and gained astonishing photographs of a city bathed in black cloud and blue light, with lightning strikes fissuring down from the clouds into Lake Michigan. In the Mojave Desert, he accessed a boneyard of decommissioned aeroplanes: climbing over barbed wire to gain entry, then hiding in the landing gear of 747s and military cargo-carriers while security patrols passed by. 'It was,' he noted drily, 'a vast playground and a long night.'

I was somewhat sceptical of Bradley at first. But as I came to know him, I grew to like and admire him enormously. He cut a broad swathe through life and he drove deep. He was generous, unpredictable, fearless, loyal – and a great deal of fun to be around.

The rest of that day Bradley and I spent together in London took us often into the capital's invisible city. We accessed the network of steam tunnels that runs beneath the Barbican. We lifted a manhole cover to drop into the course of the Fleet, one of London's so-called 'ghost rivers', with the aim of reaching the Fleet Chamber, a Bazalgettian structure near the river's outfall into the Thames. And in a north London park, we crawled under fencing, pulled away a heavy iron lid to reveal a shaft in the grass, and descended a rusted black ladder into darkness.

Twenty feet down we flicked on our head-torches – and whistled at what we saw. Dozens of brick archways extended in series away from us, the dips between them holding wide rungs of still water. The iterated forms of the archways and the reflections of the water created the illusion of infinite regress. The echoes of our whispers bounced back to us. We had entered a mid nineteenth-century reservoir, built to serve as a water depot for London and now drained almost dry. The once-drowned structures were still intact, the brick as clean as if it had been built yesterday. It possessed the functional elegance of major Victorian infrastructure, and was as beautiful in its way as the Roman cisterns at Misenum and the Basilica Cistern in Istanbul.

We walked the reservoir end to end and side to side, our voices booming. Above us in the shadows hung the ceiling vaults, made of tens of thousands of yellow-brown bricks. At the far end of the reservoir we sat for a while. Bradley smoked and set music going: a drum-and-bass track called 'Stresstest' that boomed off the bricks. We got out just before midnight. There were scattered clouds, underlit pink and orange by the city's light, with stars visible between them. Three figures moved slowly through the trees to our east, scanning the grass with yellow beams, looking for something lost.

After that first day together, Bradley and I became good friends. His pursuit of a number of the 'ghost stations' of the London Underground, and the track-trespass that this required, as well as a number of other incidents, placed him in the sights of the British Transport

Police, who decided to make an example of him *pour encourager les autres*. He was arrested, his flat was searched, his computers and phones seized, and eventually he was placed on trial on a charge of conspiracy to commit criminal damage. I acted as a character witness during his trial – which concluded with Bradley being granted conditional discharge and facing no further charges, a public relations disaster for the Transport Police, and six-figure legal costs for the taxpayer.

Bradley and I made a number of exploration trips together, and while planning these trips we communicated by postcard, on the grounds that this open form of correspondence – readable by anyone who cared to pick our postcards up and flip them over – was the most secure way to be in contact, given the authorities' interest in Bradley. No security agency still steams open letters or reads people's postcards; instead they watch text and WhatsApp conversations, and packet-sniff emails.

Travelling with Bradley both deepened and heightened my sense of landscape, and of the built environment in particular. We found our ways into many strange sites and places. As well as a daredevil adventurer's streak, Bradley had an archaeologist's interest in contemporary forms of obsolescence, and a natural historian's interest in how the wild returned to abandoned places.

One night we set out to climb a transporter bridge in Newport, ascending the supply staircase and then inching out along its trunk-thick cables strung above the dark river. With a young explorer who called himself Darmon – who specialized in accessing high-security underground sites in high-risk territories including Russia and China, who had been beaten up by authorities in both countries, and who sourced his interest in the underland to the Roman coins his father, a farmer, had ploughed up in his fields on the banks of the upper Thames when Darmon was a child – we climbed a twelve-foot iron gate to access the jackdaw-haunted ruins of an abandoned Victorian castle that spread across acres of hillside above the Irish Sea. We tended to sleep out on these trips, under hedges or farm trailers – or just not to sleep at all. I came to associate time away with Bradley with adrenaline, alcohol and extreme fatigue.

One trip with Bradley took us into the abandoned slate mines of a Mid Wales valley. We accessed the mine by a narrow adit, which led us to the top of a quarried cliff. I set up a top rope in the darkness, and we abseiled down the cliff to its base. From there a tunnel brought us out at the base of a great flooded chamber. Black water lapped the slate at our feet – and from seventy feet above us, through an opening in the rock, a shaft of golden sunlight lanced down into the chamber like an annunciation beam.

What that sunbeam gilded was far from sacred, though – for through the same opening that let in the light had been pushed, over more than forty years since the closing of the mine, hundreds of wrecked cars. Generations of locals looking to dispose of their defunct vehicles without paying a scrappage fee had driven the cars up the hill and through the gap.

The result was an avalanche of vehicles, a carchive, a slewing slope of wrecks that dropped into the chamber and continued into the black water as far as we could see. The oldest cars were the furthest down, and on the lowest stratum, a blue Cortina estate was poised as perfectly as a glacial erratic atop the moraine, with a moss-green Triumph Herald both its pivot and its point of repose.

~

The approach to the Salle du Drapeau – the Room of the Flag – is the only time I feel real fear in the Parisian catacombs.

It is early evening in the upper city by the time we get close to the room. On the surface, people are leaving offices, walking home through dusk streets, boarding trains and buses, stopping for drinks in bars.

Down in the invisible city we are heading north-west along a tunnel with no side turnings, the ceiling of which is dropping steadily lower. I walk with bent neck, then with hunched shoulders, then I have to lean at the waist, and then at last I have to drop to my knees and can only crawl forwards.

Ahead of me, past Lina, the tunnel seems to cinch to a dead end. I wait for Lina to admit that she has at last led us the wrong way.

Lina says nothing. The yellow of the limestone ahead glows in her torchlight. She shrugs off her pack, pushes it behind her, loops one of its straps around one of her ankles, and then eases herself head first into what I can now see is a tiny floor-level opening, perhaps eighteen inches high, where I thought the tunnel ended.

My heart shivers fast, and my mouth dries up instantly. My body does not want to enter that opening.

'You'll need to pull your pack along with your toes here,' says Lina. Her voice is muffled. 'And from now on don't shout or touch the ceiling.'

Fear slithers up my spine, spills greasy down my throat. Nothing for it but to follow. I lie flat, loop pack to foot, edge in head first. The clearance above is so tight that I again have to turn my skull sideways to proceed. The clearance to the sides is so scant that my arms are nearly locked to my body. The stone of the ceiling is cracked into blocks, and sags around the cracks. Claustrophobia grips me like a full-body vice, pressing in on chest and lungs, squeezing breath hard, setting black stars exploding in my head.

Drag-scratch of bag behind me, pain already in the leg to which it is looped from the effort of pulling it. Movement is a few inches at a time, a worm-like wriggle, gaining purchase with shoulders and fingertips. How long does this tunnel run like this? If it dips even two inches, I'll be stuck. The thought of continuing is atrocious. The thought of reversing is even worse. Then the top of my head bumps against something soft.

Ahead I can just see, by cocking my neck back, that the underside of Lina's rucksack is jammed against the dipping edge of a block in the ceiling. The pack is jerking around, trying to get free, she must be hauling at it with her leg, but it looks as if it could loosen the block at any moment, bring the ceiling down.

'Easy, *easy*!' I shout and she shouts back, telling me not to shout. *Pop* – and the bag comes free, slithers on.

I shuffle forwards towards the pinch when suddenly – *what the fuck?* – I can feel the stone around me, the stone that encases me, the stone that is measuring me up like a coffin, *starting to vibrate*. A faint shudder at first, but clear and now growing in strength and noise. The ceiling, the unstable ceiling, is humming with tremor. The vibrations are passing through the stone and my body, then on into the stone beneath me. The rumbling rises to a thunder and I can hear clacks and clicks among the rumblings and I remember the spectre-architecture, the faint grey outline of the upper city on this page of the map: train lines arcing in, joining like tendons and running together into Montparnasse station.

These are *trains above us*, we are *directly underneath* the Métro and overground lines, and it is decades of train-judder that has left the ceiling unstable here. I want to shout but mustn't, want to retreat but can't, so I just keep inching forwards, stone dust in the mouth, finger-scrabble against the rough rock, hauling the bag behind, all in silence, just the rumble of the trains rising and falling away, heaving breath, drumming heart, and then after five minutes of that sick-making fear, the space widens and lifts and then we can kneel again, and then we can stand, and then we can walk and then we are close to the Salle du Drapeau.

~

A flooded tunnel leads towards a chamber. Orange light on the water, washing and rocking although the water itself is still. Cries come from through the doorway, and there is the sound of music: the Jam's 'Going Underground', growing in volume, booming down the tunnel. I smile in recognition at the music, bridge onto ledges on either side of the flooded tunnel, and reach the doorway. It opens into a high-sided room, the roof twenty feet or more above us. The space above makes my head feel as if it is helium-filled, floating. A big tricolour flag is painted high on one of the walls. And there are people standing up to greet us: embraces for Lina, shakes of the hand for me and Jay, welcoming smiles for us all.

We have found our way to a different kind of *Wunderkammer* here, one filled with music and hospitality. There is a table spread with food and drink: fruit, baguettes, wheels of Brie and Camembert, bottles of spirits, cans of beer. A boxy CD player sits in the middle of the table, wired up to two small speakers.

The Jam changes to Bowie's 'Underground'.

'*Ça c'est le cataboum!*' says one of the strangers, pointing at the music box, nodding in time to the beat.

White fairy-lights are strung around the room. It is all deeply surreal – as if we have stumbled into a postmodern mead hall, far underground. A plastic glass of vodka is pressed into my hand and I knock it back gratefully. Burn in the belly, and the time in the train-rift instantly softening around its edges. My glass is refilled with brown rum from a label-less bottle. I catch myself grinning. I feel grateful for this place, for the juxtapositions of the catacombs, tilting from terror to warmth in the twist of a tunnel.

Introductions are made. There are two French cataphiles who go by handles I do not catch, and a Canadian named T, who is an old friend of Lina's, works as an au pair during the day and comes down into the catacombs often at night. All three are wearing Indiana-Jones-style leather hats, and one of the Frenchmen has a whip in his hand.

Bowie changes to 'Underground' by Ben Folds Five. Everyone cheers.

We eat more, drink more, talk more. Hours pass. I mostly listen, relaxing after the day's exertions, pinching myself at the weird sub-cultures of this underland, reflecting on the bizarre cultural recyclings that it calls out.

Much later, Lina, Jay and I set off to find a sleeping place. We reach a zone called the Bunkers. A wide tunnel-avenue is lined with a series of hooped semicircular chambers with reinforced ceilings. They are of Second World War origin, says Lina, shelters adapted to resist bomb-fall. Early in the occupation they were used by the Resistance; later they housed high-ranking SS and Wehrmacht officers when the raids came from southern England. Now they make ideal dormitories for tired cataphiles. There, with a bunker-chamber each to ourselves, we settle down. The distant passage of trains vibrates the walls.

Sleep takes time to arrive. Lying there with bedrock in all directions, I wonder at what will remain of our cities as the Anthropocene unfolds over deep time – the stratigraphic markers that will endure in the rock record. Over millions of years, the inland megacities of

Delhi and Moscow will largely erode into sands and gravels, to be spread by wind and water into unreadable expanses of desert. The coastal cities of New York and Amsterdam, those claimed soonest by the rising sea levels, will be packed more carefully into soft-settling sediments. It is the invisible cities – the undercities – that will be preserved most cleanly, embedded as they already are within bedrock. The above-ground structures we have built will collapse to form jumbled urban strata: medleys of concrete, brick and asphalt, glass compressed to a milky crystalline solid, steel dissolved to leave trace impressions of its presence. Below ground, though, the subways and the sewerage systems, the catacombs and the quarry voids – these may preserve their integrity far into a post-human future.

~

Two days later, we are ready to leave the invisible city. The original plan was to exit by means of a laddered manhole, which Lina has been told is presently unwelded. Its nickname is 'the *Chatière* of Death', which doesn't endear it to me. But the directions Lina has received as to its whereabouts are vague, and we can't locate it.

So we return to our point of entry. Hours of tiring travel through tunnels from the far north-west of the system. Lina plots a long way round that circumvents the crawl space leading to the Salle du Drapeau. We see no one else in the course of our traverse. Once we pass a stretch of tunnel wall on which dozens of hand stencils have been made with spray cans in acid green, ice blue, nuclear yellow; punk echoes of prehistoric cave art. We come back through the Carrefour des Morts, and back at last down the Bangra, in which the water level has risen noticeably since we first crossed it days earlier.

'It's been raining up there,' says Lina.

I remember the thunderclouds building as I approached Paris, the rain-veils trailing over the landscape. We reach the access hole and climb, one by one, up and out into the railway tunnel.

After days of confinement, its arched ceiling seems huge as a ballroom. The air is free of stone dust. Away to our left is a familiar arch of light. We crunch back down the track. The arch grows and brightens. Green fringes it, hanging down in long lianas, and green is a new colour again.

'Look at the butterflies,' says Lina, pointing, and there are dozens of gold butterflies filling the air of the arch, but as we approach they turn into falling acacia leaves spinning down from unseen trees, gilded by afternoon light.

The upper world moves into view. A pigeon glides, stiff-winged, across the arch-framed sky. Steep sides of the cutting show themselves, acacia branches leaning in from the banks to drop their butterfly leaves.

We stop at the point where the light meets the shadow, look up, and there is the impossible sun, soon to drop below the buildings that rise above the cutting sides. We speak quietly to one another. Our hair is pomaded with sweat and stone dust, our skin is pale. The air here in the open smells of cucumber and smoke. A woman is hanging white sheets out on the balcony of one of the apartments high above us.

I hear the first bars of Emil Gilels playing the Brahms Piano Quartet No. 1, one of the few pieces of classical music I know from even a scatter of notes. The notes drift down along with the leaves to gather in the cutting, and I think I am dreaming the music, but the others can hear it too and it is extraordinary to me that someone should be playing this now, here.

We walk on. Two teenagers, a boy and a girl, sit on a substation box under an acacia, long brown legs kicking as they chat and pass a joint between themselves. They nod as we pass, and we nod back.

We scramble up the cutting side, through the hole in the chain-link fence, out by the doorway marked '*Interdit d'entrer*'. On a street corner three turns from the doorway, a woman stops us to ask if we have come from *en bas*, from 'underneath'. Yes, we say, we have.

6
Starless Rivers
(*The Carso, Italy*)

Starless rivers run through classical culture, and they are the rivers of the dead. The Lethe, the Styx, the Phlegethon, the Cocytus and the Acheron flow from the upper world into the underland – and all five converge in a welter of water at the dark heart of Hades.

The waters of Lethe are the waters of amnesia, from which the shades of the dead must drink in order to forget their earthly existence. The Greek word *lethe* means 'oblivion' and 'forgetfulness'; it is countersaid by the Greek word *aletheia*, meaning 'unforgetfulness', 'unconcealment' and also 'truth'. By means of the Lethe, Aeneas is able to travel to meet the ghost of his father – one of the many souls that throng the flood – in the great *katabasis* of Book VI of the *Aeneid*.

Charon, the ferryman, carries souls of the newly dead across the Styx; he requires, for safe passage, an obol, or coin, to be placed on the lips of the deceased in order to pay for transport to the underland.

The Phlegethon is the river of heat, of flaming fire and boiling blood, which is thought to flow in coils and spirals, descending into the depths of Tartarus, the abyss of the damned.

The Cocytus is the coldest of the five, the river of lamentation, scoured by freezing winds, hardened in places to ice. Where the Cocytus runs, its currents call out constant cries of pain as they tumble over rapids and swirl around bends.

The Acheron is the gentlest of the starless rivers, the river of woe, over which Charon also plies his trade. It runs so deep into hell that at times it is made synonymous with it, as when Juno says in the *Aeneid*, '*Flectere si nequeo superos, Acheronta movebo*', 'If I cannot get the gods above to change their minds, I will appeal to the River of Hell.' Freud took this line as the epigraph to his *The Interpretation of Dreams*, itself an exploration of the currents and flows of a psychological underland – the starless rivers of the id that rush beneath the sunlit uplands of the conscious mind, here and there surging powerfully up.

The reason that classical literature runs with rivers dipping into darkness is geological: so much of the landscape in which that literature was lived and written is karstic in nature. Karst – from the Slovenian *kras* – is a topography formed by the dissolving of soluble rocks and minerals: principally limestone, but also dolomite, gypsum and others. Karst is vastly rich in its underlands – and it is also a terrain where water refuses to obey its usual courses of action. Karst hydrology is fabulously complex and imperfectly understood. In karst, springs rise from barren rock. Valleys are blind. A river can disappear in one place and appear in quite another, where it is given a new name by its new neighbours. Lakes appear that have no watercourse running into them and no watercourse running out of them – filled from beneath as the water table in the karst rises or drains according to season and weather (the 'vanishing lakes' described to the Royal Society of London in 1689 by Johann von Valvasor, a native of what is now Slovenia). Sinkholes and shafts pock karst landscapes like gaping mouths, making karst dangerous to traverse by night or in snow. Below the surface – if karst can be said to have a surface – aquifers fill and empty over centuries, there are labyrinths through which water circulates over millennia, there are caverns big as stadia, and there are buried rivers with cataracts, rapids and slow pools.

In karst country, there may be violent collapses of the land, as in Taipei where circular sections of road disappear, as if a monster has stomped its foot down at an intersection. The distinctive topography of karst has developed its own distinctive languages of form and absence: a 'doline' (English) is a funnel-shaped sinkhole; an *abîme* or *gouffre* (French) is a water-worn shaft that can plunge thousands of feet; a *cenote* (Spanish) is a collapsed sinkhole, often flooded; an *okna* (Slovenian) is a point where water has worn a passage through rock that can be seen through, as if creating a 'window' in the stone.

Guizhou and Yunnan in China; the Nullarbor Plain of Australia; great areas of North America, including much of Florida; the Yucatan Peninsula in Mexico; the White Peak, the Mendips, the Yorkshire Dales and the Forest of Dean in England; the limestone gorges and highlands of central and southern France – these are all karst landscapes. In the Philippines a tidal river runs for more than twenty-four miles below the karst, six miles of which are navigable by boat. In Waitomo, New Zealand, a subterranean river is lent light by the constellation of glow-worms, *Arachnocampa luminosa*, that live on the cave roof and speckle its stone with galaxies of blue stars amid the stalactites.

And where the north-east of Italy borders Slovenia, there rises a long elevated plateau of limestone known in Italian just as Il Carso, the Karst. Far beneath the wind-scoured, sun-beaten rock of the Carso runs a river called the Reka in Slovenian and the Timavo in Italian – a river of rapids and meanders that in places flows more than 1,000 vertical feet below the light.

~

I come to the Carso from Mantua. In a crypt below the cathedral of that city is kept the Grail, a relic containing Christ's blood, caught where it ran from the spear wound he sustained during crucifixion. Twice in the history of Mantua the Grail was buried and lost, twice it was unearthed. Now it is kept in the cathedral crypt, in an iron strongbox with eleven different locks, each lock openable by a different key and each key kept by a different cleric.

From Mantua, I cross three rivers to reach the Carso.

The Adige is a silver-grey snake, steaming in the heat. Its currents speak as lazy spirals. Steam coils up from the meanders where the sun boils it. Two storks beat their way westwards. Hops and honeysuckle in the hedgerows. Graffiti on the walls. A man pedalling a dusty road on a bicycle too small for him, so that his knees poke out at sharp angles. Brown earth and a sense of sea out of sight to the east, there in the sharpening of the light.

The Piave is heavy with silt from the mountains, moving with a pewter simmer, more stone than water to the eye. A sense of high peaks out of sight to the north, there in the darkening of the sky. Maize fields. Wild acacia groves in the lost ground under flyovers. Pale doves lifting in flocks off turned brown earth. Abandoned factories with pantile roofs, buddleia filling the window frames. Farmhouses lost in ivy. Everything wearing the heat like a cloak.

And the Isonzo, which marks the approach to the karst. Round pebbles of limestone, blue water that seems to glow from inside itself and the white of a flock of dozens of egrets moving eastwards over a green-rowed vineyard.

Near the Isonzo I leave the train at a small station, where no one else gets on or off. Waiting for me at the end of the platform, waving, is Lucian. Lucian and Maria Carmen live high on the karst above Trieste, in a house on a sinkhole's edge.

'Come!' says Lucian, embracing me. 'How wonderful to see you here at last!'

We drive past the fourteenth-century castle of Duino, perched on its pale stone headland above the Gulf of Trieste, where Rilke began writing his mystical *Duino Elegies* in 1912. Later, in Switzerland, after recovering from depression triggered by the First World War, Rilke would complete those elegies in what he described as a 'limitless tempest' of creativity. There he would also begin his great work of the underland, *Sonnets to Orpheus*, dedicated as a *Grab-Mal*, or 'grave-marker', to a young woman called Wera Koop, who had died at the age of nineteen. 'Ancient tangled deeps / all founding root,' begins the seventeenth sonnet, 'the hidden seeps / never to be sought.'

Not far beyond Duino we turn uphill and start the ascent to the plateau of the Carso. Lucian's little car snorts with the effort of scaling the switchbacks that climb the limestone from sea to high ground.

'You'd think she'd be used to it by now,' says Lucian, leaning forwards and patting the dashboard affectionately.

The roof tiles of the older houses we pass are weighted down with chunks of limestone. 'We're in the path of the *bora* here,' explains Lucian laconically, waving a hand at the stone-tamped roofs. 'It's the wind that rushes down off the peaks – a katabatic wind, gravity-fed. It blows at up to 200 kilometres per hour round here. It can drive people mad, set dogs howling for days, and take the roof off a house like a tin opener. Though I must say it's very good for drying clothes in its milder moods.'

Maria Carmen meets us at the door, and immediately flings her arms around me.

'Robert! *Il Professore!* Welcome to our house!'

The porch smells of pomegranates. Maria Carmen holds me at arm's length, inspects me, releases me. She is Argentinian. She wears red and black by preference, and her favourite creature is the roseate spoonbill, followed closely by the flamingo and the scarlet ibis. She mistrusts formal qualifications, preferring to judge people on their capacity for empathy. She and Lucian fell in love in the middle of their lives. Now they live together on the karst with a silver-grey cat called Raffy.

Lucian is a translator, devoid of ego, and generous to the point of impracticality. He has such kind eyes. He is quadrilingual in Spanish, French, English, Italian and moves between these languages without hesitation, like a train smoothly switching tracks on the points. He once sailed around the Horn, and was part of a number of expeditions to Patagonia. His own boat is in dry dock, and needs work, but if he ever finds the time and money to replace the buckled teak deck, his dream would be to sail her to the foot of a scarcely climbed Patagonian peak of 3,000 feet, and ascend it from sea level through a belt of tangled southern beech and swampy undergrowth that he expects to pose more of an obstacle than any glacier. He has the charts of the tip of Patagonia pinned above his working desk: channels and island groups, there for daydreaming when the word-grind gets too much.

'He *has* to give to others,' says Maria Carmen to me in a whisper one day during my weeks there. 'It is what he has to do – but he does not think about himself enough.'

Maria Carmen works in social care. 'She receives *so* little back, but goes on giving so much,' Lucian confides to me one day while we are out walking.

Lucian is a nineteenth-century explorer stuck in a twenty-first-century economy, Maria Carmen is a natural altruist in a needy culture, and together they are two of the gentlest people it is my good fortune to have met.

Lucian and Maria Carmen's house faces south-westwards towards the Adriatic, but the sea is just out of sight, visible only as silver light above the stands of oak and pine that flourish on the downslope of the Carso. There is an orchard of apricots, in which yellow crocuses bloom.

The house is cool, its windows shuttered against the heat, its roof tamped down against the *bora*. The bookshelves are glass-fronted wooden cabinets, filled with climbing, caving and sailing books in several languages. We eat lunch in the shade of an oak tree: tart Slovene apples, hard cheese, and potatoes that Maria Carmen has grown. Wild cyclamen flower on the sloping edges of the sinkhole. We throw our apple cores over its edge.

'The sinkhole gets hungry,' says Lucian.

Raffy the cat curls around my ankles like mist.

'I am not really from anywhere,' says Lucian after we have eaten, 'but I suppose I am most of all from the Carso.'

His father served as a young tank commander at the time of the Normandy landings. 'He got to France two weeks after the initial landings – and in some ways he had the time of his life: he was nineteen, in charge of a tank and spoke fluent French. You can imagine how the locals welcomed him!'

After the war, Lucian's father was posted to Trieste, where he met a young Italian woman. They married the year after in London and he moved into the long-established family business of making briar pipes. Their holidays were taken in Trieste and Lucian grew up walking the Carso, beginning slowly to fathom its secrecies, both dark and light.

'What I learned as a child is that you have to watch where you put your feet round here,' Lucian says. 'Metaphorically as well as geologically. There is much violence in the past of this region – and so little of it is spoken about. Rivers disappear and so do stories, only to rise again in unexpected places.'

Lucian has been working for years on a history of the Carso and its depths, a text which he seems to regard as potentially limitless and almost certainly unfinishable. 'It's taken me the best part of two decades to come to realize how little I know about what this landscape hides,' he murmurs, rather more to himself than to me.

Dog-rose tangles through the understorey of their garden, blossoming pink and white. Bees swim in the blossoms. I think of the strange lines Rilke wrote to the translator of his *Elegies*: 'We are the bees of the invisible. Frenziedly we gather the honey of the visible, to gather it in the great golden hive of the invisible . . .'

The air is stony. Birds skim between oaks.

'The Carso is, to my mind, the archetypal "underland", to use your term,' says Lucian. 'Here we have caves, 10,000 caves, in which humans have lived, worshipped, healed, killed, sought protection from one another and from the world, wrought terrorism, and dug for ice. In prehistory, people built forts here – but they also retreated literally into the hillsides. The Romans constructed cave temples dedicated to the underground god Mithras. And you've also – you'll be delighted to hear – come to one of the mouths of hell: the Romans declared an entrance to Hades nearby, at the point where the River Timavo dives underground at Škocjan.'

He pauses.

'Fast-forward to the nineteenth century, and a booming Trieste, made a free port by Maria Teresa, but critically short of water, sent out a series of expeditions to try and locate the missing river and supply the city. They did find it – but buried far underground. During the First World War, both the Austrians and the Italians burrowed into the limestone here, digging trenches and enlarging caves for use as hospitals, munitions dumps, and so on – and not just here, but across the Julian Alps and the Dolomites. The same during the Second World War. And in the course and aftermath of that war, too, both sides repaid their sufferings with awful interest, killing and pushing enemy troops and supposed collaborators into sinkholes, or *foibe* as they are known round here.'

He frowns.

'We have cave systems here with living glaciers in them, we have caves containing an unmentionable species of blind orange beetles – the *Anophthalmus hitleri*, which is menaced by extinction because of its popularity with neo-Nazi collectors – and we have caves in which wine is left to rest with, it seems to me, mostly indifferent results.

'And here, the earth itself is tidal. Truly! The rock here reacts to the draw of the moon, just as the water of the ocean does. Gravitational attraction pulls and then releases the limestone; the Earth's crust has spring tides, neap tides. Of course, they are tiny compared to marine tides. The tidal range of the sea can be up to sixteen metres; the tidal range of limestone up to two centimetres. Nevertheless, here the underworld surges and relaxes beneath your feet without you feeling it. There are symposia of earth tides held at Trieste University.'

The Adriatic glitters in the sky.

'And perhaps above all we have the fascination, no, the *obsession*, with mapping the complete flow of the Timavo – which they sometimes call here the River of the Night.'

~

The Timavo rises as the Reka in pine forests on the southern side of Mount Snežnik, the Snow Mountain, on the border of Slovenia and

Croatia. Its waters gather in the flat farmed valley land around Illirska Bistrica, and then wander, in idle half-mile loops, over impermeable flysch bedrock until, at the village of Škocjan, the flysch meets limestone and – in a geological conjuring trick – the Reka disappears.

The Škocjan canyon, where the Reka plunges into the underland, is a site of exceptional force. Here, over millions of years, water has cut one of the largest known underground canyons in the world. The river crashes through an immense arch in a limestone cliff and drives on through collapsed sinkholes hundreds of yards in diameter – their air filled with microclimates of mist and spray, their vertical sides offering nesting grounds for falcons and growing grounds for oak saplings and pink cyclamen – until the vigour of its fall cuts a tunnel steeply down into the limestone, just as the limestone begins its own rise up to the plateau of the Carso. The Reka-Timavo runs underground for around twenty-two miles before emerging again near Duino and debouching into the Adriatic, mingling fresh water with salt.

'The Timavo River flows from the mountains, falls into an abyss and then, after flowing about 130 stadia underground, springs beside the sea,' wrote Posidonius of Apamea, around 100 BC. The Škocjan 'abyss' was sufficiently famous to be marked on both the Lazius-Ortelius map of 1561 and Mercator's *Novus Atlas* of 1637. Systematic exploration of the river's hidden extent began in the late 1830s, in part to address Trieste's thirst for drinking water. A well expert called Ivan Svetina pushed into the canyon at Škocjan, reaching what he described as the third waterfall – and in so doing he began the first golden age of the Timavo's exploration, lasting until 1904.

These early attempts to follow the starless river were industrial in nature. Safety paths were chiselled into the stone of the canyon sides, spidering up the cliffs – vertigo-inducing even to look at from beneath – so that escape would be possible if the water level began swiftly to rise in the canyon. Boats were used to enter the further reaches, but boats were risky – hard to bring back against the current, liable to be overturned. The chambers, cataracts and channels were christened as each was reached – Hanke Channel, Martel's Chamber, the Rudolf Hall, the Müller Hall, Dead Lake, the Silent Cave – until, in 1904, further progress was stopped for almost a century by a siphon: a fully flooded tunnel that, the explorers discovered, was too long to be swum on the reserves of a single human breath.

The next breakthrough did not come until 1991, when underwater respiration technologies, and the development of cave diving as an ultra-high-risk pastime, meant that the route could be forced further. In September that year, two Slovenian divers were able to swim a siphon close to Dead Lake, which opened into a wealth of new passages and chambers, where the Timavo both ran as a river and pooled as a lake. Now diving teams travel from around the world each summer to attempt to push further along the buried waterway from those points where it can be accessed. They establish base camps in the darkness, waiting days and weeks for optimal conditions – then dive into the ink.

'The Timavo is a dream,' remarks one of its young contemporary explorers, a member of the Adriatic Speleology Society called Marco Restaino, 'a dream we are trying to bring into being metre by metre.' That dream produces obsession in its votaries, who are known as *grottisti*. These groups of *grottisti* are competitive with one another, but also understand that in order to realize the shared grail-quest – full continuous mapping of the Timavo's route and flow – they must cooperate and piece together their knowledge.

There are a few places on the Carso plateau where it is possible to reach the starless river from the surface. Almost all of these access points involve serious caving. Almost all are 'owned' by differently aligned groups of explorers and cavers, who control local access to the Timavo, and whose relationship with its flow mixes cartography, adventure, science and a compulsive kind of dreamwork that would surely have fascinated Freud (who visited one of the great caves close to Škocjan, where his attention was unsurprisingly caught by the tumescent stalactites and stalagmites, and also by the subconscious of the grotto-keeper, Gregor, who lived in this profusely phallic underland and had named each stalagmite – 'Cleopatra's Needle', 'the Eiffel Tower' – for a place or object about which he had been told by visitors to his cave).

One of the places where the Timavo can be reached is a collapsed sinkhole in the beech woods near the village of Trebiciano. There a narrow water-worn shaft drops 1,000 vertical feet from the base of the doline in a continuous passage, just wide enough at its narrowest points to admit a human body, down at last to a cathedral-sized chamber through which the Timavo rushes. It was in part to attempt a descent of the Abyss of Trebiciano – as it is comfortingly known – that I had come to the Carso.

However the Timavo is reached, the work of exploration is dangerous, difficult and dark. After heavy rain, the Timavo can flood at times up to 200 feet above typical height, killing anyone caught in a chamber or tunnel, or driving air under huge pressures up the shafts that drop to the river. Despite the efforts of the *grottisti* over more than two centuries, only around 15 per cent of the underground flow of the Timavo is presently known.

Contemplating the activities of these Timavo mappers – most of whom are men – it is hard not to see in their devotions and their rituals something of the practice of a religious cult, with the starless river their occult god.

~

'I want to show you a strong, sacred place – very much part of this region's underland,' says Lucian one morning.

We set off on foot through sloped scrub from a road-head near an abandoned farmhouse, a mile or so in from the sea. Thorns snag our ankles as we walk. We crush wild marjoram and thyme underfoot, loosing their scents. Grasshoppers crackerjack away from each pace. Lizards skitter off, tails drifting over the dust behind them. The air vibrates with heat. There is no footpath but Lucian works his way confidently uphill, contouring south-east as we climb. We cross a train track, its rails glinting. Not far short of the treeline Lucian leads us to what seems, amid that barrenness, a green oasis: acacia trees and grass growing from a shallow sinkhole in the hillside.

'Few people know this is here,' says Lucian. 'I like the fact it lies within plain sight of the train line and the main Venice–Trieste road, but is invisible to all but a handful of those who pass it.'

We push between two of the trees that together form a gateway to the site, follow a set of stone steps down – and there in the base of the sinkhole is the entrance to a cave. At the threshold are several carved limestone pedestals and column bases, one of which is part of the living rock.

We enter what is unmistakably a votive space. Two long central stone benches or altars span the cave's width, with single cubes of stone placed between them. On the sides of the cave are two carved limestone reliefs; both show a human figure grasping a bull with one hand and with the other driving a knife into its chest.

'What *is* this place, Lucian? What does all of this mean?'

'This is a Mithraeum – an underground temple devoted to the god Mithras,' says Lucian. 'Mithras was the god of the legionaries, little known in the pantheon and now hardly remembered, I think. He was born from rock – a true deity of the underworld in that sense – and the cult of his worship took place in subterranean spaces all over the empire. This was one of them, and it was probably used for more than 300 years, until it was abandoned around 400 AD. When they first excavated this place they found hundreds of coins and dozens of oil lamps and jars.'

We sit together on one of the benches. Flies dance where the light falls over the cave mouth.

'People still come here faithfully, as it were,' says Lucian. 'I once found a wooden box of coins, some of them very old, tucked behind a stone at the back. I left them, of course. They'd vanished by the time of my next visit.'

Mithraism was a so-called mystery cult that spread across the Roman Empire from the first to fourth centuries CE, standing as a provocative counterpoint to early Christianity, which perceived in it a 'diabolical counterfeit' to their own emerging rituals. Its mysticism is itself mysterious, for so little survives in terms of sources to help us illuminate its beliefs and practices. What we know of it is largely reverse-engineered from the inscriptions and artworks found in Mithraic temples, and from the fleeting references that exist in classical literature.

We know that Mithraism was centred on Rome, but that its temples existed throughout the empire, all the way to London – where in 1954 the remains of a Mithraeum were discovered under Wallingford Street, in what is now the basement of the Bloomberg building. Among the items found there during the excavation was a miniature gladiator's helmet carved from amber.

We know, too, that Mithraism was an underground cult in several senses. Politically, it kept itself secret and out of view, its initiates greeting one another with encrypted signs of recognition. Theologically, it worshipped a god who had emerged from the rock itself. Topographically, its distinctive temples were almost all located below ground: the cellars of houses, natural caves or specially built vaults; sacred chambers known as *spelea* (caves) or *crypta*.

Sitting there with Lucian, I understand what he means by this as a 'strong place'. People have been stopping here to rest and to make offerings for nearly 2,000 years. Many of its earliest visitors were legionaries, travelling back from a distant conflict zone towards Rome or home, or leaving Italy for a distant posting. Such people surely would have been in need of faith.

Lucian and I rest companionably in the cool, listening to the landscape's undersong: clack of the train track, road-hum below it, buzz-saw of grasshoppers from the scrub.

'Mithraism was a soldier's religion, and a male religion,' says Lucian. 'Only men could become initiates.'

Thinking of the diver-explorers of the Timavo as modern Mithraists, striving in their below-ground sanctums, seeking new spaces, new disclosures, I am reminded of the historically gendered nature of the underland. As far back as the classical *katabases*, it is often men who descend heroically to the underworld to retrieve women who have been trapped, taken or lost: Orpheus seeking Eurydice, for instance, or Heracles following Alcestis. Mythologically, the underland is often a place in which women are silenced or pay brutal prices for the mistakes of men. Ariadne helps Theseus negotiate the labyrinth but is abandoned by him and then, in some stories, killed by Artemis. Creon threatens to entomb Antigone alive to punish her for burying her brother, Polynices, and to disempower her politically; she hangs herself in desperation. Hades imprisons Persephone and then compels her to return annually to his domain, even after she is saved by Demeter.

Yet there are also shining modern counter-examples – women who are rewriting these ancient archetypes with courage and expertise. The Dark Star expeditions in Uzbekistan, exploring a system that may yield the deepest known cave, have been pioneered by female cavers, who cross subterranean lakes and rifts filled with blooms of blue ice. Female palaeoanthropologists have led the Rising Star expeditions in the Bloubank dolomites of South Africa, excavating early hominin burial sites. Each of these women has had to pass through an opening less than a foot wide in order to access the fossil remains, and the group has become known as the Underground Astronauts. The microbiologist and caver Hazel Barton collects microbes in extreme subterranean environments in order to research antibiotic resistance; tattooed onto her left bicep is a map of the Wind Cave in Dakota, the site of much of her research. She is drawn by the unknown as much as any modern Mithraist. 'When you're in the cave, you . . . get a feeling for what it felt like to stand on the moon for the first time,' Barton says. 'You're the first person ever to see it. There are very few things that give you that sense of exploration any more, where you can go and find unknown land that people didn't know existed.'

Lucian and I leave the cave. The sun falls hard as bronze plate upon us. On the coast below is the colourful sprawl of an industrial harbour zone, with yellow blocks of shipping containers and a series of red cranes leaning over the water.

'It's a shipyard specializing in cruise liners,' says Lucian. 'They turn ships out like Fiat Pandas down there.'

Rasp of grasshoppers, burr of bees, scent of herbs. We walk on towards the tinfoil sea.

~

The Timavo is only one of many starless rivers and flooded underlands to have beckoned people into them, sometimes fatally. 'A peak can exercise the same irresistible power of attraction as an abyss,' wrote Théophile Gautier in 1868, and the reverse is also true.

The fallen angel of French speleology is a man called Marcel Loubens, who was seized from a young age by what the British caver James Lovelock called 'a passion for depth . . . He wanted to go deeper and further into the rocky heart of the earth than any man had been before.' Under the tutelage of the father of modern French cave exploration, Norbert Casteret, Loubens led numerous mid twentieth-century explorations in the Pyrenees, which were at that time considered 'the Himalayas' of the caving world.

In the summers of 1951 and 1952 Loubens took part in the expeditions to descend the chasm of Pierre Saint-Martin, a shaft of water-worn limestone that, from its modest mouth in the western Pyrenees, drops more than 1,100 feet to its base. The Saint-Martin chasm proved to be the entry point to what was thought then to be the deepest-reached cave system in the world – a series of chambers leading down at last to an underground river – and it became the focus of intense speleological activity. In 1952, to speed up the movement of people up and down the shaft, an electric winch was devised and cemented in place at the mouth of the chasm.

Loubens was one of the most committed explorers of Pierre Saint-Martin, and he volunteered to make the first winch-powered descent of the shaft himself. He clipped himself onto the wire, backed over the edge of the chasm, and called out a farewell to Casteret – '*Au revoir, papa*' – as he disappeared from view. Then the winch lowered him down the shaft, and he saw the blue circle of sky dwindle from disc to dot, until its watching eye winked out. The sides of the shaft were in places polished glass-smooth by the action of water.

Loubens made it safely to the base, and spent the subsequent five days underground, leading the exploration of the further reaches of the system, down towards the starless river, astonished by what he and his companions were discovering. 'The show has hardly begun,' he said to his friends as he prepared to be winched back up.

Loubens was around thirty-five feet up when the clip that held him to the wire buckled. He cried out as he slipped from the line, fell, and then smashed into the boulder field at the base of the shaft, bouncing for more than 100 feet from rock to rock.

When Loubens's companions reached him, he was barely alive. Great efforts were made to rescue him, but his injuries were so many and so severe – among them a broken spine and a fractured skull – that it proved impossible to move him. He died thirty-six hours after first falling.

Loubens's friends on the surface used an acetylene lamp to burn onto a nearby rock the words '*Ici Marcel Loubens a vécu les derniers jours de sa vie courageuse*', 'Here Marcel Loubens passed the last days of his courageous life.' Those still at the base of the *gouffre* buried his body beneath a pile of boulders, and marked the site with an iron cross covered with luminous paint. Loubens had fulfilled his own wish to find his resting place deep underground.

Two years after Loubens's death, on 12 August 1954, a young Belgian priest called Jacques Attout volunteered to be lowered to the bottom of the Pierre Saint-Martin. Using a medicine chest as his altar, and with Norbert Casteret as his server, Attout celebrated Mass in memory of Loubens. He later recalled the service, in what has become among the most celebrated passages of caving literature for its convergence of theology and geology:

> Never again shall I celebrate such a Mass in a setting that was so closely united to the Divine Sacrament . . . In this vast cave we must have looked more like insects than human beings. And yet – our souls were on fire. We were so far from our surroundings, or if we sensed them at all it was because they had lost something of their material quality and become vast and luminous.

Loubens's avid striving after underland knowledge is not, of course, a recent invention. Classical sources record the use of pine cones or wooden cups as marker objects – floated into disappearing streams and rivers in the karst, their reappearance watched for elsewhere – in order to trace the submerged flow-patterns of the landscape. It is in the modern period, though, that these practices of deep-mapping have reached their most extreme and hazardous expressions.

In the Picos de Europa of northern Spain, forty years of expeditions have been spent trying to forge the connections that would result in the completion of the Ario System, which in theory might extend almost 6,000 vertical feet. The project – shared across generations of cavers from many countries – is known as 'The Ario Dream', and its aim is to create the world's deepest through-trip, whereby one might abseil into a chasm amid mountain peaks, and then emerge several days later into the twilight of a gorge. The Ario System is so extensive that its exploration requires expedition-style caving, where base camps and advance camps are established far underground as sites where equipment can be cached and sleep can be taken in tents – just as mountaineers attempting Everest move between successive camps while they gain height. Cave-diving skills are key to the Ario expeditions, for the further reaches of the system are flooded. Divers push on into the blackness with fine margins for error – often turned back by chokes or deadings-out – entering unmapped areas of the mountain's interior that are referred to, in an echo of nineteenth-century imperial cartography, as 'blank space'. George Mallory famously answered the question of 'Why climb Everest?' with 'Because it's there.' Extreme cavers jokingly modify Mallory's answer when asked why they risk their lives for an ultra-deep cave system, with the words 'Because it's not there.'

Among the driving ambitions for many of these speleonauts are connection and completion: to prove through-flow and join-up. In *The Darkness Beckons*, Martyn Farr tells the story of the four years that the cave divers Geoff Yeadon and Oliver 'Bear' Statham spent trying to connect Kingsdale Master Cave and Keld Head in the Yorkshire Dales: two chambers a mile and a quarter apart, linked by a series of submerged passages. The route became known as the Underground Eiger in recognition of its severity. Visibility in the very cold water was poor due to its silt content, and there were few air pockets where the men could surface to swap oxygen cylinders. Early in their exploration of the system, Yeadon and Statham found – and recovered – the body of a diver who had died there five years earlier. The two men finally completed the traverse successfully on 16 January 1979 – a remarkable achievement in desperate conditions. Eight months later Bear Statham took his own life in his pottery workshop in Sedburgh. He put on a full-face diving mask and regulator, connected it up to the gas supply for his kiln, then lay down on his sofa and died.

Many of the longest submerged systems are entered via modest pools that rise in open ground. There is one such system entered via a small lake called the Blautopf in Germany; another in central Norway known as the Plura that has claimed the lives of two divers. And in the Northern Cape of South Africa, on the fringe of the Kalahari Desert, is Boesmansgat, or Bushman's Hole. There, what seems little more than a pond offers entrance to a flooded chamber 885 feet deep.

Fewer than twelve people in history have dived below a depth of 790 feet using scuba equipment. Such ultra-diving exerts an awful toll on the bodies of those who survive, including lung damage and hearing loss, and the fatality rate among those attempting such depths is high. In 1994 a young cave diver called Deon Dreyer died deep in the Boesmansgat system. His body was not located until ten years later, embedded in silt on the chamber floor. Painstaking plans were made to retrieve his corpse, in order to bring closure for his grieving family. But the lead diver on the retrieval dive, a British man called Dave Shaw, became tangled in his own safety cord while seeking to place Dreyer's body in the silk bag he had brought down with him for that purpose. Shaw's breathing and heart rate increased as his anxiety rose. Dreyer's neck had become softened by his decade in the water, and when Shaw sought to move Dreyer's head, it loosened from his body, then detached completely and floated past Shaw, turning to gaze at him through blackened goggles – the moment caught on Shaw's head camera. Shortly afterwards Shaw himself succumbed to asphyxiation brought about by carbon dioxide build-up.

Four days after Shaw's death, divers returned to the cave. To their amazement, they found Shaw's body floating near the roof of the chamber, with his torch hanging beneath him, still on.

Illuminated in its beam was Dreyer's headless corpse. Shaw had – after death – achieved what he set out to do and retrieved the body of his predecessor from the blackness.

For years I could only understand these pursuits of shadowed water, blind rivers and terrible depths as fierce versions of the death drive – fiercer even than what drove the most fearless mountaineers. The language of extreme caving is often openly mortal and tacitly mythic: stretches of passageway 'dead out', one reaches 'terminal sumps' and 'chokes', the furthest-down regions are known as 'the dead zone'. But over time I saw that – as with extreme mountaineering – there was another aspect to the *thanatos* at work. Divers and cave divers often describe their experiences in terms of ecstasy and transcendence. 'I have had such beautiful moments in the water,' says the British diver Don Shirley, who dived below 790 feet in Boesmansgat. 'You are absolutely, completely in a void, like being in outer space . . . You get to the point where there is no God, no past, no future, just now and the next millisecond. It's not a threatening environment – just total serenity.'

The free-diver Natalia Molchanova similarly described her time below the surface as self-dissolving. Molchanova was one of the first people to free-dive the Blue Hole, a 390-foot-deep sinkhole in the Red Sea containing 'the Arch', an opening in the sinkhole's wall that runs through to open ocean. More than a hundred free-divers and scuba divers have allegedly died in the Blue Hole, drawn into its depths by complex longings. Molchanova dived the Blue Hole on a single breath, safely: an astonishing achievement. But on an August day in 2015, she dived recreationally off the coast of Ibiza to between 100 and 130 feet – a shallow dive for someone of her rare abilities and experience. She did not resurface, and her body has never been recovered.

'I have perceived non-existence,' Molchanova wrote in a poem called 'The Depth':

The silence of eternal dark,
And the infinity.
I went beyond the time,
Time poured into me
And we became
Immovable.
I lost my body in the waves
. . . Becoming like its blue abyss
And touching on the oceanic secret.

Only once in the course of my years in the underland did I approach a flooded labyrinth, and there I had an experience that helped me comprehend a fraction of the serenity of which Shirley spoke. The labyrinth ran under the centre of Budapest, on the Buda side of the river, and I entered it in the company of a Hungarian geologist, caver and climber called Szabolcs Leél-Őssy. Budapest is built in part on limestone, and its invisible city contains both mine networks and cave systems caused by the upwelling of warm, dissolving water. On a hot summer night with insects singing from the street trees, Szabolcs and I slipped through a gap in a heavy steel gate, unlocked a door set into bedrock, followed a tunnel that had been blasted into the limestone, and emerged into a flooded cave chamber below the city. The chamber – which was more than 450,000 cubic feet in volume – was the access point to a submerged network of tunnels below the city. It was from here that, for years, cave divers had set out to map Budapest's underwater maze.

Szabolcs and I lowered ourselves into the water at the edge of the chamber, and we floated there convivially for a night-hour in that lost space below the city. When I recall the experience now it feels dream-like. The water, rising as it did from far within the earth, was a steady 27°C. I had the sense of great depth opening below and around me in the darkness, but felt no vertigo, only occasional swoops of the spirit. The water was uncannily clear and my limbs moved in it as though they were another's.

'Here,' said Szabolcs at one point, 'I am at peace within the rock.'

Our conversation was occasional. There were long periods of silence. I have rarely felt more relaxed than in that amniotic space.

'Before we leave, you should see the true entrance to the maze,' said Szabolcs. He sculled across to a distant wall of the chamber. I followed. 'Now,' he said. 'Sink, and open your eyes. The water will not harm them.'

I took a series of deep breaths, lifted my arms above my head, joined my legs, expelled the air from my lungs in a rush of bubbles, and slowly sank. At a depth of ten feet or so, the weight of water building on skull and skin, I fanned my hands to keep myself steady, and opened my eyes. Pressure pushed gently against my eyeballs. Ahead of me in the water was the black mouth of a tunnel entrance, leading away into the rock, more than wide enough to engulf me, its stone edges smooth. The pull of the mouth through that eerily clear water was huge. Just as standing on the edge of a tower one feels drawn to fall, so I experienced a powerful longing to swim into the mouth and on, until my air ran beautifully out.

~

High on the Carso, deep in beech woods: Lucian and I approach the entry to the Abyss of Trebiciano on foot through the forest. Cicadas hiss in the acacias. Long-tailed birds I cannot name cut over the path. My body crackles with nerves about what lies ahead, beneath. I am also fascinated by what I might see and where I might reach. In one pocket I have the whalebone owl. In the other I have the bronze casket, in case the abyss discloses itself as the place for its disposal.

Sergio is waiting for us in the woods. We smell him before we see him: tobacco smoke in the air, and then Sergio himself, leaning up against the wall of a hut. I guess him to be around seventy years old. He is short and broad-shouldered, he wears a flat cap and smokes a briarwood pipe, and he is both gatekeeper and guide to the abyss.

Sergio first descended the abyss as a young man growing up in the post-war Carso. That experience gripped him, and the river at the base of the abyss became his lifelong obsession. For fifty years he has been involved in the mapping and exploration of the Timavo.

'How many times have you been down the abyss?' I ask Sergio.

He shrugs, considers the question. 'Maybe . . . 400 times?'

'Why?'

The question puzzles Sergio. He thinks for a while. Lucian helps translate his answer.

'For many years, there was nothing else to do. Also, for eighty years after its discovery in 1841, this was the deepest known cave in the world. Now we study it, come to know the river and its . . . behaviour. Our work here is not considered important by the authorities or by scientists, but still we go on. Here in the abyss we make . . . romantic science.'

He smiles. Then he says, '*Allora*,' and leads us into a hut.

On the walls are nineteenth-century aquatints of the region, and orange caving suits hung on pegs. Banks of monitoring instruments bleep quietly to themselves. Sergio unrolls a cross-section sketch-map of the Carso, flattening it on the desk. My chest tightens as I look at it. It shows the path of the Timavo under the limestone, from the point at which it cuts into the earth at Škocjan, to its outflow in the Adriatic. The abyss is marked, and Sergio follows it with his finger: a twirling, jinking line dropping down and down through the stone to reach what looks like a substantial chamber, within which runs the Timavo.

'*Allora*,' says Sergio. He is a man of few words, and most of those words, I learn, are *allora*: 'now', 'let us begin'.

We leave the hut and walk through woods of sweet chestnut and beech. It is cool in the shade. We climb to the rim of a wide, earthed-up doline. The doline is filled with spindly trees from its base, some of them forty feet high. They have few radial branches and their canopies form a sea-surface far above us, casting everything in green light that reminds me of Epping's pollard groves. From the doline's rim, a path winds down past blocks of limestone, to a brick hut at the lowest point of the crater. The hut is built over the entrance to the abyss.

It is, Sergio explains, relatively new. One day after a period of heavy rain some years previously, he came to the doline to find the previous hut in pieces. All four walls had been flattened by a blast, and the roof had been blown off. At first he thought someone – an opposing caving club, perhaps – had detonated a bomb inside the hut. Then he realized the true cause. The Timavo had risen with such speed that the spate-water, ascending the doline faster than the air above could escape it, had caused the hut to act as a pressure prison until, like an overfilled balloon, it had simply burst.

Sergio opens the hut door, and shows me to what seems like a shower cubicle, but with no visible shower head. The flooring is brown textured tiles, there because at times the Timavo still comes boiling up this far in its fury. The tiles make the clean-up job easier.

There is a hatch set into the floor, close to one wall.

'*Allora*,' says Sergio, lifting the hatch.

My stomach flips. It is another door into the dark, another portal to the underland – this one leading to a water-carved pipe of stone that runs through rock to a wild river. The usual fears rush at me like bats, flocking and tangling.

'See you on the other side,' says Lucian, who has decided to stay above ground.

We begin our descent. Progress is made by ladder, platform and down-climbing. Many of the ladders are missing rungs. There are places where I must swing out on a single stanchion, then grope down for footholds, the shaft falling away beneath me, sucking at me. I clip on and off to what safety points there are. Then come little stances, lateral passageways, sections of tight shaft. The sense – now familiar to me – grows of the surface becoming distant, other-worldly, of the rock building in mass and depth.

Sergio moves slowly but steadily, each step and drop and squeeze familiar to him. I can hear his lungs wheezing ahead of me. Mud lines on the walls mark the high-water levels of the Timavo in its different floods.

I cannot now say how long the descent took. An hour? Two? Time was irrelevant because there was nothing to keep it except for the hammer-beat of heart and the heaving of lungs.

Far into the descent, Sergio stops, looks back up at me, and holds a finger to his lips and a hand to his ear. I can hear nothing.

'Quiet,' he says. 'Very quiet.'

I breathe as lightly as I can, hanging by one arm and with my legs braced against either side of the shaft. And then, yes, I can hear it – a distant roaring sound, a hum of white noise, rising up the shaft towards us, washing at our feet and ears.

'The river,' says Sergio.

We push on down, the roar growing in volume. The shaft takes one of its abrupt sideways jumps, we squeeze around a corner, and the floor of the tunnel falls away again beneath us in a natural trap-door that gives onto pure blackness. Sergio gestures that I should lead.

'*Allora*.'

He points down at the door into the dark. I turn to face the rock and lower myself through the gap, my feet pedalling for holds below. I have an awareness of great space around me, startling after the confinement of the shaft. The roaring sound is motorway-loud now. Something, a surface, is coming up to meet me in the darkness. I jump back and thump, softly, into sand.

Black sand.

A *dune* of black sand, with gold grains amid the black. And dunes beyond that, rolling away.

Sergio appears beside me.

Eyes adjust to the space, head-torches probe for information. Rock above and behind me, curving away overhead. Black sand dunes curving ahead of me, rising up to my left, falling away to my right.

Boulders, *huge* boulders, embedded in the sand to our right but not to our left. The roaring coming from somewhere far to the right, and the air full of sand, fine black sand, that we breathe in and that swirls slowly in the light-beams.

My light-beam meets stone in the distance – the opposite wall of this vast chamber. I look upwards and around: the ceiling domes into the dark, and near its apex I can see the shadowed entrance to a shaft of some kind, impossible to reach from the ground, a single thick stalactite hanging from the rock by its mouth.

We are terranauts and we have dropped through the roof of this chamber onto another planet – dropped into an underland desert of fine-grained black-gold sand. I shake my head in wonder and fear. Sergio stands quietly beside me. He has seen this place do this to people before.

Then he reaches up to switch off his head-torch, and I do the same, and for a few minutes we stand there on that soft sand, in that thick dark. Around us, intensely, is the mystery of Mithras, the god of stone.

Then Sergio strikes a match to light his pipe, and the dark organizes itself instantly around that tiny bright fire. The smell of tobacco spreads. The pipe-bowl glows. Sergio waits, then smokes with pleasure and patience.

'*Allora*,' he says after a while. 'To the river.'

I lead, navigating by sound and slope. We move between the dunes of black sand, ascending at first to the centre of the chamber in order to outflank the crags that fall away to our right. The landscape through which we are passing is, I realize, only a temporary iteration of a dynamic terrain. These boulders are shifted and these dunes are re-sculpted by the river each time it floods. We trudge down the face of a black dune, and then pass through a narrow cleft between limestone boulders fallen from the roof, each twelve or more feet high.

Clink of my karabiners against the rock. Sergio's rasping breaths. Hush of footfall in fine sand. Stone dust in the light of our torches. The growing noise of the river. A moon landing. The night ascent of a desert peak.

The sand changes character abruptly, darkening and dampening. This is the river's most recent high point. We pick our way through a boulder field, slipping on wet sand to a small cliff.

The noise is now so loud that we can hardly communicate. There is a cleft in the cliff and I lower myself through it and climb down onto hard-packed silt-sand and here is the starless river – a living river, whole and forceful, pouring out of an arch of rock to my left, bending towards me where it has cut a bay, and then cornering again and disappearing to my right, thunderously over rapids.

The sound of this starless river is like none I have ever heard. It has volume. Its volume has hollowness. Each sound has its echo, and each echo its interior.

I drop my pack. Sergio leans up against the rock, tamps fresh tobacco into the bowl of his pipe, relights it. The beam of my head-torch gives depth to the water, which is silver and silty, and – *my God* – I see there are creatures in it, white forms moving through the silt-clouds in the slower water of the bay. The rock arch of the tunnel out of which the river tumbles has – like the mouth of the Budapest labyrinth – an incredible draw, and I feel the urge to swim here in this starless river with the white forms. I tell Sergio I plan to do so, and begin to undress. He looks at me for a while, judging how to respond, then he simply shakes his head once, firmly.

I cannot swim like a fish but I wish I had an owl's eyes to see in the dark, or that I could somehow send my sight upstream and downstream from here, to the mouth of hell at Škocjan and to the blue water of the Gulf of Venice. I understand that this is not the place to leave the bronze casket; it is a site of transit not of storage.

I step down to the edge of the water, to the bay where the white forms move, my torch-beam probing the water, and as I approach the edge the forms shrink back and away from me. I kneel and drink two stony mouthfuls of the starless river, and wash my face of the fear-sweat of the descent.

I clean my karabiners of mud in the starless river because I want the snap-gates to work sharply on the ascent. I think of the winter floods here – the river rising in volume so massively that from that rock arch ahead it fills the chamber, lifting the sand in a billowing black water-cloud, compressing the air and forcing it up through the shaft down which we have come, up which we will return.

There is an iron stake driven into a notch in the rocks by the bay. Sergio comes close, shouts into my ear above the roar, and he tells me that a team of French divers were working from here recently, that they spent a week down here in the chamber, pushing further upstream each day until the risk became too great. The most distant point they reached was almost 1,000 feet upstream from where I stand: a triviality, an immensity. I am awed and bewildered at the thought of their persistence. 'Conquistadors of the useless,' Lionel Terray once called climbers – but this is another order of futility altogether.

'*Allora*,' says Sergio.

We climb the dune-slope back to the point where we dropped through the roof of the chamber. There, by the chamber wall, is a small yellow inflatable rubber dinghy – a Marine 285 with two plastic oars shipped tidily inside it, like something you might buy from a beachfront shop.

Sergio flashes his head-torch beam along the dome of the chamber, settling on the shadowed shaft in its summit I saw earlier.

'When the cave floods,' he says, 'the explorers . . . seek the higher parts. They . . . sail this' – he nudges the dinghy with his foot. 'They float up, catch the stone, and they climb the chimney.' He nods upwards at the cave ceiling.

He shrugs. 'It is very dangerous. They do not want to fall. Of course they must . . . know the flood, so that it does not fill this place and kill them.'

He shrugs again.

'Still they do it.'

Pause.

'I have escaped . . . when the water is rising. It pushes you up. It is very powerful, like being in a . . . storm.

'*Allora*,' Sergio says for the last time, and moves towards the trap-door in the rock up out of the chamber, and we climb up to the beeches and the bees of the invisible, where Lucian is waiting for us. My eyes are wild as I clamber from the hatch.

'You look as if you've been on another planet,' Lucian says.

~

Over the days that follow, Lucian and I track the course of the Timavo above ground and below: an overland dowsing of this subterranean river. We follow it to where it shows itself and to where it dives back down. It is livelier than any river I know in its disregard for the usual rules of behaviour, and in its happiness in darkness. At the end of each day, sleep feels speleological: a nightly descent, a resurfacing each morning.

Close to where the Timavo first plunges underground, we walk to Mušja jama, a 150-foot-deep fissure in the limestone into which more than a thousand Bronze and Iron Age artefacts were thrown over a period of around 400 years, from the twelfth to the eighth centuries BC. It is clear from the archaeological record that the fissure was a major sacred site, and that people came here from as far afield as central Italy and the Pannonian plain, bearing objects of power – socketed axes, spears, swords, helmets, drinking vessels – which were broken or burned before they were ritually cast into the abyss.

Another afternoon Lucian takes me to the springs of the Timavo, where the river gushes green out of the rock into a parched land of scrub. The springs astonish me, as springs always do. This is water that fell first as rain on high ground, then made the long journey through the underland, to emerge here – filling pool after pool with its energy and colour, before tumbling away seawards.

Life has gathered around the springs. Groves of cypress and pine cast shade. Damselflies jewel leaves. The air is scored by birdsong. Emerald frogs plop into the water from the bank.

An ancient basilica was built here some 2,000 years previously to mark the spring site. Water flows through its narthex and nave. Water is part of its architecture of worship. A votive Roman capital stands above the channel, and its banded dedication reads 'To the God Timavo'.

'They've dived near here, of course,' says Lucian, indicating the stone arch out of which the Timavo surges, 'trying to force a way upstream underwater from the cave just up there. They couldn't get very far, but about eighty metres down they found stalactites in flooded chambers that are now well below sea level, but filled with fresh water from the pressure of the river system.'

We sit by the edge of the springs, take our shoes off, hang our feet in its coolness. I think of other spring sites I know: the power of everyday miracle they share, and the sense of the earth's interior that they open into. The Wells of Dee on the Cairngorm plateau. The spring sites I had seen in the Occupied Territory of the West Bank. And Nine Wells Wood, less than a mile from my home, where a circle of springs flow from the chalk.

'There is, surely, a power of peace to springs,' I say to Lucian.

Lucian shakes his head. 'Not always. This, here, was a front line during the White War – the First World War, Rob,' he says. 'The battle raged right through where we are sitting. It was a death zone.

Countless men died here: the springs themselves were swapped back and forth. None of these trees around us is older than a century, because they were all clear-cut to open fields of fire.'

Two nights later Lucian, Maria Carmen and I go down at dusk to the Adriatic coast, near the Duino Castle, close to the point where the Timavo has its final outflows at sea level. The stones of the beach still hold the day's heat. They are smooth and pale. Some are tinged with purple and have the patterns of fossil plants pressed into them. A white yacht trundles towards Venice on the night breeze.

The moon is low and full in the sky. It has risen early. The tides of the earth move imperceptibly below us. Lucian and I wade out and launch ourselves in. Salt in the mouth, the sea soft and warm to the touch. I turn shore-parallel and stroke north towards a rocky headland. The moon is a silver tunnel mouth.

Then I am startled to feel, writhing around my legs, the cold currents of another kind of water. It is the blue fingers of the starless river, born as snow on the Snežnik, plunging underground to crash through its dark chambers and black rapids, and then at last surging out here to surface under the moon. It is a moment of wonder that will be matched and more on the converse by what Lucian and I will find up in the mountains.

7
Hollow Land
(Slovenian Highlands)

We almost pass it by.

Late afternoon, late summer: harvest time in the mountains to the north of the Carso. Smell of woodsmoke, meadow. Wooden cabins with steep eaves speaking of heavy winter snowfall. An old man sitting in a chair drawn up to a western gable end, eyes closed, catching the last of the sun. Long-handled scythes leaning against walls, cut grass on the blades. Cyclamens in the shade, purple fungi poking through leaf litter under the beeches. Apple trees here and there, lit by small yellow fruit. The land's surface dimpled with grassed-in sinkholes. It is one of the most peaceful landscapes through which I have ever walked.

Then we follow, because we are curious as to where it leads, a side path that turns away from the open ground of meadows and cabins, curving gently through beech and oak, and then angling up, the trees thinning in number but growing in height, poplars now, their leaves hissing in the wind.

We walk the path in innocence because we do not know what is at its end, and through the poplars we can see golden reefs of cloud massing out over the sea, black on their undersides. The sun is warm on our faces, the rich smell of the meadow grass is thickening to rank – and then there is the first of the marks, cut deeply into the pale bark, and there is the edge of the chasm.

In front of us a sinkhole drops into blackness. Its sides are buttresses of grey limestone, softened by moss. The mouth of the sinkhole is twenty feet across at its widest point. To look into it is to feel the beckoning lurch of an unguarded edge. From the upper slopes of the mouth grow sapling beech, perched on ledges, leaning over the drop. Ferns flourish in stone niches.

Gouged into the trunks of the bigger trees around the sinkhole are swastikas. Some are old, because the bark has begun to heal them. Some are fresh; made that year, perhaps, or the year previously. The wood in the cut-lines is still pale. Some of the swastikas have themselves been scored through by knifepoint. The bark is a conflict zone of mark-making.

Nailed to the trunk of a beech near the lip of the sinkhole is a metal sheet, two feet or so high, and blotched with algae. Written on it in black ink is a long poem in Slovenian, entitled 'Razčlovečenje'. At the bottom of the poem is scrawled the word 'PAX'.

'Its title means something like "Dehumanization", or "Becoming Unhuman",' says Lucian quietly. 'My Slovenian isn't good enough to read the rest properly.'

He points to the last line of the text, which has been added with an asterisk: a postscript to the main poem.

'This, though . . .' He pauses. 'This is a curse of some kind. A curse or warning against anyone who might try to destroy or harm the poem.'

The warning has not been heeded. Parts of the poem have been scratched by blade or stone, in an effort to erase the words. Other words have been written across its text – and they in turn have been scored through. In a top corner, another swastika has been cut into the metal, bright and fresh.

I feel a sudden horror reaching up and out of the sinkhole to coil around my heart. Something terrible has taken place here, and continues to reverberate.

'Look,' says Lucian, pointing north through the canopy. There are thunderheads over the peaks now. Rain is drifting in heavy ropes far to the west. There is a distant sense of fury. Out over the ocean the gold light has become a slick yellow.

What happened here? The mouth of the chasm says nothing. The trees say nothing. Leaning over the edge of the sinkhole, I can see only darkness beneath me.

~

Earlier, Lucian and I leave his house in the Carso and travel north towards Slovenia, where the limestone rucks into sheer peaks and deep river valleys. Visible to the north are the spires of the Julian Alps, a towering limestone range where some of the most severe fighting of the so-called White War – the series of battles at the border of Austria-Hungary and Italy – took place between 1915 and 1918. There is, Lucian says, a peak high in the range he wants to reach that has been tunnelled out by war – as so many of the mountains across the front were hollowed during the conflict, for the purposes of taking shelter and of giving death.

At the Julians our plan is to separate, and from there I will walk east for three days into Slovenia, over the shoulder of Triglav, the highest mountain of the region, and down to the blue lake of Bled – though the forecast is for snow on Triglav, which would make such a journey on foot hard. Before reaching the Julians Lucian wants to show me some of the higher ground of the Slovenian karst, where extensive beech forests shelter wolves and bears, and where, Lucian says, a cave system contains an extraordinary presence.

Maria Carmen and I embrace as I leave their house on the Carso. I thank her for all that she has given me. She holds me at arm's length by the bowl of dried pomegranates in the porch.

'Robert, you are . . . a, a, a *bellissimo animale*!'

'Maria Carmen, that is about the nicest thing anyone has ever called me,' I say. 'If I had business cards, I'd print that on them as my profession. Thank you, from the bottom of my heart.'

As we drive north, gaining height on switchback roads, I ask Lucian if I was right to have heard her comment as a compliment.

'Oh yes,' he says, 'the highest of compliments. Maria Carmen regards animals as far more impressive than humans. To her, heart and kindness are more important than any honorifics or degrees.'

We follow the shore-road of a lake called Doberdò. It is completely dry: a grassy meadow through which bare limestone shows in places, several acres in extent.

'Doberdò is what I think in English is called a "turlough",' says Lucian, 'an intermittent lake that wells up from underneath and within the rock when the rains come and the water level rises – but that drains dry in the summer months.'

The roads are lined with cypress trees, planted in commemoration of the dead of the armies that fought here in the course of two world wars. The trees have elegant boles shaped like candle flames, burning green.

'Neither of the world wars has ever really left this region,' says Lucian. 'Brush fires in the Vipava Valley caused unexploded First World War ordnance to go off last summer. You could hardly have a better metaphor for the politics of this area.'

We pass through Nova Gorica, a border town. 'TITO' has been sprayed twice on the exit road in blue paint, reversed on either side of the central line so that it can be read by drivers coming in both directions.

The road rises to a pass and then drops down to a bridge over the Isonzo. The Isonzo runs bluer than any river I have ever seen. It is the blue of Cherenkov radiation, beautiful and chilling.

Lucian pulls over in a lay-by above the bridge.

'A century ago it would have been death to try to go from here to there,' he says, pointing to the bluffs of limestone that rise to either side of the bridge. I realize that the rock of the bluffs is unnaturally textured: perforated with squared-off holes and portals.

'It's Swiss cheese, this stone,' says Lucian. 'Warrened by war. The high ground is a honeycomb of gun emplacements, access tunnels and chambers. The low ground is all trenches and foxholes. They burrowed into the mountains; they made a *war-machine* of the landscape. You'll see even more of this from the First World War when we get up into the Julians, where the snow was heavier and the fighting, if possible, more desperate.'

I have, again, a powerful sense of this landscape as one in which geology both produces and confirms ways of feeling. Here in this hollow terrain of the karst, historical memory behaves like flowing water, disappearing without warning, only to resurge under new names, in new places, with fresh force. Here in this topography of cavities and clandestine places, dark pasts get hidden, then brought to the light again.

We have entered a contested frontier region – part of the Julian March, the name given to the border zone of what are now Italy, Slovenia, Croatia and even Carinthia. Cultures and languages have mingled productively here, but this is also where groups perceiving themselves to be of different ethnic or national identities have visited appalling persecutions upon one another. Traces of conflict are still scored into the physical terrain (trenches, mass graves, monuments), archiving and perpetuating a modern human geography of violence and displacement.

We climb higher. The reflected light off the sea still silvers the sky to our south. There are cheerily coloured beehives, ranged in rows in fields set back from the road. Wild-flower meadows, small vineyards.

We cross a wide pass between high peaks. Beech and pine have thickened in the lower ground, and I can smell resin in the cooler air. A sense of mountain communities and forest wilderness grows. The extent of the woods here makes a nonsense of human frontier lines. Beeches march over borders.

Dapple of shade. Pools of light. A glade, a meadow, a hut. Caves everywhere visible in the cliffs, hidden in the forests. Dips between trees where sinkholes have collapsed, filled, regrown. Big slides of light tilt down hillsides. The upper ridge of one mountain is holed by a window-like gap running right through it, the ancient relic of a long-vanished river. Through the window I can see blue sky and clouds, framed by stone: a surrealist's canvas.

Where the treeline breaks against cliffs, distant beech trunks stand smooth against the rock. Two winters previously, western Slovenia was struck by a severe snowstorm that coated millions of trees in ice, leaving them so heavy that their root systems were unable to bear the weight of their frozen canopies. Millions of trees died, toppled by the burden of their own crowns.

Another valley is sheer-sided on its eastern flank, with 400-foot-high white cliffs rearing almost from the road. Plumb in the centre of one is a cave mouth, and out of that mouth roars a silver river which plummets to a plunge pool at the cliff's base. Rainbows drift in the spray.

I have never seen a feature like it. It defies all the usual geological and fluvial rules. Rivers are not meant to issue from the centre of cliffs. But then the earth is not meant to have tides, and mountains are not meant to have windows – and caves are not meant to grow glaciers.

We find the glacier-cave sunk high in the mountains, where the beech trees grow to sixty feet or more and the canopy is so thick we can hardly see the sky. We follow a thin contouring path, knotty with tree roots, through the forest. The air is heavy, hot.

As we walk, Lucian explains the glacier's existence to me, but I can hardly believe what he is saying to be true. A flowing river of ice at this height, in this heat? There is no lying snow for many miles around.

'This cave system is a mile in length, nearly 400 metres deep, and it perforates an entire mountain from one side to another,' says Lucian. 'The free movement of wind along the cave system, combined with the chill of the rock itself, keeps the temperatures within the system well below freezing. Snow gathers in the cave mouths in the winter, blown deep into it by the northerly winds, and – hey presto! – over thousands of years, the snow becomes a long thin glacier, winding within the mountain.'

The ground begins to fall away to the left of the path. We are soon on the edge of a huge doline, perhaps 150 feet across. The far side is near vertical, but our side slopes at fifty degrees or so, and a thin track switchbacks down into the chasm, where a cave mouth gapes.

With each hairpin of the path, the air chills around us. I have never before experienced such a precipitous temperature gradient. Thirty degrees Celsius at the doline's edge becomes twenty-five degrees Celsius within sixteen vertical feet, and so it drops as we drop, and though we pushed first through tepid air, soon an evening cool is around us, and then as we approach the cave mouth, 100 feet down, the air prickles cold as metal in the nose, our breath feathers in front of us and then we pass into a fine silver mist – the breath of the glacier itself.

The steep temperature gradient produces a steeply raked ecology. The trees shrink in size with each turn of the path, from towering beeches to bonsai pines near the cave's belly, clinging on in the arctic temperatures. By the entrance to the maw, where the temperature rarely rises above freezing, there is only moss and lichen matting – a low polar tundra. The smells at this depth are wholly different from those of the Carso and the forests; instead of heat, herbs, resin and stone, here is moss, winter and ice.

Lucian and I scramble down a short slab of rock, cross the threshold of the cave and step into darkness. I glance back up through the mist to see a crescent of blue sky still visible, meshed with beech branches. I remember the arch of light left behind as we entered the Parisian catacombs. I sense movement in the cave's far corner: a creature of some kind, big and powerful.

Cold is burning my ears and fizzing in my teeth. Underfoot is a hard crust of rocks and debris – lichen, twigs, bones – fallen from the sinkhole's sides, oddly fixed to the touch. Then between two lying branches I see a gleam of blue-black metal. I kick at it with my toe, feel my foot skid. Not metal – ice.

'We're on it,' I call out. 'Lucian, we're on a glacier! It exists!'

Lucian doffs an invisible hat, mock-bows.

Treading carefully now, we move on towards the furthest reach of the cave. The mess of the threshold thins out, and we are walking on blue-white ice that slopes away and down to the corner where the creature lurks.

The creature is a sinkhole in the ice – a vertical shaft cut down into the glacier by the action of meltwater. The ice slopes towards the sinkhole and so too does the light, as if pulled into it. We approach it with care – this black hole set within blue-black ice – conscious of our unsteady footing, of how easy a slip would be. A few yards from its edge we stop and regard it briefly, shivering and chilled.

Returning to the rock slab down which we have earlier climbed, we hear a call.

'*Živjo!* Hello! You want some help?' A man is at the top of the slab and he reaches an open hand down to help us up the final tricky moves, one after the other. A woman is standing on flat ground above the slab, wrapped up against the cold in an ankle-length sheepskin coat. Her chest bulges and squirms, and then a little poodle pokes its head out from between the lapels of the coat, yapping at us.

'That's a fine hot-water bottle you have there!' I say.

'We keep each other warm!' she replies, stroking the dog's head, laughing.

An eagle spiralling far above looks down through the sunlit green-gold canopy, past tall old beech trunks, past the lichen that hangs in wisps from the lower branches, past the gentians that bloom blue amid the leaf litter, down the sides of the sinkhole, past the sloped band of tundra and the bonsai pines, to where Lucian and I stand talking with the man, the woman and her poodle in the mouth of the ice cave, all of us laughing now.

~

It is late afternoon, elsewhere in the upland beech forests, when we come to the place of horror.

We pass through the meadows by the wooden cabins, we follow that footpath up through the woods, past the trunks of the trees with their swastikas scored into them, and we come to a halt at the sinkhole's edge, where the text on the sheet of metal is nailed to the beech, as the clouds build over the sea.

Between 1941 and 1945 the limestone of southern central Europe – from the Cansiglio plateau below the Dolomites, across and down into what was then Yugoslavia – became the site of a brutal conflict. In April 1941, Yugoslavia was invaded by the Axis powers. It was captured and trisected, with Italy occupying southern Slovenia and Ljubljana, Hungary annexing the Prekmurje region, and Nazi Germany taking northern and eastern Slovenia. Germany and Italy soon began ethnic-cleansing activities in their new territories, deporting, resettling, driving out and killing thousands of Slovenes.

In response, partisan groups began to form across the Julian March and beyond with the aim of resisting the occupations. These anti-Fascist resistance groups – nicknamed 'the woodchoppers', and becoming increasingly left-wing in affiliation as the occupation continued, until formally declaring Communist alignment in March 1943 when they united with Tito's partisan army – largely took the forests of the karst as their fortress and battleground. They fought *nel bosco*, in and with the woods. The British and Americans, aware of the power of these partisan troops, began to invest arms and intelligence in their operations. Among the officers sent to support the partisans was Fitzroy Maclean – later famous as the author of *Eastern Approaches* (1949), an account of his time with the resistance in the Yugoslav mountains – and John Earle, Maclean's liaison agent with the Slovene and northern Italian partisans.

The high karst was perfect for strike-and-retreat partisan tactics in occupied territory. Dense forest cover meant ground activity was hard to see from aircraft. Steep-sided valleys and the prevalence of sinkholes made it difficult to move heavy vehicles off the main roads and tracks. Ambushes could be planned on the narrow mountain roads, with attackers firing down onto vehicles before melting away into the woods again, pursuit almost impossible. The ubiquity of natural caves, and the readiness of the limestone to be enlarged into tunnels and chambers by blasting and excavation, made it the ideal geology for a guerrilla war. Weapons stores, sleeping places, even field hospitals were established in the rock, with sly systems of tunnels used to disperse woodsmoke from underground fires, so that the smoke did not rise in a column and betray a position.

From the summer of 1942, seeking to counteract the growing partisan threat, Italian authorities started to create their own 'anti-Communist' militia among ethnic Slovenes, named first the 'White Guard' and then – under Nazi command – the 'Slovene Home Guard'. A brutal civil war developed in the forests and the villages of the karst, aligned chiefly along Fascist–Communist divisions, but also inflaming hostilities between the partisans and Catholic activists in Slovenia. Nationalism, religion and revenge tangled terribly together. Large-scale reprisal killings began to be exercised upon the civilian populations, as well as between fighters.

The worst phases of these reprisal killings came in two waves: in the autumn of 1943, after the Italian surrender, and then during the notorious *Quaranta Giorni*, or Forty Days, of the Yugoslav administration of Trieste, following the fall of the city to New Zealand troops in early May 1945. During these terrible periods, geology and atrocity intersected: the landscape of the karst – which had served the partisans so well in terms of shelter and concealment – was repurposed for mass murder.

Sinkholes, caves, ravines and mineshafts throughout the limestone regions of Venezia Giulia and Istria became the locations of individual executions and group killings, carried out predominantly by Communist partisans but also by Fascist militias. Civilian and military victims were transported to the edges of sinkholes, and there were pushed alive, wounded or dead into these chasms in the limestone. In some cases, victims were bound to one another by barbed wire. Others were buried in scooped graves in the forests. The caves and glades of the karst filled with hundreds, perhaps thousands, of bodies. These extrajudicial killings are known today, especially by Italians, as the '*foibe* massacres' – from *foiba*, meaning a 'sinkhole used for killing'. The bodies of those executed are still being disinterred in the shallow soils of deep woodland, or down in the sinkholes, where cavers will occasionally encounter human bones, bullets, rusting wire.

History itself possesses its own burials and exhumations. The history of the *foibe* killings is still today heavily contested, not least because for decades it was deeply submerged. In the years after the war ended, a strategic 'good-neighbour' policy emerged between Italy and Yugoslavia that encouraged the forgetting of the atrocities. Italian politicians seeking to rebuild a united Italy saw little benefit in focusing on the crimes carried out by partisan troops on both sides. Yugoslav leaders rejected evidence of the existence of Communist atrocities, preferring to emphasize the suffering under Fascism experienced by Slavs, and to align their cause symbolically with the ultimate atrocity of the Holocaust. While the consequences of the partisan war played themselves out damagingly in individual and family contexts across the Julian March, in public discourse it was mostly relegated to *la politica sommersa*, 'submerged politics'.

It is primarily over the past three decades that the *foibe* killings have surfaced again in the public sphere, becoming an intensely controversial subject in the region. For Slovenes and those broadly on the left, the details of the *foibe* killings are seen to have been lavishly exaggerated by the right for purposes of propaganda and political leverage. For Italians and those broadly on the right, the *foibe* atrocities serve as a convenient shorthand for all the reprisal killings, jailings and deportations that happened to Italians during and after the end of the war – and they stand, too, reflexively, for the ways in which the post-war Communist governments of these areas disposed of the history of these persecutions. The language of this ongoing debate is riddled with subterranean imagery, literal and metaphorical. Images of light and dark, burial and exhumation, concealment and revelation run through the discussions: historiography and topography tangle with one another. The numbers and identities of those who died in *foibe* vary considerably, with the numbers given often depending on the political alignment of the researcher in question. In all cases, at stake is what Pamela Ballinger – in her major study of 'the terrain of memory' at the borders of the Balkans – refers to as 'autochthonous . . . rights', meaning the battle for the right to claim authentically to 'belong' to a given area of land, rock and soil.

The *foibe* have also become focal points for contemporary right-wing and fascist groups seeking to stoke popular patriotism and fire anger against perceived left-wing influence in government. The sinkholes have become sites of ritual return for Italian nationalists and exiles. Commemoration marches are held, ending at *foibe*. Swastikas and other marks or mottos are often inscribed at the site. Priests perform annual services of remembrance. Bones of the *infoibati* (those killed in the *foibe*) have been displayed as versions of sacred relic. At the most notorious of the *foibe* – in fact a mineshaft – near a village called Basovizza/Bazovica in the north-east of the Carso, a few miles from Trieste, two contrasting monuments have been established: one commemorating those killed by Yugoslav partisans in the shaft, and one commemorating the 'heroes of Bazovica': four Slovenians shot in 1930 for anti-Fascist activities. The mineshaft at Basovizza/Bazovica was sealed in 1959, in a ceremony carried out by a Catholic priest and attended by 2,000 people, because the deposition of explosives at the time of the killings made the subsequent safe exhumation of the victims' bodies impossible. Because detailed scrutiny of the contents of this *foiba* is unfeasible, it has remained a void – susceptible to multiple projections of claim and belief. More hopefully, the village is also now home to the Elettra Sincrotrone, an international research centre that involves people from all the neighbouring countries and affiliations: it, too, is underground.

Basovizza/Bazovica has become, more even than the other known *foibe*, an example of what Pierre Nora calls '*lieux de memoire*', 'memory-sites': places in a landscape where the meanings of history are most actively created and contested. The matter of the *foibe* continues to resist closure. By keeping these sites 'open', so the history of the past continues to wound the present.

~

High in the Slovenian beech woods, as a storm grows to the south-west, Lucian and I have found our way to the edge of a *foiba* now known as *Grobišče Brezno za lesniko*, the Wild Apple-Tree Shaft Grave. The details of what happened here, as at all the *foibe*, remain unclear and highly disputed. At some point in May 1945, between forty and eighty people – some of them Italian police, some of them

Slovene Home Guard, some of them civilians – are alleged to have been marched through the trees, along the curving track that Lucian and I have also followed, to this chasm. Here they were either killed at its edge and pushed in, or pushed alive into its depths.

The swastikas cut into the tree bark have been made recently by right-wing protestors, who march to this *foiba*, as they do to others, in order to protest the killings and to commemorate those who died here. The scorings-out of the swastikas have been done by objectors from the other side. And the poem has been written by someone chiefly in memory of the victims, lest they be left voiceless by the battles of claim and counterclaim.

Later, a Slovenian friend will translate the poem for me. I should have warned her about where I had found it, what it might contain. I did not anticipate the powers of horror that the text possesses:

Dehumanization

But despite it all, they were people like you and me.
Who are you? The living thrown into the madness,
Killed with clubs and stabbed,
Here crucified and no cross for you.
But O, you humans,
Your bones in the bottomless pit,
They were people like you and me,
Killed in the golden freedom.
As you pass by, stop for a while,
Think of your wrists bleeding in the dark night,
Barbed wire wrapped around them,
As they, cursing, goad you on,

Beaten, naked, a corpse still living,
You can hear the blows of the rifle butts,
The screams, the groans, the terror turning into the sweetness
Of approaching death.
The fear, the pain, are vanishing,
The footsteps echoing towards the void.
In the bottomless pit countless numbers of them lie,
But despite it all: they were people like you and me.

PS: A curse be upon anyone who might attempt to erase this record.

Imagine yourself as victim, the poem orders its readers. Think yourself into the skin of another human, for then – sunk into a different being – you will surely find yourself unable to inflict suffering. It is as unsettling a text as I know: the vividness of the scene of execution it conjures, the curse it threatens as protection against its own erasure. The poem at once challenges and charges its reader, both forbidding and demanding response. Above all, it is a poem about compassion – about feeling as another feels. To the poem's author, the darkness of the 'bottomless pit' represents the utter failure of empathy that characterized the war in those regions, as it must of necessity characterize war at all times and in all places.

~

Apple trees by the roadside, their fruit yellow as lamps. Steady lift of the land. Wide river valleys, and pale limestone peaks rising higher to either side. A vaulted blue sky, strong sun gleaming off stone. We are passing through a mountain paradise, but we drive in silence. The *foiba* has shaken me deeply, and it has shaken Lucian too, I sense, familiar though he is with the hidden violence that this landscape contains.

Birches on the turn, now, leaves seething sulphur. Bindweed flowering white in the hedges. A southerly breeze sets the poplars shaking. Air cools as we gain height. The air brightens. *The shadow past is shaped by everything that never happened. Invisible, it melts the present like rain through karst . . .*

What is the relationship of beauty and atrocity in a landscape such as this? Is it possible, even responsible, to take pleasure in such a place? What had Anselm Kiefer written? *I think there is no innocent landscape, that doesn't exist . . .* I recall Kiefer's paintings of German forests: tall-trunked, shadowed woodlands that bewilder and entrap the viewer, their trees often nourished by the cruelty that has occurred among them. Kiefer's Europe carries an immanent history of guilt and pain. Pines grow tall on bones. Kiefer longs for – but disdains as futile – a soteriology whereby our sins might be absolved by the earth's own stigmata.

The true peaks of the Julian Alps are now beginning to show on the horizon: a Gothic dream-range. Limestone summits spiral up to towered tops. Structures of hollow and fold replicate up and down the scales, from ridges and valleys to the water-scores on a single boulder. Matter shifts appearance, changes place. It is hard to tell cloud from snowfield from pale rock-face.

I think of W. G. Sebald's writing about landscape and the relics of violence; how his narrator in *The Rings of Saturn*, walking the tranquil but chronically militarized coastline of East Anglia, becomes preoccupied to the point of 'paralysing horror' by the combination of an 'unaccustomed sense of freedom' in the landscape with 'the traces of destruction, reaching far back into the past, that were evident even in that remote place'. I remember taking a friend to the former nuclear weapons test site of Orford Ness off the Suffolk coast – where Sebald had also been – and seeing her weep uncontrollably on its shingle bank by the brown waves of the North Sea. The state violence latent at the Ness had caused the surfacing in her, unbidden, of memories of a cruel relationship in which she suffered for several years. *The violent event persists like crushed glass in one's eyes. The light it generates, rather than helping us to see, is blinding.*

We are up into the heart of the Julian Alps now, and at a turn in the road, where it crosses a river bridge, we see an elderly woman sitting alone on a shingle beach by the water. She is in a wheelchair, pushed out among the shore boulders. She wears big dark glasses, dark amber in tint, and her legs are wrapped in a green blanket. Her hands are placed together on the blanket, and she is looking without moving into the whirling blue water of the river. It is not clear how she has arrived there or how she will leave, but she seems to be at peace by the current.

Dissonance is produced by any landscape that enchants in the present but has been a site of violence in the past. But to read such a place only for its dark histories is to disallow its possibilities for future life, to deny reparation or hope – and this is another kind of oppression. If there is a way of seeing such landscapes, it might be thought of as 'occulting': the nautical term for a light that flashes on and off, and in which the periods of illumination are longer than the periods of darkness. The Slovenian karst is an 'occulting' landscape in this sense, defined by the complex interplay of light and dark, of past pain and present beauty. I have walked through numerous occulting landscapes over the years: from the cleared valleys of northern Scotland, where the scattered stones of abandoned houses are oversung by skylarks; to the Guadarrama mountains north of Madrid, where a savage partisan war was fought among ancient pines, under the gaze of vultures; and to the disputed valleys of the Palestinian West Bank, where dog foxes slip through barbed wire. All of these landscapes offer the reassurance of nature's return; all incite the discord of profound suffering coexisting with generous life.

A mile or so up the river valley from where the woman sits watching the water a stream tumbles down to the main river from a side valley. This is marked on the map as the 'Rio Bianco', the White Torrent, and it is to be our path up to the high tops where war was waged exactly a hundred years earlier. We set off from the road-head along a thin track through the beech woods that fringe the stream. The path itself is lightning white where it has been worn to bedrock. It strikes up and on between the trees.

Holes in the trunks of the beeches hold micro-gardens of moss and ferns. Dwarf pines spread between the boulders of the stream-bank. Harebells, gentians and edelweiss star the understorey. Little trout flick as quick shadows in the bigger stream-pools. Towering above us are scree-slopes and bone-white summits jagging several hundred feet up from the ridge line. Can we really be going up there? Always to our left is the Bianco, pooling and splashing. It is a mysterious and wilful presence, a fine companion for our ascent on that hot day, and soon I can resist its invitation no longer.

'Lucian, I'm going to use the stream as my way up.'

'Enjoy. I'll stay dry, I think, and I'll see you up in the cirque.' He gestures up to cloud level. 'Head towards the junction of valleys, and then hook left and up. You'll find yourself in a big, flat-bottomed corrie, with a small bivouac hut bound to the rock by iron cables. We'll meet again there in, what, three hours? Four?'

He wanders on into the woods, and I clamber down to walk the stream.

Sunlight blazes off stone. I hop from rock to rock, climb the bigger boulders, clamber up the faces of plunge pools, and where the stream runs deep and wide I just wade through it, relishing the snowmelt bite of the water on feet and shins. There are water-worn humps of limestone as smooth as skin. Little spill-pools have their own white-sand beaches a few inches wide. Each new section of the stream poses a different puzzle of ascent.

It is a beautiful stream for the whiteness of its light, and it is a strange stream because it tricks. In the stiller pools, the water is transparent to the point of seeming absent, and more than once I stop to dip a hand to prove the water is still there.

The challenge, really, is to keep moving at all, for each pool is inviting as a place to wait and jabble, and each side-stream beckons its own following. At last, in a waterfall-fed basin of polished limestone twelve feet wide, its lower rim giving a view across the valley to a pillowy peak beyond, I swim. It is a natural infinity pool, and I wallow in it for five minutes or so, letting the waterfall pummel my back until it numbs.

I proceed lazily upwards, boulder-hopping, stopping, each rapid luring me up and on, each pool detaining me, until the sides of the gorge become high enough that I risk getting trapped. So I climb out using tree roots for ropes. Seven chamois watch – with the feigned disinterest of true voyeurs – a near-naked man dressed in little but a rucksack clamber over a gorge-edge into a glade, and there re-clothe himself.

From the glade it is up again, the path switchbacking past a hut in a clearing, the trees diminishing in size as height is gained. Purple scabious remind me of the chalklands of home. As I ascend, towering beeches shrink to ten-foot-high mature trees, then to a wide scrub forest, forked with different paths. The scrub is of pines and glossy-leaved pin oak, and the trees stand first at head height, then at shoulder height, then at waist height, then they are gone altogether – and I am out into ground scoured clear of trees by altitude and avalanche.

Bare rock, shrill whistles of marmots echoing, the peaks pressing closer around. Rock towers rising, their forms continued above by the white-bossed thunderheads that are building in the sky, and continued invisibly below by the chasms and cave systems that descend from the landscape's surface.

Flocks of finches sweep across the pines below me, vanishing with a flutter amid the leaves. I climb up through a boulder field to reach the corrie's belly and there, lashed to a flat boulder by steel cables to brace it against ferocious winter storms, is the bivouac hut Lucian mentioned. It is little more than a metal capsule. I open the front door. The space is just high enough to stand up in. Six bunks, three on either side of the space. Stacks of blankets neatly folded on the beds, two full jerrycans of water. A life-saving outpost. But where is Lucian?

I lie down to wait on a promontory of grass near the hut. Warm wind. Pincushions of alpine flora. Clouds, rock, marmot whistle, happiness. Raven cries hexing off the cliffs. Clink of stone-fall, hoof-hit of ibex – *ibex!* – just twenty yards away. The hum of something like silence. The corrie is horseshoed by a great curling wave of limestone, rising to peaks and falling to sharp notches. I know our eventual destination lies to the west, but have no idea how we will reach it.

Half an hour later, Lucian emerges over the lip of the corrie, hot but cheery. I passed him somewhere in the scrub maze without realizing it. We eat apples and drink river water there by the shelter.

'In winter the snow drifts to fifteen, twenty feet here,' he says. 'This gets buried.'

'This place lifts my heart,' I say to Lucian. 'Thank you for bringing me here.'

'I'm glad, Rob,' he replies. 'There was war here too, sadly, although you wouldn't necessarily know it by looking around. They burrowed through rock and scaled cliffs to get at the enemy. But here more men were killed by the winter conditions than by bullets.'

Across the Dolomites and the Julians, retreating glaciers have begun to disclose their contents from the conflict a century earlier: rifles, crates of ammunition, unsent love letters, diaries and bodies. Two teenage Austrian soldiers have surfaced from a glacier in Trentino, lying top to toe beside one another, each with a bullet wound to the skull. Three Hapsburg soldiers melted out of an ice wall, hanging upside down near the peak of San Matteo at an altitude of 12,000 feet. *The problem is not that things become buried deep in strata – but that they endure . . .*

From the refuge, we begin the proper climbing. Up towards a notch in the ridge on a tongue of scree: two steps up, one step back. Sugary snowfields to cross, punching through with each step. Hard, hot, private work. Helmets on now to guard against rockfall. We reach the notch. It is an extreme place. We sit astride it facing one another, as if upon a horse, for here the rock is a spine a foot or so wide. To the south is a huge fall-line, dropping several thousand feet to the white ribbon of limestone that marks the path of the Isonzo, its water shining blue even from this height amid the dark green pines of the valley.

Ahead of us the ridge leads off in peaks, fins and drops. These are the Cime Piccole di Rio Bianco, the Little Peaks of the White Torrent, and their traverse is made possible only by the cables and brackets of the *vie ferrate* – the 'paths of iron' – that have been bolted into place. Lucian and I get into our harnesses. My karabiners still carry the silt and mud of the Abyss at Trebiciano, and at the sight of them my mind flies to that dark chamber, some 7,000 vertical feet below us.

'That's the Canin,' says Lucian, pointing to the other side of the valley. A humped, slumped white mountain, with what look like – but cannot be – vast snowfields descending from its whale-backed summit, gleaming in the light and pocked with holes.

'The Canin is a true karst peak. You can see how the limestone behaves differently. The type we're on is more friable, sharper. The Canin is more of a bread loaf in outline, more like the moon in texture. You have to imagine it in cross-section, too. It's honeycombed with natural cave systems. There are caves with their entry points on the slopes of the Canin that descend for almost two vertical kilometres.'

'A mountain has an inside,' Nan Shepherd wrote in her great study of the Cairngorms, *The Living Mountain*, and it took me years to understand what she meant with respect to that granite range, which appears so outwards-facing. Here in the Julians, though, Nan's proposition seems merely a statement of the obvious. These are hollow mountains, lightless peaks, which everywhere turn in on themselves in the form of valleys and caves.

We are about to begin the traverse of the Cime Piccole when we hear a sustained peal of thunder, rolling in from the north-west.

'This is *not* good timing,' I say to Lucian. 'We're attached by metal karabiners to metal cables, with metal ice axes sticking up from our packs, on the exposed ridge of a mountain, with thunder and lightning incoming.'

'Well, we could wait it out back in the corrie,' says Lucian, 'or we could race the storm and hope it slides past us, or doesn't reach us until we can take shelter in one of the tunnels.'

We race the storm. A two-hour sprint against the thunder. Pinnacle after pinnacle, ticking them off. I remember it in shutter-clicks and sharp shards. Hot rock under the hands. Drops pulling at us. First, second, third tops. Adrenaline, bloodied fingernails, lactic in legs and arms. We are alive in the world, happy to be alive in the world, and the thunderstorm slides slowly past several miles to our north.

The cables of the *via ferrata* interweave with the conflict infrastructure of the First World War. We balance up trembling timber steps hammered into the rock a hundred winters previously. We use rusted iron ladders to cross notches in the rock. We reach the slopes of the ninth pinnacle and there ahead of us, somehow, is a tunnel mouth – gapingly dark in that sunlit upper world. It has been blasted and hacked fully through the pinnacle, and during the war it must have been one of the safest places in this deadly conflict zone, secure from ordnance, lightning and avalanche.

We step into the tunnel, grateful for the respite it offers from the wind, and the shelter it can provide if the storm swings our way. We walk on into the mountain. The tunnel dips for sixty feet, takes two corner turns, and then we are in darkness so complete we must turn on our head-torches. On again, dropping into a lower level by a rusty ladder, helping each other down with raised and reached hands.

Light rises and we round a corner to find a gun window cut out of the limestone, offering lines of fire across the valley towards the Canin. The rotational iron circle on which the gun was wheeled to east and west is still set into the stone of the emplacement. A recoil space has been cut back into the interior wall. In this confined space, the detonation noise of each shell-fire would have been shattering. The men who worked this gun must have lost their hearing almost immediately.

Light shows again at the turn of the tunnel – a doorway of light now. Having passed through the phases of that hollow peak – light, dark, light, dark, then light again – we are at the end of the ridge, and above a scree-slope that falls away towards a col. I think suddenly of *Le Passe-Muraille*, the figure in the catacombs who walks through walls.

I run the scree, skidding down to green sloping pastures seamed with paths made by chamois and by people. Patches of old yellow snow lie in the shadows of the peaks. A mile or two away I can see a hut, perched at the point where the ground drops thousands of feet to the valley. It holds the promise of rest, food, company. Thoughts of war tumble away. Clouds pass rapidly over the sun, casting an occulting light over the landscape.

~

The hut is presided over by seven-year-old Theresa and her white cat, Luna. Theresa's father is the custodian, though he keeps to the back room. Theresa's mother is nowhere to be seen. Theresa makes the pasta for dinner, and comes out to greet us with flour on her face, carrying Luna under one arm like a rugby ball. She speaks to me in Italian and I to her in English, and neither of us can understand the other but it doesn't matter at all.

Seeing Theresa, my heart aches for my own children. I haven't seen them in almost two weeks. The darkness of these beautiful landscapes has leached a little into me, shadowing the edges of sight and spirit. I want to be with them, to make them safe.

The hut is a reliquary of the White War. Its windowsills are lined with the debris of death, picked up by walkers over the years. Shell fragments, bent bayonets, bullets, boot buckles, helmet spikes and chinstraps, and a shell-casing peeled back like a banana skin by the blast. It is a grim museum of slaughter.

There is a small library of books, many of them concerning the war. I sit on a wooden bench, reading about what happened here. There are black-and-white photographs of the fronts that shifted across these mountainsides, the men who fought here. Tunnel after tunnel, portal after portal cut into the stone of the peaks. Men in shadow, looking over to enemy cliffs perforated as the flank of a cruise liner. To get *inside* the mountains was the only way to shelter from the killing avalanches, from the killing cold, and from the killing enemy shells and bullets. These Alps became weaponized peaks, their topography forcibly reorganized by the imperatives for cover and concealment. Shell-fall alone lowered the height of one mountain by twenty feet. The theatre of the White War extended from the summits of the peaks, through their hollowed-out interiors, down into the caves of the slopes and valleys.

I am reminded again of Eyal Weizman's study of the landscape architecture of the Israel–Palestine conflict, *Hollow Land*, and its proposal of 'elastic geography', whereby space is to be understood not simply as the backdrop for actions of conflict, 'but rather the medium that each . . . action seeks to challenge, transform or appropriate'. Weizman mapped the 'elastic geography' of the West Bank and Israel: the attempts made by erecting walls and fences hermetically to seal areas of territory, the nonsense made of such sealings by the tunnels that were dug under these barriers by Palestinians in order to smuggle people and weapons, and the arcs described by rockets fired out of Gaza by Hamas militants. He wrote of the reconceptualizations of space undertaken by both sides in the conflict: the way the disputed terrain ran vertically from the militarized airspace far above ground level, all the way down into the competition for control of the water aquifer that lies deep in the limestone, sunk thousands of feet below the West Bank. Weizman's name for this fluxional space is 'hollow land', possessing as it does 'a complex architectural construction . . . with its separate inbound and outbound levels, security corridors and many checkpoints. Cut apart and enclosed by its many barriers, guttered by underground tunnels, threaded together by overpasses and bombed from its militarized skies, the hollow land emerges as the physical embodiment of the many and varied attempts to partition it.'

Something comparable happened in the Julians during the White War. In this 'laboratory of the extreme', new types of warfare were developed, and new transformations of space occurred. Mountains were seen no longer as solid structures, but as honeycombs that could be opened, the interiors of which could be traversed, the walls of which could be walked through. The landscape itself had become actor, agent, combatant. In the Second World War, as Lucian and I saw at the *foiba*, it would be differently instrumentalized into a means of execution.

Theresa brings Luna over to see me. She drops him onto my lap, then holds him by his ears and kisses him full and hard on the mouth. Luna yowls in protest and digs his claws deep into my thighs. I yowl in protest and dig my fingernails deep into my palm. Theresa wanders off, pleased with the result.

We share the hut with four Triestini. They are regulars, two couples who come up here often from the city to spend time together: ski-mountaineering in the winters, climbing and caving in the summers. They fold us into their conversations, share stories of the mountains. One of the Triestini is a broad-shouldered man, bearish of build. He wears an orange fleece and a blue bandana, and his hair is sweated tight to his scalp. He explains without bravura that he is an extreme caver. I am surprised; he seems quite the wrong shape for such a specialism. I do not convey this thought to him. He gestures over at the Canin.

'There are some of the deepest caves in Europe from surface to lowest point,' he says. He comes to sit with us, points out on our map the entrances to the caves.

That night there are lightning flashes far in the distance, illuminating the Canin. Lucian and I go out onto the balcony to watch the show. We can see its pitted limestone plains in the scorch-light. They look like the asteroid-cratered surface of the moon. It is unearthly and beautiful.

We watch the storm, time the gap between each lightning bolt and its accompanying rumble of thunder.

'You can hear stags bellowing from down in the valley a little later in the year,' Lucian says after a while. 'Such a sound – haunting and violent. It drifts up from below and echoes around the cirques.'

Later the storm reaches us, and the rain strikes the tin roof like bullets.

~

We wake to calmness and a miracle.

A cloud-sea fills the landscape below us. The valleys are fjords and we are islanded. As we watch, the cloud surges slowly upwards, roiling higher until I have the illusion that we are sinking – an atoll shuddering down into white water. Green of the pines shows amid eddies of mist and pinnacled peaks; a Chinese scroll painting unrolls before us.

We set off to the west on a thin path that cuts between sheer cliffs above and sheer drops below. We pass in and out of the cloud-sea as the path moves. Where waterfalls drop from the cliffs above, we must crouch and run through them, meltwater clattering on head and neck.

Feline tracks on a snow patch. An ibex glimpsed in the distance. A pair of mating black salamanders in a moist clasp on the path's white stones, their long toes and fingers pressed ardently against one another. More embrasures, more tunnels, every cliff riddled with them, the whole mountain range a hive, really, a terrible war-hive. *We are the bees of the invisible . . .*

A flock of choughs dips far below us with pinging cries. Two chamois flee on the gallop, then stop on a boulder, looking back over their shoulders at us. Trenching is still visible in the rocks and soil, grassed over. We walk freely across a side valley, the openness of which once meant death. Coils of barbed wire are sunk in turf and stone.

We drop off the high path and take a contouring line down into the cloud, which gathers us into its white world. At a stand of wild raspberries we stop to eat. The berries are tart in the mouth. On and down from there for hours, and as we fall the sun rises to burn off the cloud.

Early in the afternoon we reach the valley floor, through which flows the young Isonzo. Here, where it passes over the karst of the Canin, its water is ice blue. I want to slip into it and float to the Adriatic. Lucian and I rest on a shingle beach by a deep pool. Trout shadows flip for flies, or hang wavering upstream. What did the mountaineer-mystic W. H. Murray say after being released from years spent in German and Italian POW camps? *Find beauty, be still.*

A fine mist rises from the water and hangs white above the river, so that the water is clearer than the air. The trees that border the river are lush with moss and lichen. This is not a rainforest but a mist-forest, and through it runs this otherworldly river. I find a flattened, rounded black stone on the shingle, and throw it into the middle of the current. The stone rocks down through the blue water to the riverbed, where it part-buries itself in the white sand.

THIRD CHAMBER

Down in the labyrinth under the old ash tree with the riven trunk, a last passage is chosen, followed. It dips fast down, twists, curls, then steadies out. A gravel beach is reached, at the edge of which dark water pools to depth. The rift-roof dips to meet the water. The only means of progress from here is into the pool and on through the flooded passageway.

So into the pool, the water black as stone and cold as snow, the cold running quick as dye to bone, no sight, no light, feel for ridges in the roof-rock and kick on, lung-air hot and red now, head-pressure rising, rising, rising . . . then up and out at last into clear air beyond, gasping in the darkness on the far side of the water. Is this what death feels like? Or birth?

Another chamber has been entered. Stalactites run from ceiling to floor. A light is struck, lifted, shifted. The chamber walls are alive with image and story – each slope of stone bearing a scene from the underland. These are scenes of haunting and afterlife, shifting across time but echoing each other.

A woman's body is prepared for burial in Thessaly in the fourth century BC. Her lips are closed by a coin bearing the head of a gorgon, there to pay the ferryman who will carry her over a dark-watered river towards the realm of death. Placed on her chest are two heart-shaped leaves of gold foil, into which metal words have been etched. Together the leaves form a *Totenpass* – a death pass or death map. The text they bear is for her to read in the underland; it gives her directions to the dominion of the dead, where she will be placed in the care of Persephone. The text warns her of mistakes made by others, who have not navigated their way to safety in the underland and are now condemned to haunt the mortal realm eternally as spectres. *You will find on the right in Hades' halls a spring, and by it stands a ghostly cypress-tree, where the dead souls descending wash away their lives. Do not even draw nigh this spring . . .*

A man walks open country west of Pennsylvania in the 1860s. He carries a silver coin in his pocket and divining rods in his hands. He walks, then stops, waits, and seems to listen. He bends at the waist to bring his ear closer to the ground. Listens again. Watches the rods, waits for their twitch. Nothing. They hang limp in his hands. He walks on. The man is a medium, a geologist-spiritualist, an oil speculator. Oil is a gift given from God, limitless in its underland abundance, divinely stored for the use of mankind. One must just know where to find it. For the oil emits 'coruscations' – atmospheric glitterings above ground, perceptible to those few people sensitive enough to detect them. The man walks on through the grassland – and the rods in his hands begin to twitch. His spirit guides have led him at last to the place where he will site one of his Harmonial wells. He stops, listens, confirms his soundings. Smiles, kneels on the ground, takes the silver coin from his pocket and pushes the coin deep into the turf. *This* is where the drill bit will bite. *This* is where the oil will rise.

The year is 1971 and a Soviet drilling rig squats on the sand of the Karakum Desert in Turkmenistan, near the village of Darvaza. Suddenly there is a crack, then a roar, and a disc of the desert floor 230 feet in diameter shatters and collapses into the abyss that opens beneath it, swallowing rock, sand and drilling rig in a few seconds.

The void migrates to the surface . . . The drilling has punctured a natural-gas cavern, the cavern's roof has collapsed, and now poisonous fumes are pouring into the upper world. The decision is taken to ignite the gas and burn it off. It is expected that this will take only a few weeks. More than four decades later, the pit is still on fire. It has become known as the 'Door to Hell' and 'Hell's Gate'. At night its orange flames light up the desert for miles around. People travel from around the world to approach its rim and sleep within the radius of its glow.

Early this millennium, on the sweltering north coast of Java, a lake of toxic mud has spread over four square miles of landscape, gouting out of a central crater from which a plume of foul-smelling gas also rises, and burying twelve villages. This mud volcano began to erupt ten years previously, shortly after a multinational corporation, drilling for oil in a Late Miocene stratum some two miles below ground, ruptured a high-pressure aquifer and opened a series of blowout vents on the surface – from which ever since has flowed this torrent of ancient, poisonous sludge. By some the mud volcano is seen as a consequence of corporate greed – an unnatural disaster. By others it is seen as an emanation of *batin* – of the submerged occult forces of the underland, of ghosts and spirits that dwell in the landscape and exist far beyond human bidding.

A great flock of snow geese, more than 25,000 birds strong, whirls across a plain in the American west in 2016. The birds have been driven from their usual flyway by a snowstorm, and they are desperate for a place to land where they might escape the wind and cold. They pass over the gleaming red-black waters of a flooded former open-pit copper mine. It seems to offer refuge, so first one goose dips its shoulders and then ten birds follow that single bird and then 10,000 birds follow those ten birds, draining down in a welter of wings and honking cries to the pit, where they settle, shrug their feathers, and drink gratefully. But the gleaming water – 45 billion gallons of it – is toxic because of the mining that has taken place here: highly acidic and contaminated with heavy metals. Thousands of the birds die and become a new surface, hundreds of floating acres of dead geese, dark-barred and white-winged, folded over one another in the pit.

The same year, a man dressed hood to toe in white stoops to pass through a narrow doorway, braced by a frame of steel, and into the darkness of a chambered tomb. The wall is made of rough concrete and is more than two feet thick; it is known as the Sarcophagus, and the space it encloses is the Reactor Cavern. The man, who has a camera slung around his neck, moves on through the cavern, his torch-beam lighting his surreal surroundings. He sees bent wedges of fallen steel, twisted girders, contorted piping, control panels slumped and dripping. This is a space that has been reorganized by a force beyond imagining. There were once seven rooms here, stacked on top of one another, *but they are no longer in the same places or in the same order* . . . A stalactite of lava runs from ceiling to floor, thicker in girth than the chest of a human, formed of molten rock, rubber and uranium: to stand by it for a few minutes is to die. The man has forty minutes at most in the Sarcophagus before he is overexposed. In what was once a control room he stops, raises his camera, and takes a slow-shutter-speed photograph.

Later, he develops the image. It should be an image of darkness. But down through the image falls a scatter of white points of dust, like static or a fine snowfall. These points are not dust, though – they are the imprints on the photosensitive film of pure energy, the radioactivity that was swarming invisibly around him in the Sarcophagus, swarming *through him*. They are the dazzling radio-autographs of uranium, plutonium and caesium – burning points of light that ghost the eye.

PART THREE

Haunting (The North)

8
Red Dancers

(Lofotens, Norway)

Looking across the bay to the northern shore – and there by the glimmering birches is a figure standing dark on rising ground, where no figure should be.

Two oystercatchers flicker over the water between us with quick cries of alarm, catching my eye as they fly.

Looking back across the bay to the northern shore – but now by the birches there is nothing there, no one at all.

~

Days earlier, sailing the Vestfjorden in worsening weather, due to make landfall at Moskenes an hour before dusk. Sunlight pools to the south, then is soaked up in shadow. Small snow-squalls blow in and blot out sight from the boat. Snowflakes hum in the air with angry speed.

Impossible islands grow to the west. I glimpse a long band of black and white; crag and snowfield between low grey cloud and high grey sea. Gleams of light on the snow, in the gullies and the shallower flanks. So much more snow than I have expected to see – and the peaks themselves even steeper and sharper than I have anticipated. The long band of land widens as we approach.

Mountains shade into view, developing out of the squalls like photographs. A scatter of red-walled, black-roofed houses. Thousands of cod, frozen hard, hanging by their slit throats in ranks from A-frames of wood, clacking in the wind. The squalls thickening to an easterly blizzard and a buzz of worry building in my belly.

Later I will remember the days that followed mostly as metals. Silver of the pass. Iron of the bay and its clouds. Rare gold of the sky. Zinc of the storm in its full fury. Bronze and copper of the sea to the south as I escape.

~

'Look out for them,' Hein says to me in Oslo. 'There are more of them there, without doubt, more figures on that shore.'

Pause.

'But first you'll need to get over the Wall safely. I only ever went there by boat, the long way round, in the summer. You'll be walking there, in winter.'

Smile.

'Have you ever thought of taking up smoking? It's never too late to learn!'

Pause. Smile.

'Smoking can be a good survival skill in a landscape like that.'

~

The majority of European prehistoric cave paintings are located in the chambers and shelters of south-west France and northern Spain. As you move further north from this region, the quantities of such art decline and their age also decreases. Above sixty degrees north, relatively little exists.

The main reason for this scarcity of painted art at higher latitudes is that much of this landscape was buried under glaciers until the end of the last Ice Age. Twenty thousand years ago, when the seventeen-foot-long red aurochs was being painted in the Hall of the Bulls at Lascaux, in what is now the Dordogne, all of Scandinavia and most of Britain and Ireland were still glaciated. As the ice slowly retreated, it left behind a shattered landscape scoured of life. Northwards human colonization of this barren terrain happened only slowly.

Geology also has a role to play in the rarity of surviving northern-latitude painted cave art. Cave chambers form the most secure gallery sites for such art, and such chambers form most naturally in limestone: Lascaux, Chauvet, Altamira – all of the most celebrated prehistoric artworks were made in and on limestone. Limestone has the added curatorial power of often running a film of transparent calcium carbonate over wall paintings, which then sets and acts as a preservative varnish, mitigating degradation of the pigments. Northern Europe is sparser in limestone than Spain and France, though, and richer in igneous and metamorphic rocks. Where caves or overhangs form in such rock types, they do so by the erosive forces of ice or seawater and as such tend to be shallower and rougher-sided. Their interiors lack the inviting canvases of water-smoothed limestone. A jagged granite cavity does not offer the same pictorial possibilities as a limestone chamber pillared with stalactites. Arctic-latitude prehistoric rock art does exist in Europe, including the astonishing concentration of work at Alta in far northern Norway, where more than 6,000 images – predominantly petroglyphs – depicting reindeers, bears, humans, hunting scenes and the aurora borealis were made between *c.*7,000 and 2,000 years ago on glacier-polished rock. But painted art – far more vulnerable to damage and weathering than incised imagery – is scant.

Some of the most striking painted rock art from these northern landscapes is found in the decorated sea caves of Norway's western coastline. To date, twelve sea caves have been discovered to contain such art, dispersed over a 500-mile distance from Nærøy in the south to the Lofoten archipelago in the north. All of these caves are located in remote areas, often on wild coasts where peaks fall sheer to the ocean. All of them have been smashed into sea cliffs or crags by the hammer strength of wave action over many millennia. Some of the caves could – at the time of their painting – have only been reached by boat, requiring hazardous navigation of the exposed coasts of islands and peninsulas.

Together these 'painted caves' contain around 170 simple stick figures, arms and legs wide as if dancing or leaping: mostly human, but with occasional human-animal hybrids, and one image of a single hand. All the figures were made using red iron oxide pigment, and all were daubed using fingers or brushes. Dating the art is challenging, but the most secure estimates – based in part on the radiocarbon dating of artefacts found in the caves, including an arrowhead of polished slate, a seagull leg bone with a drilled hole that may have been used as a flute, and a great-auk amulet – place the making of the figures to somewhere between 2,000 and 3,000 years ago.

These painted figures are, then, Bronze Age peri-Arctic artworks, made in some of the harshest country in the world by hunter-gatherer-fisher people who moved along an isolated coastline, surviving there only by the gift of the Gulf Stream's warmth. Their lives would have been short, arduous, and might reasonably be thought to have left little space for the creation of art.

Yet the red dancing figures exist.

One of the remotest of these painted caves lies near the western tip of the Lofoten archipelago: the island chain that extends for almost 100 miles into the Norwegian ocean at a latitude of around sixty-eight degrees. The cave's modern name is Kollhellaren, which translates roughly as 'Hole of Hell', and it lies near the tip of the island of Moskenes, on its uninhabited north-western coast.

There are two ways to reach Kollhellaren. One way is on foot, over what is known as the Lofoten Wall – the precipitous ridge of peaks that runs down the centre of the island chain, and that in winter can be traversed only by means of a small number of passes. The other way is by boat, rounding the tip of the archipelago, and passing through the notorious Moskstraumen, one of the strongest whirlpool systems in the world, about which Edgar Allan Poe wrote his 1841 short story 'A Descent into the Maelstrom' – in which the whirlpool is figured as the portal of a tunnel leading to the core of the Earth. The Old Norse name for a maelstrom was bluntly pragmatic: *havsvelg*, 'ocean-hole' – a hollow in the sea into which everything flowed.

Two entrance points to the underland, therefore, lie close by one another: a mouth of rock and a mouth of water, locked off by fierce mountains and fierce seas.

The people who created the art of Kollhellaren more than 2,500 years ago took considerable risks just to reach the site of their making. Before even entering the cave-space, they had to cross powerful landscape thresholds.

~

Winter has returned to the Lofotens when I reach them. Arctic gales have blown from the west for four days the previous week, stripping the windward slopes of loose snow, and dumping it in the east-facing gullies of the Wall as storm-slab. Avalanche risk has risen from low to moderate, and is due to keep on rising: 'Storm-slab avalanches likely in east and south-east facing aspects, with triggering possible by high additional loads upwards of 300 metres.' This is not the avalanche forecast I want to hear for my plan of reaching Kollhellaren and its red figures on foot.

There are only two points close to Kollhellaren where I can cross the Wall in winter. Both present difficulties in these conditions. One is a gully that cuts up below an adze-shaped peak named Mannen. The other ascends a slabbed shoulder of peak. I consider them on a map. The gully is much steeper, but will probably hold less snow. The shoulder is a less severe slope, but looks more avalanche-prone. I decide on the gully. I like gullies. They hold you. You feel you're likely to fall less far. They're more comforting than ridges or shoulders, even when they're more dangerous.

The night before I set out for Kollhellaren, it snows steadily from dusk. I am in a tiny village called Å, at the very end of the road that snakes almost the full length of the Lofoten archipelago. Beyond Å there are only lakes, peaks and sea. I am staying with a retired fisherman called Roy. Roy broke his pelvis and leg falling onto a winch from the dock at Å six years earlier, after thirty-eight years of fishing. He accepted a state pension and retired early to take photographs.

'You should not go over the Wall,' Roy says that evening. 'It's not the right time of year. There is nothing over there on the west side. No houses, no people, no cell-phone signal. Just the cliffs and the sea. And snow. Why do you want to see Kollhellaren anyway?'

I think of trying to explain: how the figures have come to fascinate me since I first heard of them years earlier. How I am trying to understand what drew their makers to that hard, strong place to leave their mark. But it seems too frail a structure of reasoning to risk exposing, just when I need my confidence most.

'I just want to see the cave and its figures, and be over there on the west side for some time,' I say.

Roy shrugs. 'There have always been Englishmen doing such things, since Slingsby,' he says.

We discuss his Indonesian holidays instead, and his relationship with an Indonesian woman that went fabulously right and then fabulously wrong for him out there. He shows me a video of the small palace of black marble and pink stucco that he built for her to house her nail-parlour business. We look at photographs: Roy straddling a moped outside the palace, with its candy-coloured horns and raked slate roof; Roy eating with his partner in a restaurant, shirtless, smiling.

That night, I cannot sleep. I open the curtains, stand by the window and watch snowflakes rushing like fire-sparks through the light of the last street lamp in the archipelago. It is an oddly peaceful sight, but I know it means the snow on the peaks and in the gullies will be building, and that the avalanche risk will be growing.

Early next morning, as I am preparing to leave, Roy rustles in his freezer and pulls out a plastic bag.

'These are five fishcakes, made with the *skrei* I caught two days before, round off Helle, not so far from where you may be going.'

My pack is already too heavy, but I push them down into the outer webbing.

~

Looking back afterwards, from the far side of the Wall – though there were other hazards there, and other wonders too – I recall the crossing chiefly as a white whirl, a dissonant combination of finely granulated decision-taking and chaotic blur.

I follow a dead-end road away from Roy's house and out of Å, a little after dawn. The fresh snow squeaks under my boots. It has fallen softly through the night, settling to a six-inch layer. Sound is muffled. The village is asleep. Mine are the only tracks on the road.

The gully rises from the head of a long low-lying lake called Ågvatnet that runs westwards from Å, draining a horseshoe of peaks to its north, west and south. Movement along the lake shore is immediately difficult – rocks slippery underfoot, under snow. The water of the lake is frozen to steel in its main reaches, clear only at its outtake where the current keeps the water moving. A riprap of stranded ice-plates has been piled on the bay shores by the recent winds. Roosting on the lee crag of a rock island in the lake's centre is a colony of gulls. Their chatter and screams are comforting in that austere valley: the convivial sound of social life. Far ahead of me, black cloud hides the peaks almost to their bases. This worries me. It will be hard to locate the correct gully.

Slow going over snow-hiding boulders and slick rock. Trip, slip, fall, hard to get back up with the pack. Four times there are small crags to be scrambled, the hand- and footholds outwards-tilting and icy, requiring delicate, orderly movement.

Then the land eases, opening to a wide bowl of open ground at the lake's head that rises gently for half a mile or so to the foot of abrupt cliffs. Scrub birch forest grows here; it is hard to force a way through. Bushwhack, posthole. Low wreaths of quick-moving white cloud conceal and disclose the terrain around. There is no sunlight. Just water-streaked rock, wind chill and the occasional rumble of small avalanches. I have a strong sense of the terrain's disinterest, which I might at other times experience as exhilaration but here, now, can feel only as menace.

In the lee of a boulder just short of where the Wall rears up into cloud, I rest and take stock. Still no sight of the peaks themselves. Small spindrift cyclones roam the slopes. I can see the starts of three gullies ahead of me, leading up into the cloud. I know from a photograph I have been sent that only one is passable: the other two lead to sheer cliffs. Avalanche debris has gathered on the run-outs of all three gullies. I am reassured by its nature, though: it consists mostly of larger chunks of snow, rather than full avalanche fans.

How to choose in this poor visibility? Left, right or middle? The left-hand gully seems to veer too far to the west to be correct. The right-hand gully looks truest, but also appears to narrow abruptly as it enters the cloud-line. I remember that I have the photograph of the gullies on my phone. I pull the phone out, try to read the image against the land itself. But the photo was taken in late spring; it shows black rock and a few lines of snow. It bears no meaningful resemblance to the blizzarded wall of white ahead of me.

Rattle of rockfall.

I choose the middle gully through a mixture of hunch and *eenie-meenie-miney-mo*, hoping I can reverse it and choose again if I must.

Crampons on, helmet on, ice axe out. I move up to the gully mouth, and dig an avalanche test pit where the slope steepens. There is clearly give in the top layer: new windslab slipping over the harder, older snow beneath. Not so good. But the volume of windslab in the gully doesn't seem enough to bury me if it does go.

So – proceed. Just.

Up into the gully proper now, the ground angling, the axe necessary. The snow in the gully's throat is deeper than I expect, thigh-deep, so that I am soon wading up in a steep white river. Little avalanches are already beginning to trigger, which make me uneasy, so I wade across to the left side of the gully, where it curves up at its edge like a gutter. There at the edge it is rockier, the snow thinner and icier, so there is less chance of an avalanche. But there, too, is a more serious drop and greater chance of rockfall. That trading-off of hazards between avalanche, drop and rockfall becomes the trick of the ascent: picking the best-fit line that optimally minimizes all three.

Time slows, swirls, repeats. Each step is hard going, the heavy pack peeling me back off the slope or jamming me into it. Spindrift hisses into my face, frets my cheeks. I murmur a mantra to myself: *Take the time that needs to be taken, take the time that needs to be taken.*

Why are you here? Why are you here? ask the rocks and the wind in reply.

Still no sight of the pass. Is this the right gully? Then a crunch comes beneath me, a sudden drop and – *bang!* – hard snow punches my lungs. I am sunk to my arms. My legs are dangling in a void of some kind. *Think, think*: a snow crevasse. I must have dropped partially through into a fissure formed where the old snow flows over a bulge of rock. I *really* don't want to drop fully through into the space beneath. I don't know its extent, but I do know that it will be a hell of a place to get out of under these conditions. So carefully I cut, pull, swim, *float* myself free, as if getting out of quicksand. A long reach with the axe to gain purchase, a couple of heaves with knees and feet, and I am out and above it, and then – *there!* – eighty feet or so above me I can see the top lip of the gully, and clear air above it. I have chosen the right path and this is the way over the Wall.

But thirty feet short of the top, the slope becomes more severe. It is heavily loaded with windslab, and a small cornice has formed at the rim – a frozen lateral wave of snow, perhaps five feet long, curling out and over the lip above me.

I don't like the look of either the cornice or the loaded slope, so I explore possibilities among rocks to the gully's left side. But the terrain is even more serious there, perhaps just fifteen degrees off the vertical. The front points of my crampons skitter on exposed granite, and with only one axe I cannot tool my way up. The fingers on my left hand are beginning to freeze where I have been jabbing them into the snow for grip. I sense a serious fall opening below me, and so I retreat from the left-side rocks, carefully reversing the ten or so moves I have made to that point, move by move. *Take the time that needs to be taken.*

So: the cornice, then. Pace by pace, diagonally up to the loaded exit ramp. Windslab falling away below me with each step, in yard-wide chunks. Avalanche a real risk with every movement now. On and up, placing each foot as carefully as if walking on thin ice over deep water, until I am just beneath the cornice. I get myself as stable in my footing as possible, kicking the front points of my crampons in deep, and then go to work with my axe on the cornice. Chunks tumble around me and away down the gully. Six or seven blows and I have cut a passage through it. I reach up into the gap, plant the axe – *thunk* – in the frozen turf of the ridge beyond, kick up and over and through the cornice, and haul myself onto the saddle of the pass with a whoop.

I lie on my back, a gaffed fish, breathing heavily, and there above me, showing through the mist, is a sea eagle, low and circling, and the queasy fear in my throat is forgotten and my heart leaps to be overflown by that remarkable bird in that remarkable place. Then I think, *It's just sizing you up as lunch* – and I laugh out loud at my stupidity and the land's indifference.

~

To reach and enter one of the painted caves of the Norwegian coast would have been a 'rite of passage' requiring 'physical and mental ordeals', writes Hein Bjerck, the archaeologist who discovered many of the painted caves and who I met in Oslo before coming to the Lofotens. The ordeals were several in number: firstly the journey to the site of the cave itself, and secondly the passage into the cave, crossing its two key thresholds: first the entrance mouth, and then the point at which light fails and dark takes over. Bjerck writes of the challenging short-term visits made by the artists as 'ritual actions', journeys to 'the outer fringe of the human world'. He notes, too, how the surviving names for cave-sites continue to emphasize their status either as performance space or as access point to a hostile otherworld: Church-Cave, Hell's Mouth, Hell's Hole, Troll's Eye.

These caves are, unmistakably, dramatic places. The Troll's Eye is a wave-smashed tunnel around 100 feet in diameter, running east–west wholly through the rock of a small island, in which the setting orange sun is framed once a year. Bukkhammar Cave is set into a sea cliff so sheer that the mouth itself can be reached only from the water: a vault visible from the sea for miles in clear weather. Solsem Cave contains an overhanging panel of rock more than 100 square feet in area, on which is painted a monumental cruciform figure. In Fingal's Cave, the most southerly of the painted sites, at the point where the cave splits into two main passages and deepens away into the rock, stands a sharp menhir, the front face of which is struck by the sun's rays for a short time twice a year. And Kollhellaren itself is an immense north-facing cruciform cave, 150 feet high at its entrance, with a 600-foot-long gallery system. During the midsummer weeks, the outer reaches of Kollhellaren flood with the yellow light of the midnight sun.

Of all archaeological specialisms, the study of prehistoric rock and cave art is among the most speculative. These acts of marking are irrefutable – but the immediate circumstances of their making are scarcely retrievable. It is hard to locate with confidence the intent or significance of individual artworks in wider webs of cultural practice.

It is possible to say, however, that the art of the painted caves of Norway is part of a circumpolar cultural presence left by the northern populations of Eurasia during what is now called the Bronze Age; other artwork from this period includes the complex of incised rock art at Bohuslän in southern Sweden. Most of this art is located at liminal places – coastlines, riverbanks, caves; sites where, as Richard Bradley puts it in *An Archaeology of Natural Places*, 'land meets sea', darkness meets light, and worlds 'come closest together'.

At Bohuslän – during the same centuries that the red dancers were being painted in the northern sea caves – a densely ritualized landscape was developed in a transitional site close to the coast. There, burial cairns were built in number on higher ground above the sea. Where bedrock was exposed, worn by glaciation to offer an ideal surface for inscription, hundreds of carvings were made. Hauntingly, many of these carvings are of footprints, which leave track-lines running down the angle of the rock into which they are cut. These ghostly prints – made by beings who are not present in any form other than the impress of their feet – seem therefore to record the passage of walkers from the barrow cemeteries on higher ground, down to the sea itself – as if spirits were leaving their tombs to make a final foot-journey to the domain of the deceased. Bradley connects the stone-prints at Bohuslän and the Norse myth that the recently dead must be helped on their journey to the otherworld by the provision of 'hel-shoes' – specially cast soles that will allow the spirit to make its journey along 'the path from the grave to the world beyond'.

The painted caves of northern Norway are all, also, clearly strong transitional sites. The presence in the painted caves of at least one figure wearing ceremonial headgear suggests possible relation with the three-tiered Saami cosmos, whereby the universe is arranged vertically into three layers – the sky, the earth and the underworld. Only shamans and the dead are able to pass between the tiers, by means of an *axis mundi* that – in the form of a river or tree – connects the upper and lower spirit realms to the living present of the middle realm. Terje Norsted and Bjerck both propose that practices in the painted caves may have been rites of passage, permitting mortal movement – through the membrane of the stone – to the cosmic underland or overland.

The paintings and petroglyphs of these extreme landscapes may also be understood as an early form of land art, whereby the specific place of making (the cave interior itself) is chosen not only for the pragmatic purposes of shelter and preservation, but also as part of a potent larger location that extends in radiating context both outwards (to the cliff, bay and coastline that contain it) and inwards (to the implied further depths, metaphysical and actual, of the cave's interior). In the case of Kollhellaren, certainly, it is hard to imagine that the cave's closeness to the Moskstraumen Maelstrom was not considered part of its power as a place of making.

Any encounter – modern or ancient – with those painted figures will therefore necessarily be shaped not only by the confrontation with the red forms themselves on the cave wall, but also by the details and moods of the landscape beyond the point of darkness – by the fall of sunlight or of snow, the temper of the sea, the gliding presence of eagle or the flowing presence of otter – and by the experience of reaching the caves of the dancing figures in the first place.

~

My laughter up there on the pass is whipped away by a wild westerly, from which the Wall has sheltered me during my ascent. It is a hostile wind, this: low gale-force in strength. Troubling, too, given that I will be on the exposed west coast for several days to come. Visibility is fifteen yards. The ground falls fast away below me into whiteness. Hail needles rattle my jacket. The drop into the mist on unsure ground is problematic, but there is no question of reversing the gully. As I begin the descent I recall the sense of doors shutting and locking behind me that I felt in the boulder ruckle in the Mendips.

This westerly flank of the ridge is slightly less severe than the eastern gully, though, and I feel comfortable moving down over mixed winter ground, as I have done often in mountains before. It requires probing and prospecting for a viable line: testing gullies, gathering information in the cloud from the way slopes and crag-lines fall in order to determine which route will dead out in a cliff, and which will take me safely lower.

I make a long traverse down the flank, losing height wherever I can, using a tongue of snow then crossing a buttress of rock to reach the next tongue, moving with particular care on slick rock and tussocked grass, steering away from what I sense to be a big drop to my south-west.

After twenty minutes of this tricky work, the cloud begins to thin. Lines show through the white, broken black and grey-green, hard to read except as abstract forms. A roaring builds in the air. A tear in the cloud – and there is the shoreline, 200 feet below. I can see white waves foaming on black boulders, a scatter of driftwood timbers – and, puzzlingly, hundreds of perfect spheres, dark orange in colour.

Half an hour after that, I make sea-fall. I drop the pack, sit on a rock, take stock. I look south-west along the shore I must now follow for several miles to reach Kollhellaren.

Black walls of wet granite fall sheer to the sea, which from here look almost impossible to traverse around. Sharp skerries out to sea. A bay of sand, then a bay of rock.

I am wet through from the gully, and the cold is beginning to soak deep into me. This is probably the most intimidating above-ground landscape in which I have been. It is a place for whatever self-reliance and composure I can muster.

Scattered all around me on the beach are the spheres. They are, I now see, hollow iron net-floats from fishing trawlers – vast numbers of them, beached and rusted, like alien eggs. Between and around them is a thick wrack of plastic jetsam, repulsive in its presence on this wild coast: plastic bottles, tangles of nylon netting, fish-crate fragments.

Far to the north-east a patch of blue shows in the clouds, and for a few seconds there is a glitter of light out on the water below. For those seconds I love that blue with all my heart, dream-dive deep into it, drown in its hue.

~

Hard, slow miles along the shore. Boulder fields, scrub forest, crag. Always the cliffs rising sheer to the east, and always the waves falling white to the west.

A pair of ptarmigan whirr off on silver wings. A snow hare stops among mossy rocks, white on green.

Bilberries, heather, moss. But no water. No fresh water. Caught between salt to the west and ice to the east, eating snow for a dry mouth.

Through a bay with rocks big as houses, navigating a canyon maze between them. Pop of wrack, kelp slicks.

Hail falling.

A boulder field so densely mossed the rock cannot be felt beneath the feet. Lichen bearding the trunks of stunted birches.

Sleet falling.

A bay of black-gold sand, bound by marram, angled from the base of ice-lined cliffs.

Rain falling, then hail again.

Forest of birch and willow with a six-foot canopy. Birch bark gleaming in the light, first buds bursting furred on the willows.

Up and over crag and boulder to a headland shoulder, each step sore now, the wind colder. Pack heavy, head heavy, throat chilled, body older.

Headland after headland, until there at last to the west is the bay and beyond it, perhaps, the vent of the cave. Green sea over white shell-sand in the bay. Guard-arms of rock curving out and around to either side, the bay water calm, though the outer ocean by the Maelstrom is in chaos.

Five sharp-coned summits rising steep from the shore to the peak called Hellsegge, each higher than the last. From the top of each flows a white plume of cloud, bent flatter to the east, and there, *there*, set low in the belly of one is the black vault of a cave.

~

Boom of waves on the offshore reefs. Two sea eagles circling, unmoved by the wind, silent. *Ping ping* of crow-cries sounding steel off the cliffs. Croak of a raven.

I reach the north side of the bay below Hellsegge – named on the map as Refsvika Bay – exhausted, excited. It has taken me more than an hour to move each mile over this exceptionally obstructive terrain.

On raised ground I find what seems a good place to pitch camp. It is exposed if the wind were to swing to the north, but that is its only serious vulnerability. Two big boulders give shelter from the west wind.

Crucially, a deep pool of rainwater has gathered in a dip in the tundra, with a white gull feather floating at the leeward end. The eastern edges of the pool are clotted with the hailstones that fell earlier. I drink handful after handful until my skull aches from the cold.

Underfoot is a layer of heather, moss and lichen, soft as a winter duvet. I lie full-length in it and sink down a foot, the heather rising up and leaning over me in a gesture I experience as a sheltering. I lie there for a while, looking up and out, feeling the anxieties of the day flow from me. Late light glints in the west of every raindrop held in the bones of the lichen, beading on the bosses of moss.

Lying there I sleep, unexpectedly, for half an hour or so. Rain wakes me, a brief squall, then the wind drops almost to nothing for the first time since I set off at dawn. I pitch the tent, and put the whalebone owl in one of its corner pockets, the bronze casket in another. I have hated the casket's weight that day, resented the burden it has added to my load. When the work of setting up camp is done, I eat Roy's fishcakes. They are the best food I have ever tasted, no competition.

In the Celtic Christian tradition, 'thin places' are those sites in a landscape where the borders between worlds or epochs feel at their most fragile. Such locations were, for the *peregrini* or wandering devouts of circa AD 500 to 1000, often to be found on westerly headlands, islands, caves, coasts and other brinks. This place, now, is one of the thinnest I have ever been.

~

That first night at Refsvika is uneasy, broken. The weather turns again. Wind clatters the flysheet. Hail showers blow through and over, spitting on the canvas. Rain falls for hours at a time. I wake for good at five from sleep to sleet; eat, drink water from the feather-pool. The waterfalls have frozen overnight, high on the cliffs.

There are two bays to cross to reach Kollhellaren, and in the first of them is the remains of a settlement.

From the mid nineteenth century until the mid twentieth century, a tiny community survived at Refsvika: a handful of houses, a handful of families. There were twenty-two inhabitants in 1900, thirty-eight in 1939. They kept cows, which cropped the grass of the thin belt of land between the cliffs and the coast. The men fished the rich waters (cod in the winter and early spring; pollock and ling at other times) off Helle. When the weather was fierce, which was often, the cows were led into Kollhellaren cave for shelter. The bay was just enclosed enough to allow safe anchorage for fishing boats, even in winter storms. The community had no electricity until the final decade of the settlement, and there was no way in or out except by boat over the Maelstrom, or on foot over the mountains, a considerable journey even in summer. For much of each winter, the inhabitants of Refsvika were locked off from the world beyond.

Between 1949 and 1951 – like many of the island communities

along Norway's coastline – the people of Refsvika were 'brought in': relocated with the aid of government subsidy to larger settlements, in this case to Sørvågen on the lee side of Moskenes Island. When the families left Refsvika, the houses were demolished, and most of their stones and timbers were carried by boat to Sørvågen, where they were used to build new dwellings.

I follow the land's curve round from my camp spot. Oystercatchers pipe alarm and scatter at my approach. Five eider ducks ride the swell out near the mouth of the bay, moving as if they are of the sea and not on it. I pass between two boulders that are covered in a yellow lichen I do not recognize.

Movement in the corner of the eye and I see that there is a family still living in the ruined settlement: four otters, sea otters, two parents and two children, loping uphill through a boulder field, their fur still slick with sea, moving fluidly between rocks, chattering and mewing but not once looking at me. I lean against the northern boulder and watch them travel, watch them flow, watch them *pour* themselves one by one into a mossed hole between boulders and vanish. I am struck with happiness to have seen them there, at home in their habitat.

I reach the first of the former houses, surviving now only as a vanishing ground-plan of stones. It reminds me of the derelict crofts and cleared villages I know from the Scottish Highlands and islands. Here, as there, moss and lichen are reclaiming the stones. Here, as there, small straight birch and slender young rowan flourish in the stones' lee. Walking on, I count the remains of twelve houses. Few are more than a single layer of stones high, and saplings step through the interior of each. I cannot guess at the resilience of the people who lived here for so long with so little. What must it have been like to be part of a community this small, in a place this hostile?

The bay itself is of coarse white shell-sand, flecked with fragments of whelk and mussel – and strewn with human debris. A doll's head, two toothbrushes, shards of plastic bottles, pots and hanks of blue rope, tangles of nylon with rusty hooks, net-webs rolled up with weed.

Something I heard an archaeologist say in Oslo about deep time returns to me: *Time isn't deep, it is always already all around us. The past ghosts us, lies all about us less as layers, more as drift.* Here that seems right, I think. We ghost the past, we are *its* eerie.

The crags are seamed with blue ice-falls. My eye is caught by a thread of green, drawing my sight on. It is a thin path, leading between the stones, running a fine line through the moor grass, joining former doorway to former doorway and then on around the bay, picked out by the bright mosses that grow on it. It is a path made perhaps a century ago, still there as a trace in the land, now kept open by otters and others.

I add my own feet to the path, thank it for its softness underfoot, for its elegance of route – and for its movement within time.

~

A summer's night 3,000 years ago. At this latitude, in this season, darkness scarcely exists above ground. Low tide, calm sea. A small group of figures follows the shore, stepping from rock to rock. The cave's mouth is vast and its lower lip sits close to the waterline.

The figures pause at its threshold. The distant roar of the Maelstrom. A sea eagle turns overhead, wing-tips near the cliffs that drop sheer to the water. The figures pass one by one into the cave – and the world changes.

Colour thins out. Yellow of the late sun ebbing. Green is gone, grey rises. Grey of the rock, streaked with brown, streaked with red. Wet sand underfoot. White of the sand. Black of the deeper shadows ahead. Smell of damp stone. A hundred feet into the cliff and the last full light falls on a pale central buttress of rock, around which the cave-space splits. It would be a good canvas – but it is too near the outer world of waves, and eagles, too near time told in its usual ways.

A passage to the right of the buttress rises straight ahead before ending in stone-fall. A narrow tunnel cuts away into the mountain to the south-west. And a high rift, taller than a human, shaped in cross-section like a teardrop, climbs up into the rock to the north-east, into full dark.

The figures follow the teardrop rift, moving up between fallen stones.

Here in the shadows, space and time spill into one another. If life exists it is the slow life of rock, it is the sea's patient exploration of the mountain's inside.

Up where the tunnel wall overhangs them, the figures halt, make their preparations. Rock is to be the painter of rock. In a cup of stone they crush haematite and mix it with spit, earth and rainwater to make a red paste.

The painting begins.

Dip of a fingertip and a single sure red line moves across the pale slope of rock, curving down in an arc that echoes the chest and one leg of a figure that is dancing, a figure that is jumping.

Dip again, reach, draw down in a curve to make the second leg of the dancing figure.

Dip again, a crossed line for the outstretched arms – and on to the next figure.

Dip, draw – single sure red lines moving over the slope of the rock, filling the slope of the stone with dancing figures.

In the shifting light of a burning torch, and the faint steady light of the distant summer sun, the figures on the rock seem here to move – to sway with the play of shadow and flame. These are presences made to exist in the dark but also perhaps to survive it.

Dip, drag, and a fingertip draws a line through time – to a summer's day in 1992.

A young archaeologist named Hein Bjerck is investigating a cave on the Lofoten archipelago's remote west coast. The weather is fine, the sea is flat-calm – what they call on the islands *transtiller*, 'oil-film still'. He and a friend have sailed around in a small boat early that morning. The cave lies under towering sea peaks. Hein and his friend are there because shell fragments found in silt on the cave's floor have been dated to 33,000 years ago. They want to dig test-pits that might reveal the details of this site's ancient human history: to see if they can track across this gulf of time something of the hunters to whom the cave gave shelter on this edge.

They anchor, dinghy to shore, haul up the dinghy and climb grass and rocks to the cave mouth.

Smell of moss, smell of stone. Pause at the threshold. Roar of waves on far reefs, the distant churn of the Maelstrom. A sea eagle turns overhead, wing-tips near the cliffs that drop sheer to the water.

The figures pass into the mouth of the cave – and the world changes. The cave twists into the cliff. Time reverses space – the deeper in they get, the younger the cave-space. The journey into darkness is a journey to the present. The sea has taken thousands of years to win each yard of stone.

Hein tilts his head and the light of his head-torch falls upon the west wall of the cave, slips off it again – *but what was that?* – flicks up again, searches, settles, finds nothing, searches again, and there, *there*, is a faint red line too firm and sure to have been authored by the rock itself, moving against the fall of the wall, too counter to gravity to be deposit from run-off, and there, *there*, is a cross-piece to match it, cutting boldly through the first line, and suddenly there, *there*, shimmering out of the dark is a red figure, a leaping red figure of a person – and another, and another.

The discovery, Bjerck will say later, is like 'a shooting star' – unexpected, undeserved and magnificent – and it leaves him with a desperate longing to experience such a moment again, once more to be the first person in thousands of years to set eyes on these figures dancing in the dark.

He begins years of travel up and down the western coasts, sailing and walking to cave after cave, an undertaking that moves from longing to addiction. He finds himself drawn both in his dreams and in his daily life into what he comes to call the 'cavescape'.

And he does find more figures, enough to feed his addiction. Red figures, always red, almost always the same simple form, leaping and dancing in the darkness of caves up and down the coasts, familiar in their shape now and yet still utterly mysterious in their making. Each time he finds them his heart leaps too and there is a collapse of time, or a coexistence of multiple kinds of time, as the figures dance and flicker in the low light.

Dip, drag, and a fingertip draws a line through time, to a late-winter's day in the now, and a man alone on the bay near the cave.

I walk the last few hundred yards to the cave mouth on ground falling sharp from cliffs to the sea. No choice but to keep tight to the apron of the cliff, though the faint threat of rockfall here means my pace is hastened. Snow banks below soaked cliffs shining black with water. Bird-cries doubling off the rock. Drop to the sea and up through bare earth and grass to the mouth.

I pause at the threshold, look back and around. Roar of waves on far reefs, the distant churn of the Maelstrom. A sea eagle turns overhead, wing-tips near the cliffs that drop sheer to the water.

The scale of the entrance is astounding. A 150-foot-high zawn. The arc of the bay, the maw of the cave: this is an unmistakably performative space, a place for meaning-making. The cave is a slip-rift, an entrance to darkness where time shifts, pauses, folds.

Quick clicks of water falling, drops curving silver to the eye from the granite far above. Lichen blotching the entrance orange and grey-green. A prickling in the shoulders as I cross the threshold.

On and down the main rift, pupils widening, the light still here but the colours already falling away. A hundred feet in, the cave becomes cruciform in its architecture. Two side-rifts cut away to left and right, separated by a white bastion of stone against which the space breaks into three. I place a hand flat on the bastion and feel the cold rush fast up the arm.

The air here thrums, sounded by sea and wind, driven into this hollow so that it turns upon itself. This space has been won by wave, by war.

I take the left-hand rift, rising up and into the rock. Yellow-white granite to the higher side, leaning away, and leaning over me now is a darker stone, streaked with red and black, broken and veined. A teardrop shape of clear light is visible back at the bastion.

I am here at last. Such a long, cold journey to reach this place. I rest against the rock behind me, let my eyes adjust to the shadow, look at the wall of granite ahead.

But there are no figures on the stone.

None at all.

I look again. Stare. Search.

There is nothing here.

All this way, all these miles, and the dancers have vanished. Were they ever there at all?

I lean back, right back against the rock behind, let the stone take my skull's weight, let the shadow settle into eyes that are tired of looking.

And when I open my eyes and look again, there is – yes, there, *there* – the flicker of a line that is not only of the rock's making. The line is crossed by another, and joined by a third, and there, *there*, yes, is a red dancer, scarcely visible but unmistakable, a phantom red dancer leaping on the rock. And there is another, and another, *here*, a dozen or more of them, spectral still but present now, leaping and dancing on the rock, arms outstretched and legs wide, forms shifting and tensing as I blink.

Their red is rough at its edges, fading back into the rock that made it, blurred by water and condensation, and all of these circumstances – the blur, the low light, my exhaustion, my blinks – are what give the figures their life, make them shift shapes on this volatile canvas in which shadow and water and rock and fatigue are all artists together, and for once the old notion of ghosts seems new and true in this space. These figures are *ghosts* all dancing together, and I am a ghost too, and there is a conviviality to them, to *us*, to the thousands of years for which they have been dancing here together.

Suddenly, unexpectedly, my head begins to tingle and then my back and my chest start to shake, and I find myself crying, sobs shuddering my body in the teardrop-shaped rift, far from another human and so close to these generous figures. The dangers of the journey to reach the dancers ebb from me, the joy of their movement ebbs into me and I cry there, surprised and helpless, deep in granite and darkness, weeping for feelings I cannot name.

The sea eagle gyres by the cliff. The waves crash on the boulders below the cave. The Maelstrom spins and unspins. *The hands of the dead press through the stone from the other side, meeting those of the living palm to palm, finger to finger* . . . Time proceeds according to its usual rhythms beyond the threshold, but not here in this thin place.

~

'Art is born like a foal that can walk straight away,' wrote John Berger, 'the talent to make art accompanies the need for that art; they arrive together.'

In December 1994 three French cavers, led by Jean-Marie Chauvet, were exploring the Ardèche gorges close to the great meander of the Cirque d'Estre. Using smoke from a mosquito coil, they detected air moving from a boulder-choked limestone fissure high on a valley side. They cleared the boulders and dug out what revealed itself as the entrance to a downwards-sloping tunnel, just wide enough for the thinnest of the cavers – a young woman called Eliette Brunel – to crawl down. Using a chisel and hammer, Brunel was able to clear the tunnel of its stone snags such that her larger companions could follow her. After thirty angled feet the tunnel dropped near-vertically towards what seemed to be a large chamber. They descended this chute and were excited to find themselves in a substantial space, later measured at around 1,300 feet in length and up to 165 feet wide. In places stalactites stood like pillars, fusing floor to ceiling. The three advanced further, sweeping their torch-beams around in wonder. It was every caver's dream: to be the first to discover a chamber of such dimensions, and to explore the systems with which it was connected.

Then Eliette cried out and all three of them stopped, astonished. Her torchlight 'flashed onto a mammoth', she recalled later: 'then a bear, then a lion with a semicircle of dots which seemed to emerge from its muzzle like drops of blood . . . We saw human hands, both positive and negative impressions. And a frieze of other animals thirty feet long.' Giant stags with fabulously tined antlers roamed the walls of the chamber, rhinoceroses fought with locked horns, a single owl perched on a rim of stone. Some of the images were incised, others depicted in red and black pigments. On a high-standing slab of rock sat the skull of a bear.

The trio had entered what would become known as Chauvet Cave, also nicknamed 'the Cave of Forgotten Dreams', and it contained the greatest gallery of prehistoric art ever discovered. An uncanny sense of immediacy haunted the space at the time of that first modern entry. Some of the paint palettes used to make the work more than 30,000 years previously were still on the cave floor, abandoned beneath the paintings they had helped create. The tapers used to light the chamber had been dropped where they were held, their black ash spilling across the limestone. Many of the walls had been scraped clean before the paint was applied or the incisions made, to increase the contrast between mark and medium.

The art of the chamber has an astonishing liveliness to it. Despite the rudimentary materials and the lack – to our knowledge – of any kind of training or tradition on which the artists could draw, the animals of Chauvet seem ready to step from the stone that holds them. The horns and cloven hoofs of the bison are painted twice, the lines running close to one another, to give the impression of movement – a shake of the head, a stamp of the foot. The horses are painted with soft muzzles and lips, which one wishes to reach out and touch, feel, feed. Sixteen lions – muscles tensed, eyes fixed with hunting alertness on their quarry – pursue a herd of bison from right to left across a wall of stone. This is, you realize, an early version of stop-motion; a proto-cinema. *Art is born like a foal that can walk straight away . . .*

Throughout the cave there is, strikingly, little foreground present – no line of landscape or vegetation on which these creatures exist. They have no habitat save the rock and the dark, and as such they seem to float free, unmoored from the world. They exist at once as exquisite anatomical drawings – and as embodiments of a world-view utterly different from our own. These animals live, as Simon McBurney memorably puts it:

> in an enormous present, which also contained past and future. A present in which nature was not only contiguous with them, but continuous. They flowed in and out of a continuum of everything around them; just as the animals flow into and out of the rock. And if the rock was alive, so were the animals. Everything was alive.

Perhaps, concludes McBurney, 'this is what truly separates us' from the makers of this art: 'not the space of time but the sense of time . . . In our minute splicing of our lives into milliseconds, we live separated from everything that surrounds us.' Certainly, the three discoverers of the cave recognized, as they stood there that first day in 1994, something of that older sense of being. 'It was as if time had been abolished,' wrote Jean-Marie Chauvet, 'as if the tens of thousands of years of separation no longer existed, and we were not alone, the painters were here too.'

The modern history of cave art is scored by such shooting-star moments of discovery, of which Chauvet is only the most luminous. The stories of another of the great unearthings vary, but one version runs as follows. In September 1940, four months after the German invasion of France, a teenager called Marcel Ravidat, out with his dog in the woods near the village of Montignac in the Dordogne, discovered a fissure in the limestone near an uprooted tree that was wide enough to squeeze through. Enticed by a local rumour of a secret hiding place containing buried treasure, Ravidat returned with three friends, and together the four young men eased into the opening, and then followed a long passage on and down into a chamber deep inside the rock. The chamber did indeed hold treasure – but not of the kind they had anticipated. For the walls of this rotunda-like space were, like those of Chauvet, covered with paintings: a miraculous bestiary of animals that, in the dim light, seemed to be moving. A frieze of thirty-six animals circled the gallery, comprising six stags, a bear, eleven aurochs, seventeen horses and a fantastical creature resembling a unicorn. More galleries led off the rotunda, and their walls also held spectacular paintings, created more than 15,000 years earlier: hundreds of horses with bristling manes, a stag with curlicued antlers, throwing its head up and rolling its eyes back as it bellows, aurochs, oxen, cats, bears, and a human with a bird's head facing a bison, bending its own neck in defiance to show its horns.

Five years after the finding of Lascaux, other people were discovering other chambers of darkness elsewhere in Europe. On 27 January 1945, Soviet troops pushing west through Poland entered the death camp of Auschwitz eleven days after the Germans had evacuated it, driving the survivors of the camp on a brutal westwards march that would kill more than 15,000 of them. In the haste to abandon the camp the Germans did not have time to destroy its infrastructure. The Soviets found the dark interiors of the gas chambers, the bodies of the dead and the dying, and the after-effects of mass killing on an unthinkable scale: hundreds of thousands of folded men's suits and women's dresses; mounds of dentures and eyeglasses; tons of shorn female hair. Through the months ahead, Soviet and Allied troops would reach and enter dozens of labour and death camps, encountering in those places evidence of the worst crimes of which humanity had ever shown itself capable. Many of the men who 'liberated' the camps and the gas chambers were never able to describe what they had seen there. In this way, the generous secrets of Lascaux – as Kathryn Yusoff writes in a brilliant essay on these doubled discoveries – 'became known just as everything visible on the surface was in darkness, illuminated only by the exploding field of destruction. In this ruptured landscape, a gift of such wealth arrive[d] to suggest the potential of the universe to be otherwise.'

The philosopher Georges Bataille visited the cave at Lascaux in 1955, fifteen years after its discovery, at the point when the nuclear arms race was entering rapid escalation, and atomic testing was being pioneered in underground chambers and desert spaces. A new order of destruction was declaring its possibilities: the annihilation of a species, of a planet.

'I am simply struck,' wrote Bataille after surfacing from Lascaux, 'by the fact that light is being shed on our birth at the very moment when the notion of our death appears to us.'

~

I stop on the threshold of the cave, stepping out of rock and into air. The rain is heavier now. The landscape comes back to itself: first brightness, then colour. Surge of water, echo of wave in the cave-space behind me. I pick my way back along the bay-line, towards the remains of the settlement.

I have a strong, strange sense of being watched.

Gulls watch me from shit-stained rocks in the bay.

What did I see in the dark? A shadow-play of pasts, events refusing sequence, the fingertip drawing its lines through time far from the well-lit world, there in the unfathomable cave. This was a place that absorbed those visitors who crossed its threshold – as it had absorbed me, another in the long history of meaning-seekers and meaning-makers in its shadows.

A sea eagle watches me from the burly air above Hellsegge.

I think of the other dark spaces I have entered in the underland. I do not know yet that I will enter one more, 400 miles south-east of here, that is perhaps the darkest of them all.

Oystercatchers watch me from the sand of the bay.

Wave-water moves between big shore boulders, surging up around my feet as if upwelling from within the earth, and there rises in me a longing to hold again those people I have loved who are dead.

Otters watch me from among the mossed stones of Refsvika.

I look across the bay to the northern shore and there, *there*, by the glimmering birches is a figure standing dark on rising ground, where no figure should be. The figure is in silhouette and does not move; it is human-like and it is facing me.

The figure watches me from the birches.

Then two oystercatchers flicker across the water between us with high cries of alarm, catching my eye – and when I look back across the bay there is nothing on the rising ground, no one there at all.

~

Early evening on my last night in the bay, and the wind falls almost to nothing. After the days of gale, the silence is astonishing. Freed from the wind's rushing, all other sounds are crisper. I sit on a flat stone near my tent.

The tops of the peaks are clear, and show their snow. The sky above them is streaked in blue and the sun glows through a haze out to sea. A windless half hour. Waves still boom out on the reefs. A calm starts to rise in me.

Then I hear a noise. It resembles a jet engine starting up. A grainy roar, building steadily in volume. I cannot place it. It worries me. The air temperature begins to drop. I see that plumes of cloud from the top of the peaks above the cave are now not streaming eastwards towards Hellsegge. They have swung southwards, are streaming inland, and they are longer. The wind is blowing again, but now from the true north. It is strong and cold, and growing stronger and colder. I understand that the roaring noise is the sound this new northerly makes as it rushes over the granite summits. The sea is already starting to chop and brawl, and it has changed in colour from grey-green to grey-black. My tent shifts and tugs on its weak moorings.

A wall of white sweeps towards me, and hailstones the size of peppercorns hiss into the lichen around me. Then small flakes of snow, then spikes of sleet.

That night there is no hope of sleep. The northerly builds and howls, and so do my worries. How will I get out of this locked-in space, this trap of a bay? The surf on the reefs sounds like bomb-blasts, detonating every few seconds.

At midnight a gust of blizzard punches the tent flat and rips all but two of its pegs free. I have no choice but to fight my way out of the collapsed tent, and then carry it whole into a soaked dip in the land, weight the corners down with rocks, and crawl back into what shelter is left.

Half-light comes at four in the morning. I am too cold to stay still any longer, hunched in the sodden canvas. I walk to a high point from where I can see the sea through the ongoing blizzard. It is a shocking sight. A hell has broken loose beyond the encircling walls of the bay. Big grey waves mosh and smash. Spray shoots fifty or sixty feet into the air where surf strikes the reefs.

Sleet blacks the sky to the north. A guillemot whirrs just above wave level, at home in this storm. Then there – *there*, can it be? – I see, out in the direction of the Maelstrom, a thin wire of light, running below the blizzard. It is a gleam of bronze, and it suggests that somewhere out beyond the storm, sun is falling on water, and this in turn tells me that the storm will end before too long – and this means that I may have the weather window I need to leave the cave and what it holds.

For a long time after those days at the cave of the red dancers, I find myself unable to shake the sense that I have left one of my selves back in the bay – left a figure on the shore. This feeling is powerfully with me as I travel still further north up the Norwegian coast from the Lofotens, to the big Arctic island of Andøya in the Vesterålen archipelago, where a battle over the sea's underland is under way.

9
The Edge

(*Andøya, Norway*)

'I have four pets,' Bjørnar Nicolaisen tells me at 69.31° N, 'two cats and two sea eagles. I feed them all together on the shore, there by the throne, with the best fish in the world!'

He gives a huge laugh, and points east through the window of his living room: snow-filled fields sloping away to a rocky beach that borders a fjord several miles in width. Steel-blue water in the fjord, choppy where the currents are running. Far across the fjord, ranks of smooth-snowed peaks gleam in the late sunlight. They are shaped more wildly than any mountains I have ever seen before. Witches' hats and shark fins and jabbing fingers, all polished white as porcelain. I cannot see a throne on the shore, though.

'Here, try these.' He hands me a pair of binoculars. Black leather-clad barrels, weathered in places to brown. Polished eye-pieces – and a Nazi eagle engraved into the left-hand barrel-back.

'Wehrmacht-issue,' says Bjørnar. 'Beautiful lenses. An officer's. When my father was dying, he asked me what I wanted from his possessions. "One thing only," I told him, "the binoculars you took from the Germans."'

I lift the binoculars and the shoreline leaps to my eyes, close enough to touch. Calibrated cross-hairs float in my vision. I pan right along the beach. Nothing. I switch back left. Yes, there, a chair of some kind – but six or seven feet tall, built from driftwood lashed and nailed together. It looks like something the ironborn of Westeros might have made.

'I take the eagles a cod or a saithe whenever I come back from a good day's fishing. I feed them by my chair, there.'

'Bjørnar, you are the only person I know who counts sea eagles among his pets.'

'I am more of a cat person,' Bjørnar replies.

'Than a dog person or than an eagle person?'

'Than a people person!'

Bjørnar laughs and laughs – a deep, explosive laugh coming from far inside his chest.

~

The blizzards of Lofoten lighten and then clear as I travel north to Andøya. My first day ends in a cloudless dusk at Andenes, the town at the island's northernmost tip. Andenes is a town of wide streets, hard winters and night-sailings. The chimneys wear chrome caps. A magpie chatters on a street light. There is a violet haze to the air, and a burning cold. Peaks carry fine ridges of snow. The sea opens away from the town. North from here is 100 miles of ocean and then the Svalbard archipelago.

The sunset is opulent, all purple and orange satins behind the peak line. Later a white moon hangs over the ocean.

The next morning I go to see Bjørnar and Ingrid. Their house is a few miles south of Andenes. It is set back from the road and faces east towards the sea channel that separates the island from the Norwegian mainland. Cross-country skis and poles are leaned up against the garage.

I ring the bell, the door is flung open and there is Bjørnar, roaring a welcome, putting out one great hand to grasp mine, and at the same time clapping his other to my forearm and seizing it tightly.

I am straight away in the grip of Bjørnar Nicolaisen, and I will not leave it for many days to come.

'Come in, come in!'

Black leather flat cap, a short white beard, a grey woollen fisherman's jumper. Sixty years old, I guess, or fifty, or seventy. Massive arms and chest. Wide-set legs. A vast smile, an even bigger laugh – and the strangest eyes I have ever seen.

Bjørnar's pupils are white-blue, so pale it seems he must be blind. They are the eyes of a seer, unsettlingly steady. He holds me still for a moment, looks me up and down, and I feel those eyes see into and through me.

Then: 'This is Ingrid!'

Ingrid is wearing red fluffy Liverpool Football Club slippers and she is carrying a baby. She smiles the kindest of smiles, and motions an apology that she cannot shake my hand.

'This is our granddaughter, Sigrid,' says Ingrid. 'Coffee is in the flask. Come, sit, be comfortable.'

A cat with tortoiseshell fur and lizard eyes yawns on the carpet of the living room. Instead of a row of flying ducks on the wall, there is a row of four brass sea eagles ascending the wallpaper in orderly flight and decreasing size. Two polar bears are tooled into the iron doors of a wood-burning stove. On decking outside the window, headless cod dry in racks, swaying in the slight breeze like wind-chimes.

Bjørnar is a fisherman, a fighter, and he understands the underland of the sea, and for these reasons I have come to meet him. In the winters Bjørnar fishes long days, from five in the morning until seven or eight at night. Winter is cod season, and it is just ending as

I reach Andøya. When the cod are running he goes out in the dark of these high latitudes, comes back in the dark, and it is dark during most of the time that he is at sea, save a few hours' light around noon.

Bjørnar fishes alone. There is no one to see if he goes overboard or the boat is swamped. The temperatures he works in can be as low as -15°C for a fifteen-hour working day. But the cod are the prize for the risks and hardship – and such a prize. The finest fish from the best cod waters in the world; cod up to seventy kilograms in weight. *Kaffetorsk* – 'coffee-cod' – as the biggest fish are nicknamed.

Like many people who have arduous and dangerous jobs, Bjørnar is uninterested in narrating his hardships. Fishing is the task, hardship is the cost, and the rewards are clear to him: he is the sole ruler of his floating kingdom of one, he earns a living, and he satisfies his profound love of the sea. He has no intention of giving up his fishing until his body forces him to do so. Life on land is, anyway, scarcely less hazardous. Fifteen years ago Bjørnar fell twenty feet between levels in a factory. He drove his wrist into his forearm and fractured his pelvis. He was in hospital, he tells me, waving his hand dismissively at my concern, 'for some weeks'.

There is something of the polar bear to Bjørnar: there in his powerful physique, his heftedness to the north, those white eyes, and of course in his name: Bjørnar, the Bear, from the Old Norse *björn*. He is an intense, intelligent presence; a person you would want fighting for you and would dread as an enemy. He is not without self-regard, but I do not begrudge him that.

There is also a strong mystical streak to Bjørnar: unexpected, perhaps, in a man whose working life compels him daily to such pragmatism and self-reliance. But – as I will learn – Bjørnar looks often *through* things: hard into them and right through them with those pale eyes of his. He looks through people, through bullshit and through the surface of the sea.

Bjørnar settles himself in a big black swivel chair near the window, from which he can keep an eye on the water of the fjord. I jounce chubby Sigrid on my knee, delighted to have been trusted with a baby.

'When I was a young man, do you know, Rob, I decided I would never leave my island of Andøya.'

'It's a rare thing these days, to stay so rooted,' I say.

'Maybe. To me it was obvious. This island has everything I need for a long lifetime, and I love it.'

He pauses.

'Yesterday Ingrid and I watched killer whales, just *there*.' He points east to the sea channel. 'Orca, a family of them. We watched them for *free*!'

Bjørnar stresses the final words of any sentence. He speaks with a plosive and idiomatically perfect English. He rolls his *r*s, he pops his *p*s and *b*s, and he adds a stressed schwa to the end of many words. *STOPPP-uh. BOAT-uh. RRRROB-uh.*

'I have been to Oslo, of course, but I never like being off this island, unless I am in my fishing boat. This island, Rob, it grew me up.'

Ingrid sits nearby. Sigrid begins grizzling, and Ingrid passes me a teething ring. I ask Ingrid about her own childhood. She tells a remarkable story. She grew up on an island so small and remote it was a two-hour boat journey to the next largest island, itself a substantial boat journey from the mainland.

'Our island was home to ten families when I was born,' says Ingrid. 'Which meant it was really home to one big family.' I think of the Refsvika settlement – Ingrid's had been even further out, even smaller.

'Oh yes, I knew every inch of my island!' she says, smiling. 'When we were young we explored, that was what we did. No one with us to take care of us except ourselves. We knew each part of that place.'

One by one the families left, though, and by the time Ingrid was in secondary school only two families remained.

'Gradually, the government made it more and more difficult for us to live there, and so we were forced to "come in" to the mainland. And that was where I met Bjørnar . . .' She trails off with a smile.

Bjørnar bellows with laughter.

'Never leave your island! *That* is the moral of *that* story, Rob! You will immediately find yourself in trouble for the rest of your days! Now, come – sit here by the table, I will bring the charts and show you where we will go together these coming days,' he says.

He lays a chart out on the table. It is dog-eared and stained with what looks like blood. It is traversed by arcing purple lines and dotted with depth marks and buoy positions. It shows the northern half of Andøya, the fjord-cut western edge of the mainland, and perhaps forty miles of open sea to the north and west of the coast. Contours indicate the shifting depths of the seabed.

'Here is Andenes, from where we will sail tomorrow,' says Bjørnar, indicating with his forefinger. 'And look, *here*!' He moves his fingertip north by four or five miles, to a point where the contour lines bunch close together and tuck back in on themselves. On a mountain it would represent a gorge cutting through big cliffs. I have a flashback to the Lofotens and my crossing of the Wall.

'Here in Andøya we call this the *Edge*,' Bjørnar says, running his finger back and forth along the bunched lines. 'Here in Andøya, we live – how do you say it – on a *bookend*. This drop-off, this *cliff*, it happens only a few nautical miles from the coast. This is why the fishing is so rich and easy here: the fish gather at the Edge, and we do not need to go far to harvest them.'

He shakes his head.

'To me the land does not *stop* when it dips into the ocean. It keeps on going and I know that land under the sea as well as I know this world above. I can see it as well as you can see *that*.' He gestures through the window at the fjord.

'It's the knowledge about what is *under* the surface that for all times has kept these coastal people and this coast *alive*.

'And here,' he continues, jabbing his forefinger repeatedly into the chart near the Edge. '*Here* in some of the finest fishing grounds in the Arctic, *here* is where they were sonic blasting, testing for oil, *here* is where those *idiots* want to place the rigs.'

~

On 15 June 1971 production began in the offshore oilfield known as Ekofisk, sited to the south-west of the Norwegian continental shelf. At that point the extent of the Norwegian oil-holdings was still unknown, but the rapid success of Ekofisk began a speculative oil-rush along the west and north-west coasts of Norway. The Norwegian government responded quickly, creating Statoil in 1972 and establishing the principle of substantial state participation in each production licence issued for these wealthy waters.

Oil is Norway's life-blood. Its system – political, infrastructural – is thickly oiled, through and through. Substantial taxes have always been applied to the income produced by oil and gas: in under half a century of operations, the oil industry has generated a national sovereign wealth fund – the *Oljefondet*, or Oil Fund – of more than three-quarters of a trillion pounds, equivalent to around £150,000 per citizen. The petroleum sector accounts for almost a quarter of value-creation in the country as a whole; almost a third of the country's total real investments are oil-based. Huge sums have been invested by both companies and government together in oil exploration and oilfield development, as well as transport, supply and support facilities.

Yes, it is oil – and the Gulf Stream – that have made Norway's modernization possible. One of the country's most distinctive features is its combination of infrastructure and wilderness. The road that runs the length of the Lofotens – an engineering miracle that connects more than 100 miles of islands, involving undersea tunnels, mountain tunnels, avalanche-sheltered high roads, and dozens of bridges – was paid for in part from the oil chest. Norway loves nature, it loves technology too, and it sees these chiefly as complementary rather than opposed categories.

But Norwegian oil is running down. Around the turn of the millennium, production from the North Sea fields peaked at 3.4 million barrels per day. By 2012 it had reduced to almost half that level, with a correspondingly diminished income to the sovereign wealth fund. The obvious solution to the dwindling production volume was – and remains – to open new oilfields. Attention turned to the North Norwegian and Barents seas. Early in the 2000s, interest grew in the possibilities of tapping the reserves that were thought to exist beneath the waters off the Lofoten and Vesterålen islands. Around 1.3 billion barrels of oil were estimated to be buried near these archipelagos. The drilling areas were in relatively shallow water, were relatively close to land, and their geology promised steady returns. They represented good oil and – compared to other drill sites being considered much further north in the Barents Sea, where Arctic conditions drastically increased extraction costs – they represented cheap oil.

However, these same seas are also home to one of the world's largest cold-water reefs, and the Lofoten and Vesterålen archipelagos are among the most astonishing coastal landscapes in the world, drawing visitors from across the globe in a highly remunerative tourism industry. The waters off the island groups are also home to the fishing grounds that have been Norway's gold for a thousand years, long before the discovery of oil. Dried cod from those fishing grounds was thought to have been carried as a staple seafood by the Vikings on their founding voyages to Iceland and Greenland. Cod is the nation's founding fish – its original wealth fund.

The issue of whether to drill for oil off Lofoten and Vesterålen has, over the past fifteen years, become a battle for the soul of Norway. The stakes are high and the forces are powerful. On the one side is a state machinery lubricated by oil money, and a population indebted to and embedded in oil culture. On the other are Norway's perception of itself as a green nation – devoted to a secular religion of nature, committed to reducing global temperature rise and to fighting climate change – and its ancient identity as a fishing nation. Point 112 of Norway's constitution declares that 'natural resources should be managed based on long-term considerations, safeguarded for future generations', and this is seen by many in the country to countermand the opening of new oilfields, especially in fragile northern waters.

During the 2000s, as the first proposals to drill off the Lofoten and Vesterålen islands took shape, so too did resistance to the plans. Those opposed to the drilling began to organize. Unlikely alliances were formed. A coalition emerged uniting national green groups (especially young people), local activists from the islands, conservationists, environmentalists and fishermen. The campaigners quickly learned how to make their case more visible. They took their battle to the capital, to the airwaves and to the newspapers. They held torchlit protest marches through Oslo. They convened public meetings by the twilight of the midsummer-night sun on the beaches of the islands under threat.

One of the people who at that time became a leading figure in the fight was Bjørnar Nicolaisen.

~

We leave for the Edge just after dawn, chugging through the breakwater sequence of Andenes harbour. Thud and splutter of the engine, a high blue sky, the sea oil-film still. Sunlight flaring green and red in the ice crystals caught in my eyelashes. Two thin reefs of white cloud to the west; otherwise clear, cold and still. Perfect fishing weather on a northern ocean.

Out past the last harbour arm. Snow-lines of peaks to east, west and south, dropping to the sea. A raft of eider on the swell, and a single cormorant perched on a tidewater mark, facing the sun, its wings cranked open into an iron cross. Then three swans steadily overhaul us, wings creaking like doors, flying north into Arctic space.

'You tell me what to do and what not to do out here, and I'll obey,' I say to Bjørnar.

He looks back at me, tilts his head quizzically. 'You obey rules? I *never* do!' A barked loud laugh. 'But for today, my rule for you is – *don't fall in*! Anything else is fine.'

Bjørnar is wearing a raccoon-skin hat. The head of the raccoon is still attached and sits immediately above Bjørnar's forehead, gazing forwards. Its body has been curled over a skullcap and stitched into place, with its tail hanging down at the back. The raccoon looks comfortable; a long-term squatter.

The raccoon's eyes have been replaced with gleaming black false eyeballs. The effect is very disconcerting. Whenever I speak to Bjørnar I find myself looking into four seemingly sightless eyes, two of them jet black and two of them ghost white.

Out beyond the breakwaters the swell comes in long, slow hummocks, moving towards and then beneath us, tilting the boat sometimes by twenty or thirty degrees to the horizontal. The compass slopes in its gimbal with each peak and trough. Bjørnar moves around the boat as easily as if it were dry-docked.

The boat is a thirty-three-footer: a Libra class, Norwegian-made. It is called *Trongrun*, 'Seabed of Tron'. Bjørnar bought it fifteen years previously from a man in Finnmark County for a million kroner. It is a hard-working space, stripped to the essentials, messy but efficient. A cockpit cabin with a sealable door for running in big seas. Two winches on the starboard side drawing two jig-lines, one fore and one aft, the aft line held clear of the propeller by a metal arm that can be swung out to starboard. Four hooks on each jig-line, with sand-eel or squid lures. It is about as simple a rig as could be – but more than enough in waters as good as these, with quotas as tight as they are.

Knives cling by their blades to a magnetic strip next to the cabin door. Lines of red and yellow lures are hooked in rows to the edge of the table in the cockpit. Bjørnar is wearing neoprene boots with grippy soles, yellow and blue waterproof salopettes, an orange jacket – and the raccoon. Every half an hour or so he takes a fresh plug of black tobacco from a tin, lifts back his cheek and slots it into place between gum and cheek, as if inserting a new bit of software.

Above the dashboard of the cockpit is a brown baseball cap: salt-marked and bloodstained. It glitters with fish scales. I tap it with my fingers. It's hard as a fossil. The fish-finder hums and updates on a split-screen monitor: a jaggy Rorschach of orange, green and white.

'The white line shows the seabed,' says Bjørnar, pointing to the monitor. 'The orange above it shows the fish.'

'And what is the orange and the green beneath the seabed?' I ask.

'That is the *underworld*, Rob! That is the *OIL* !'

We roll on over those blue hills.

'Now,' says Bjørnar later. 'Now we are going out over the Edge. Here the land, how would you say it, *plummets* away beneath us.'

I feel my stomach drop, and have a sudden recollection of moving along in the drift tunnels of Boulby Mine years previously, passing underneath the threshold of the coastline and then out below the North Sea.

A plume of gulls follows us, crying into the wind. Bigger, longer waves, the boat riding them out. The spire of the Andenes lighthouse thinning with distance. I have the casket with me, think about slipping it overboard once we have crossed the Edge. There will be few deeper places.

'As a fisherman,' says Bjørnar, 'you have to be able to see through the water. When you are out here, *you* can't see anything. But I? *I* can see the shape of the landscape underneath us: there are bumps and valleys and mountains down there, and streams flowing, and fish moving in those streams. To be able to imagine this, you have to use your brain at the same time as watching your machinery, talking with friends on the radio here' – he taps the two-way – 'and sometimes it's heavy waves, freezing cold, and we have to turn the boat up into the wind. Yes, fishermen must be multitaskers!'

He booms with laughter – and then stops smiling.

'We face death every morning to bring food to those *idiots* onshore,' he says, jerking a thumb over his shoulder. 'Politician-idiots. The people who want to blast open this seabed for *more oil*.'

A kittiwake among the gulls now.

'Cod were here long before the oil was found, and they will be here – if we let them – long after the oil is gone. Cod fed the Vikings on their journeys, and they feed us now. When man has become so crazy that he is willing to offer his food in order to get more rich, more oil, then the craziness is complete and we have no hope any longer.'

Bjørnar's fight with big oil began in the spring of 2007, when the Petroleum Directorate – the government agency responsible for regulating the oil and gas resources on the Norwegian continental shelf – arrived in Andøya. The Directorate had already reached out to marine biologists and to fishermen's unions in the north of Norway, preparing the way for their campaign of suasion in Andøya and Lofoten. Now they wanted community endorsement for their plan to open up new fields beyond the Edge. Among the evidence they showed in favour of their plans was data gathered by seismic mapping.

Seismic mapping is a means of seeing the marine underland. A specialist ship carrying a low-frequency, high-volume air gun fires sound pulses into the water. These pulses are powerful enough to penetrate some distance into the sea floor, before reflecting back upwards to where they are recorded by seismic sensors dragged on long cables behind the ship. The blasts can occur at intervals of under a minute, for weeks or months at a time. Barely audible above the surface, they fathom the seabed. But the sound blasts also travel for hundreds of miles laterally below water, sending thunderclaps sideways through the ocean. Seismic surveys are used not only by the oil industry, but also to target deep-sea sedimentary sections that are suitable for unravelling the nature and causes of past climate change, so that models for future climate change can be tested and refined. Most survey boats now carry specialist observers who watch for cetaceans and order firing to cease if they are seen, and who advise on how best to schedule blasting to avoid migratory patterns. Controversy and uncertainty nevertheless surround the technique, especially concerning its effect on whales, dolphins and other marine life.

A public meeting was called in Andenes, where the Directorate representatives laid out what they were framing as a 'consultation' with the people of Andøya about the possibilities of further oil prospecting, including more seismic blasting.

Bjørnar checks the lures on the jig rigs as he speaks.

'I remember sitting on a chair there, listening to the first people talking. And I was thinking – it's *over*. It's already been planned, all *fixed*. The testing is already happening. This consultation is to me a performance, how do you say – a *sham*! It's *over*! They are coming for the seabed, and to destroy our livelihoods.' He pauses.

'And at the same time I was thinking, I would like to see myself grow old, and maybe when I am old I will sit in a chair unable to move myself, and know then that I didn't do anything to stop this. So I thought to myself, I must begin this fight now, *today*!'

He speaks with confident sentences and long silences, the memories clearly vivid to him. He checks the last hook, lets the rig fall, and fixes me with his unsettling gaze.

'Rob, I have a gift for – how would you say – seeing what the future will be.'

Looking into those white eyes, being looked through by them, I do not disbelieve him.

Bjørnar began his campaign against the plans, even as the oil companies continued their seismic blastings. He fished and he fought. He was elected secretary of the local fishermen's union, which gave him a political authority he then wielded for access and voice. He knocked on doors up and down the islands. He took to the newspapers, writing about the dangers of blasting and drilling. He activated the old Norwegian allegiance to cod, and set it against the new Norwegian allegiance to oil. He challenged oil company representatives to debate. He used satire, mocking them and their plans in print and broadcasts, and he challenged the hard science of their claims.

'Maybe my main strategy was *delay*,' says Bjørnar. 'Time was working for the resistance and the people. I knew this. If you get things delayed, new information comes in – and the new information is working usually against industry.'

He is telling his story faster now, speaking in a torrent, hard to interrupt or question. His mood flickers as he talks: big smiles, big laughs, then pulses of sadness and loss. There is an aggrandizement at work, too, but I hear this not as bragging or chest-beating, rather as an echo of the self-heroizing that was necessary for Bjørnar to fight his battles, and to absorb the damage that he personally sustained.

Six months into his campaign against big oil, Bjørnar broke down. The strain was too much. He was found at his keyboard one day by Ingrid in something like a fugue state. He spent weeks in a psychiatric ward. When he came out, he spent three months rebuilding himself. Then he began the fight again.

Thump of the engine, roll of the big swells. Two fulmars in the seabird plume now, the kittiwake gone.

'I will tell you a picture I had in my head when I came back from that first trance,' says Bjørnar. 'It felt like I was standing out on the peninsula furthest away from shore there' – he points back at the spined peaks of the Andøya coast in the distance – 'with my boots in the sea, my front turned to the people onshore, fighting against mankind, and with the Edge waiting to claim me. This was the image that surfaced from my subconscious at that time. It was that crazy. Can you imagine?'

The boat is locked on its autopilot course. Bjørnar has stopped all work with the fishing rig, focused wholly on telling his story. The *Trongrun* noses north-west, wallowing on the hummocks. He has braced himself against the wheelhouse and is looking unblinkingly at me. The story is gushing from him under great pressure now.

'Slowly, though, others joined me on the shore. More and more environmental organizations came up here and joined us. Individuals came together to protest.' He spreads his arms wide, then curves them around in a gesture of gathering. 'My project was always to build all these organizations into one big *army*!'

'"Coexistence", Bjørnar,' I say. 'You learned your tactics from big oil!'

He laughs. 'Together, yes, we were coexisting, the resistance, we were making history, starting to turn the tide back against these big men. We raised so much noise out here. They were going to *take* it. They were going to annex the area at that time. But we *stopped* them.

'The season of the seismic surveys was from May until September. They blasted for three years and I fought them for three years. My brother died in Finnmark of cancer, and my sister near Paris also died of cancer. They blasted for three years, and for three years I was brought away to the mental hospital once a year, in a trance. I was blown out.'

Shriek of the gulls, mew of the kittiwake.

'I do not regret those years of the battle and the trances, Rob. I learned from them, though they were hard on all of us, there is no doubt about that. During those years, even my fisherman son, he looked at me as a stranger. I could not have done this without Ingrid. She is such a strong woman, so very strong. Always there behind me. Taking care of my family . . .' He trails off. I nod. Even from my short time with them, it is clear to me that Ingrid is a person of exceptional surety and subtlety; bedrock to Bjørnar's torrent, calm to his storm.

The pressure is ebbing. He speaks more slowly now.

'The tide was turned, though. Minority parties in coalition blocked the drilling. It was a victory for us. And we have only grown stronger. Now the majority in Norway says *no* to oil. So much has happened because of the oil battle and those years. The youth is coming back to the fishing, moving back to this way of living, to the rural areas. All of the nation has been watching this fight up and down the coast.'

The aftershocks of those years were disastrous both for Bjørnar's health, however, and for the undersea world.

'Since the seismic testing,' says Bjørnar, 'everything has changed here. You know the fish that we are going to catch today? They disappeared. Before the blasting we could get up to 3,000 kilograms of fish just with these jigging lines in only one day. That was one of the reasons I bought this boat.' He pats the wheelhouse of the *Trongrun* affectionately.

'But in the first year of the blasting, the saithe disappeared, and only in 2015 did they start to come back. Six years after the last blasting. The whales, too, it affected them. The orca left also. And we began to see sperm whales in the fjords, where they had been driven by hunger.'

He pulls back on the throttle, setting the engine to a gentle idle. He puts his hands together in a wry gesture of prayer, bows smilingly towards me.

'And now, let us fish.'

~

The night before Bjørnar and I set out in the *Trongrun*, I stayed up reading Edgar Allan Poe's 1841 story 'A Descent into the Maelstrom', about the whirlpool that lay just off the Lofoten shore; the whirlpool I had seen and heard during my days in the bay of the red dancers, the drill hole of which was thought by some – including Athanasius Kircher, author of the early modern epic study of the underland, *Mundus Subterraneus* (1664) – to penetrate the earth and resurface in the Gulf of Bothnia.

Poe's story opens with two men near the summit of Hellsegge, the blunt peak that rises to the south of Refsvika Bay. The men are sitting on the brink of 'a sheer unobstructed precipice of black shining rock', looking at the distant island of Værøy. One of the men is a nameless visitor to the archipelago; the other is a Lofoten native from Moskenes, who has a head of strikingly white hair.

When the two men first gain their observatory, the ocean beneath them is a 'wilderness of surge' with 'something very unusual about it'. The visitor experiences an apprehension – a disturbing sense of something part-glimpsed. Then comes a sound, loud and gradually increasing: a moaning, like that of a 'vast herd of buffaloes'. Quickly the sea changes its textures: currents of 'monstrous velocity' begin to run, and the sea is seamed and scarred into a 'thousand conflicting channels', which gradually resolve into a multitude of small whirlpools. These vortices vanish and then, 'very suddenly':

> [a] circle of more than half a mile in diameter [appeared]. The edge of the whirl was represented by a broad belt of gleaming spray; but no particle of this slipped into the mouth of the terrific funnel, whose interior, as far as the eye could fathom it, was a smooth, shining, and jet-black wall of water, inclined to the horizon at an angle of some forty-five degrees, speeding dizzily round and round with a swaying and sweltering motion, and sending forth to the winds an appalling voice, half-shriek, half-roar.

'This,' stammers the narrator, who, feeling the mountain shake beneath him from the force of the water's rage, has thrown himself to the ground in panic, 'can be nothing else than the great whirlpool of the Maelstrom.' So it is, and into its maw, the white-haired islander tells him, have been sucked over the years whales, pine trees and countless boats. Even a polar bear was once caught within its traction, consumed by 'the abyss of the whirl'.

Poe's description was, of course, nautically preposterous. He had never been to the Lofoten Islands, nor had he spoken to anyone who had seen the Maelstrom. His vision of the Maelstrom was constructed from legend, hearsay and the cartouches of sea charts. The image of a funnel descending to the bed of the ocean and beyond bore no relation to reality. The Maelstrom resembles neither a tidy double spiral, nor a dipping hole in the ocean with blackness at its core. Rather it is a churning field of water, roughly circular in outline, ranged over a diameter of a mile or more. Within that rough circle the water stands up in waves, and from that rough circle – like the arms of a spiral galaxy – wandering lines of foam track the incoming tidal currents that make the Maelstrom.

Poe's surreal vision of an irresistible, helical vortex does, though, speak to the draw that whirlpools – from plughole swirls in a bathtub to cosmic black holes – exert upon the imagination. Such structures captivate us because of the distant tractions they exert, the event horizons they establish. Their victims are trapped before they are even aware they have been caught.

In Poe's story, the islander proceeds to tell the narrator about how he and his brother, out fishing, once became ensnared by the Maelstrom. As their boat was pulled towards the vortex, the man said, he found himself peculiarly calm, his terror giving way to an odd kind of fatalist love: 'I became possessed with the keenest curiosity about the whirl itself. I positively felt a wish to explore its depths, even at the sacrifice I was going to make.' Spinning furiously, held in the power of the Maelstrom's centrifuge, the boat slowly slid down the slopes of the black-sided shaft. 'It appeared to be hanging, as if by magic,' recalls the islander, 'midway down, upon the interior surface of a funnel vast in circumference, prodigious in depth and whose perfectly smooth sides might have been mistaken for ebony but for . . . the gleaming and ghastly radiance they shot forth.' The spray thrown out by the vortex formed a moonbow of light above the whirlpool – an unearthly crescent hovering over this underland portal.

Poe's story partook of widespread nineteenth-century fascination with the idea of an actual global underland to which certain entrance points existed, giving access either to a wholly hollow planet or at least to substantial interior space. A subgenre of subterranean fiction flourished in the 1800s, in which the Earth's crust and mantle were frequently imagined as riddled with tunnels, often leading down to a habitable core. In 1818 an American army officer called John Cleves Symmes began to advance as fact his belief that the Earth was structured as a series of concentric spherical shells, with vast openings some 1,400 miles in diameter at either pole. Symmes argued for the need to make an expedition to the North Pole to descend into these spheres and explore their potential for resources and habitation.

That expedition was never made, but in an early science-fiction work entitled *Symzonia: A Voyage of Discovery* (1820), purportedly written by a 'Captain Adam Seaborn', a group of travellers descend into the Earth's centre via the North Pole, where they do indeed discover an inner continent. Poe extended Symmes's theories in his 1838 novel *The Narrative of Arthur Gordon Pym of Nantucket*, then in 1864 came the most famous of these fantasies, Jules Verne's *Journey to the Centre of the Earth*, in which his explorers enter an Icelandic volcano, descend to a depth of eighty-seven vertical miles, sail an underground sea, and emerge from the crater of Stromboli off the coast of Sicily. The following year, Lewis Carroll published *Alice's Adventures in Wonderland* – originally titled *Alice's Adventures Under Ground* – a very different kind of expedition to the subterrane.

These hollow-Earth fantasies persisted and mutated in the twentieth century. In 1923 the Russian mystic and painter Nicholas Roerich staged a Himalayan expedition with his philosopher wife, Helena, to seek the entrance to the city of Shambhala, which would lead in turn to a 'Hollow Earth Kingdom'. They travelled on horseback from Darjeeling on their futile quest, carrying an American flag fluttering from a Mongolian spear, and probably aided by Soviet intelligence agents. After 1945, disturbingly, a post-Nazi geo-fantasy emerged of caverns in the Earth's crust into which Hitler and his closest allies had supposedly escaped from their bunkers during the final Russian assault on Berlin, and from which Aryan power might in the future resurge.

That night in Andøya, I came to think of Poe's story as a premonitory oil-dream. In it, the Maelstrom operates both as a kind of boring drill and a means of seeing the seabed where it lies bared at the base of the vortex. Often Poe describes the water of the Maelstrom in oily terms: it turns 'smooth', 'shining' and 'jet-black', it 'gleam[s]' like 'ebony'. Like oil it is both fatal and miraculous – and like oil, it re-sequences time.

Poe's story and others like it speak in part of the mid nineteenth-century dreams of the 'oceans of oil' that were imagined to exist under the earth. These narratives advanced the Holocene delusion of a planetary interior containing inexhaustible wealth and energy – a delusion that still characterizes expansionist oil discourse, nearly two centuries after Poe was writing. 'We need new acreage to explore, and want to step up our exploration activities,' Norwegian Statoil declared the autumn before I travelled north. Several months later the Australian oil and gas giant Karoon announced its wish to open up new fields in the Great Australian Bight on the grounds that the area held 'underexplored Cretaceous basins'.

The Deepwater Horizon catastrophe of 2010 was partly a consequence of pushing deep drilling to its limits in an effort to open new territories. On 20 April that year, forty-one miles off the south-east coast of Louisiana, the borehole of a semi-submersible oil rig burst. The ensuing blowout at rig-level killed eleven crewmen and ignited a fireball that could be seen onshore. Two days later the rig sank, leaving the well gushing from the seabed at a water depth of around 5,000 feet. Two hundred and ten million gallons of oil escaped into the Gulf of Mexico, rising on the ocean as a slick that was visible from space. At sea level the oil devastated marine life. Tar-balls, rolled by the waves, gathered in their thousands on the coastline. Striped dolphins leaped through floating slicks. It would take until the autumn to cap and seal the well successfully, such that it could be declared 'effectively dead', but the consequences for the ecosystems and communities of the Gulf persist today. Deepwater was a rare laying-bare of the darker operations of the global extractive industries. One of the agreements tacitly made by consumers with these industries is that extraction and its costs will remain mostly out of sight, and therefore undisturbing to its beneficiaries. Those industries understand the market need for alienated labour, hidden infrastructure and the strategic concealment of both the slow violence of environmental degradation and the quick violence of accidents. Deepwater violated that agreement shockingly, manifesting a substance on which most modern human life depends but that few people encounter in the raw.

After returning from Norway, I would learn that the Moskstraumen Maelstrom had become literally enabling of the oil industry. In the 1980s a man called Bjørn Gjevig – an antiquarian scholar, professional mathematician and amateur sailor, who seems as if he must have been invented by Poe, but truly exists – became fascinated by the hydrodynamics of the Maelstrom. Using data gathered in part while sailing close to the whirlpool, Gjevig began to model the maths of its currents. When oil was discovered off the Lofotens, he realized that his data had gained application: oil companies would need to understand such ocean forces in order to construct rigs that could withstand 'destructive currents of the kind found in the Maelstrom'.

At the climax of Poe's story, the human body loses all volition and becomes a kind of drift-matter, helpless within the 'destructive currents'. The fisherman and his brother are drawn steadily deeper into the vortex. The fisherman realizes that he has entered a giant grading-machine, which weighs and measures the objects that have been pulled into it – and moves the heaviest and most irregularly shaped items to destruction at its base.

In a wondrous flash of intelligence, he understands that in order to survive he must, counter-intuitively, leave the apparent safety of his heavy fishing boat, and lash himself instead to a lighter wooden barrel. Unsurprisingly he cannot convince his brother of the wisdom of this course of action, so he is left with no choice but to abandon both brother and boat. The barrel onto which he is lashed rises slowly to safety, as he predicted. But the fishing boat with his brother spreadeagled on the deck is dragged down to its destruction.

All of these nineteenth-century hollow-Earth texts read, now, both as beckonings into and warnings of the void. All are Anthropocene works *avant la lettre*, about longings to gain access to the Earth's wealthy interior. They foretell the arrival of the extractive industries in all their gargantuan force. They portend the establishment of the immense infrastructure that has spread across the Earth, dedicated to retrieving from the underland the raw materials it holds, creating petro-scapes from the burned-back wastelands of the Niger delta, to the flaming oil wells of the Middle East and Houston's sprawl of refineries and silo tanks. Our modern species-history is one of remorselessly accelerated extraction, accompanied by compensatory small acts of preservation and elegiac songs. We have now drilled some 30 million miles of tunnel and borehole in our hunt for resources, truly riddling our planet into a hollow Earth.

~

The *Trongrun*'s killing space was stripped back and simple: a zinc trough bolted to the starboard side of the boat, and a removable wooden cutting board that lidded the trough. Bjørnar ran the jig-lines off a winch. The press of a button brought the line up, the winch creaking with the weight of fish.

Tick of the winch. Click of the jig. Bjørnar peers over the edge. Silver shapes swimming up into view, finding focus, then thrashing to the surface. Bjørnar holds the line clear from the boat with one hand, and with the other gaffs the fish, each in turn, jerking them up and into the trough with single, practised movements. A shake of the lure to free the hook, and each fish falls to flap in the trough, their orange swim-bladders extruding out through their mouths like party balloons. They are saithe, similar to the pollock and coalfish I've caught off British shores before, but huge: seven, ten, twelve pounds. A strong white line runs along the mid-length of the flank of each fish, like the line on the fish-finder; black-copper above the line, and a bronze-brown below it. They are gorgeous even in death.

'My grace at the table,' says Bjørnar, 'when at home we are eating fish that I have caught, is always this: "Fuck! We don't know how lucky we are!"'

After each haul of fish, Bjørnar lowers the lines again. Once they are down he takes a red-handled knife from the magnetic strip, pulls each fish out with a hooked finger in its under-gills, turns the fish on its back, and with a quick jerk and slash cuts its throat and breaks its neck at the same time. Blood dribbles on to the deck beneath the trough.

'That's a sharp knife, Bjørnar.'

He looks at it as if it were a stick.

'This is not a sharp knife. Later you will see a sharp knife.'

Kittiwakes, fulmars, black-backs picking up the scraps. Creak of the winch, slosh of the scuppers as Bjørnar hoses the deck clear of blood.

A cod comes up among the saithe: malty-brown dotted flanks, barbels, snow-white belly.

'You should have seen the cod in the wintertime. They make

these saithe look like sardines. They have just gone this fortnight. My eldest son is following them to North Cape right now. I had one this year – thirty-two kilograms!'

The gaff is now red with blood, gobbeted with flesh. A curious fish comes up, slender of body, with large iridescent scales that rainbow in the sunlight, and wide, flat eyeballs. In the sunlight its pupils, adjusted to the dark of the depths, have dilated to the size of bottle caps.

'A beautiful fish, no?' says Bjørnar. He does not say its name. He shakes it off the gaff onto a metal tray. Its upwards eye has been burst by the hook, and as I watch it slowly fills with ruby blood. With its rainbow scales and jewelled eye, it resembles a Fabergé model that might spring into clockwork life.

My mind is pulled northwards to the Svalbard archipelago, 100 miles beyond our wallowing boat, where the Global Seed Vault has been constructed: a billion-dollar storage site sunk beneath the permafrost for the preservation of biodiversity, readying for a future in which variety has been depleted by extinction and genetic modification. I think of the seismic charges detonating underwater, of the oil ships dropping their drills to the seabed, of the Deepwater Horizon blowout, and of our species' instinct to open what has been sealed without thought for the consequences.

'Let's go home and *eat*!' says Bjørnar after we have caught thirty fish or so. He revs the engine, pulls the boat's nose around, and with a chuckle of contentment sets course for the Andenes lighthouse.

~

We moor up back at the dock. It is cold in the shadow of the deckhouse. Oil rainbows the water around the boat.

'*This* is a sharp knife,' says Bjørnar, prising a yellow-handled knife from the magnetic strip.

He reaches into the vat, takes a fish by its tail, slaps it down onto a wooden board criss-crossed with knife scars. He hooks a finger into the gills to hold the fish steady, then cuts down from the head in a filleting stroke along the flank. He seems barely to lean in with the knife, more just to place it – and the flesh falls away out of respect for the blade's edge. Down and along to the tail, flip the fish, repeat on the reverse. Flip the fillet, strip the skin, peel and rip. Yellowish-white flesh, soft as putty, faintly translucent. Head and skeleton into the harbour, fillet into a bucket of water.

A man in a hat with furred ear-flaps walks down the gangplank, stops by the boat, nods to Bjørnar, looks me over.

'Aha! This is Sven,' says Bjørnar. 'An old friend. We've fished together many times.'

They talk as Bjørnar works: about the fishing, the prospect of the seismic blasting starting up again, and the cod's recent departure north to Finnmark.

'I will follow the cod tomorrow,' says Sven. 'I will be perhaps two, three weeks away? I still have quota. Maybe I will look for lumpfish also.'

Lumpfish are caught for their roe, a cheap caviar. They are cut open and the red roe is scraped out.

'I always make sure to cut the throats of the lumpfish before I take their roe out,' says Sven modestly, as if confessing a substantial charitable donation.

'Some "environmentalists" say we shouldn't kill the lumpfish for its roe,' he continues. 'But the rest of the fish is not for eating. Just two patches of meat on its cheeks – so we take the roe, cut out the cheeks, and return the rest of the fish to the system of the sea. It feeds that system. They don't understand that the sea needs feeding, just like us.'

Bjørnar grunts. 'I expect to be returned – how do you say, *in my next life* – as a lumpfish. So I always cut their throats before I take their roe, just as I would like to have my own throat cut before I have my roe taken.'

'Do as you would be done by,' I say. 'The golden rule of reincarnation.'

~

Early that afternoon we eat saithe with butter and potatoes while the cat with lizard eyes watches us from a corner. Ingrid ladles the chunks of saithe from pot to plate. Bjørnar thumps the table with both fists, and says grace: 'Hell! Thank you for the fish in the pot!'

'That's a more polite version of grace than the one you quoted in the boat,' I say.

Bjørnar laughs and thumps the table again. 'One language for sea, another for land!'

After lunch, Bjørnar takes me around the island. He wears his raccoon hat again, and we carry the Wehrmacht binoculars with us. As Bjørnar drives and talks, I begin to understand something of the ancient and contemporary complexities of Andøya. Ecologically, it is an island of four zones: peaks, peat, marsh, beach. The glaciers have bulldozed it flat on its east side, but left it mountainous on its west. Much of the island is open to all-comers, but other areas are controlled by NATO, enclosed in high perimeter fences. I am strongly reminded of the Isle of Lewis in the Outer Hebrides: the peat, the remoteness, the openness – and the same attractive potential for industrial exploitation and military colonization.

'You know, Rob,' says Bjørnar as we bump down a side track on the west coast of the island, 'if there is a blowout at one of these proposed rigs, well, it will destroy this coastline. The Gulf Stream is pumping in and out of all the fjords. The oil is going to be spread everywhere. A blowout at Lofoten will spread oil all the way north across here and up to Finnmark County. The Gulf Stream will be the oil's conveyor belt.'

What Bjørnar fears is a version of 'solastalgia', the term coined by Glenn Albrecht in 2003 to mean a 'form of psychic or existential distress caused by environmental change'. Albrecht was studying the effects of long-term drought and large-scale mining activity on communities in New South Wales when he realized that no word existed to describe the unhappiness of people whose landscapes were being transformed about them by forces beyond their control. He proposed his new term to describe this distinctive kind of homesickness. Where the pain of nostalgia arises from moving away, the pain of solastalgia arises from staying put. Where the pain of nostalgia can be mitigated by return, the pain of solastalgia tends to be irreversible. Solastalgia is not a malady specific to the Anthropocene – we might consider John Clare a solastalgic poet, witnessing his native Northamptonshire countryside disrupted by enclosure in the 1810s – but it has certainly flourished recently. 'Worldwide, there is an increase in ecosystem distress syndromes,' wrote Albrecht in an early paper on the subject, 'matched by a corresponding increase in human distress syndromes.' Solastalgia speaks of a modern uncanny, in which a familiar place is rendered unrecognizable by climate change or corporate action: the home become unhomely around its inhabitants.

Bjørnar spots a sea eagle on the shore. The side track takes us closer to it. We drive slowly past a row of wooden houses near the beach. I watch the eagle through binoculars. It is perched on a kelpy boulder. Its four-foot wings hang around it like an oversized cloak.

There is movement from one of the houses. A finger twitches back a curtain and a face looks anxiously out at us.

'Why does that man look at us like that?' Bjørnar asks, puzzled.

'Bjørnar, I've read enough Scandinavian crime fiction to know that we are behaving quite a lot like murderers. Two men in a big black car and dark glasses, one wearing a dead raccoon on his head, the other scanning lonely houses through a pair of binoculars. That person should be forgiven for feeling worried.'

The booming laugh again. 'You are a *good* man, Rob.' He drives on. The face at the window disappears.

Blue hues to the snow now. A wooden swing on a beach shifts in the wind. Purple shadows creep over eastern peaks. Sea eagles pluck a dark carcass far out on a frozen lake.

~

In the days that follow, a northerly wind builds. We are unable to go out fishing, so I take to climbing in the mountains on the west of Andøya, then returning to Bjørnar's house in the afternoons and evenings.

The weather stays clear. The days blaze with metal light: silver off the snow, gold in the sun, iron in the shadows. Star-filled nights freeze the snow hard. It is -10°C in the forests at noon. The wind makes moving cyclones of snow grains, far bigger than any I have seen in Scotland or the Alps. They roam the windward slopes of Andøya's peaks. Some are hundreds of feet high. I watch them across the valleys, abruptly changing direction and speed, their tops whipping around like trees in the wind.

One day I ski up a valley over deep snow, through scrubby birch woods to the base of a mountain shoulder. I stash the skis and continue on foot, punching through the crust with each pace. It is difficult, exciting going. The snow holds an archive of print trails: snow hare, fox, raven. The wind rattles my skin, presses on my eyes. A fifty-foot snow-devil wanders towards me, hits me with a hiss that rises to a crackling roar, then roams off across the slope in silence. I feel as if I have been passed through by a ghost. On the plateau the wind has sculpted extraordinary structures in the snow. Rime ice grows in feathers on boulders. Cloud shadows slip over peaks to the west. A raptor hunts the birch woods below me in the valley. It is one of the most pristine places I have ever been, though I know this to be an illusion. I sit in the lee of a crag, grateful for its wind-shadow.

Returning across the plateau, I meet and follow the line of my own outward footprints. The wind has already whittled away the loose snow around my prints so that they are starting to stand out from the snow – as if time were reversing, and what was pressed down below the surface is now rising up.

That afternoon I go down to one of the beaches on the north-west of Andøya. A skerry shaped like the dorsal fin of a shark stands a few hundred yards offshore. Hundreds of seabirds wheel around it. It is low tide, and the sand of the bay is strewn with jetsam, almost all of it plastic. Here, as in the Lofotens, the density of human debris is shocking. Fishing buoys, toothbrushes, bleach bottles, tangled fishing nets, thousands of unidentifiable shards.

I feel sick as I walk the wrack-line and its litter, appalled by the contrast with the plateau, implicated by my part in the scene. This was once all oil too. Oil – the 'monstrous transformer' – is in all of these things, vital to the manufacture of the plastics that we first synthesized only a century ago. I think of the photographs I have seen recently of hermit crabs on the remote Pacific atoll of Henderson Island; one crab had taken a plastic doll's head as its shell, another an empty tub of Avon night cream. Plastic is the substance that has served as our most perfect container – and that now overwhelms our systems of containment. The substances we have made are relentlessly accumulating around us, forming a very present past. Over the last two centuries, and especially the last fifty years, our mass production, consumption and disposal has created 'an empire of things' with its own unruly material afterlife, 'a swelling topography of scrapped modernity', as Þóra Pétursdóttir and Bjørnar Olsen write, 'which despite ever more effective regimes of disposal is increasingly confronting us with its pestering presence'. Nuclear waste waits in vitrified flasks for the underground tombs into which it might be buried. Seas and coasts thicken with plastic trash. Carbon dioxide accumulates in the atmosphere. I recall Don DeLillo's laconic, lethal one-liner from his novel *Underworld*: 'What we excrete comes back to consume us.'

These surging, multifarious substances of the Anthropocene are what Timothy Morton calls 'hyperobjects': entities that are impossible for us to perceive in their dispersed, 'viscous' entirety, and of which we find it hard to speak. Our accumulative activities have even produced a new type of rock called 'plastiglomerate' – a hard coagulate that contains sand grain, shells, wood and seaweed, all held together by molten plastic produced by the human burning of beach rubbish on campfires. Plastiglomerate was first identified by geologists on Kamilo Beach in Hawaii; it has been proposed – due to its durability and distinctive composition – as a plausible future Anthropocene strata horizon marker. Plastiglomerate is surely an emblematic substance of our epoch. It is made by a stickiness that picks up and coagulates other entities, it is born of an untimely new geology that practises a kind of sampling and remixing, and it holds both the natural and the synthetic in grotesque hybridity.

Perhaps stickiness is one of the defining experiences of the Anthropocene as it is lived, I think there on the beach. Each of us is implicated in the effects of the epoch, each of us an author of its making and its legacies. In the Anthropocene we cannot easily keep nature at a distance, holding it at arm's length for adoration or inspection. Nature is no longer only a remote peak shining in the sun, or a raptor hunting over birch woods – it is also tidelines thickened with drift plastic, or methane clathrates decomposing over millions of square miles of warming permafrost. This new nature entangles us in ways we are only beginning to comprehend. As with the sticky strands of self-tightening silken plastic that drift down from the helicopters of the 'New People' at the end of John Wyndham's premonitory novel *The Chrysalids* (1955) – originally titled *A Time for Change* – the more we struggle to distance ourselves from the Anthropocene, the more stuck we become.

~

'Come, Robert, we will walk together one more time, it is your turn to sit in the throne!'

Good Friday on Andøya: my last day with Ingrid and Bjørnar. We have all eaten together: cod's tongues, cod steaks, saithe fillets, big pink-skinned potatoes that you peel on the fork.

We walk down to the shore, treading carefully on the sheet ice that lies over the sloping fields, placing our feet full and flat. The wind from the north is flayingly cold. It bites my ankles, burns my shins. Our breath is steel wool.

At the water's edge stands the driftwood throne. By its side a small standing stone has been raised and set deep into the ground.

'My god is the god of stone,' Bjørnar says with a quiet smile. 'I don't need any other god.'

Then he roars again, bellowing with laughter and patting the arm of the throne.

'Come! Macfarlane! Sit here and be King of Andøya for a few minutes!'

The throne's legs and back are made of birch trunks as thick as my wrist. Its back and base are nailed racks of driftwood, stripped of their bark. Its arms are two limbs of driftwood. It is perhaps eight feet high and its seat is four feet off the ground. It is a chair you mountaineer into.

I sit in the throne, and look across the fjord. There is a cheeping and a whirr of white wings – a blizzard of snow buntings flies past us and over the waves.

'This is where I leave the fish for the eagles,' Bjørnar says, pointing to the rocks in front of the throne. 'And when the killer whales come, we see them in the near channel. They are moving between one hunting ground and another, always very sure of where they are going.'

Just along the shore from the throne is a rusting vertical pipe, sticking six feet up out of the shoreline. Three plastic bottles are beached by it.

'What is that, Bjørnar?' I ask.

Suddenly he seems tired, sad. His eyes are rheumy. He works his jaws silently, as if they have become stuck together, his mouth gummed up. He does not answer, then says quietly – as though he has not told me before, as if he is saying it to himself or to the wind – 'They blasted for three years, and I fought them for three years. Now they are coming back. It is all returning.'

Then he says, 'Enough, Rob. We need to go no further. It is too cold.'

We walk carefully over the icy fields to the house.

That afternoon I play with baby Sigrid, bouncing her on my knee while I hum the William Tell overture and 'Bye, Baby Bunting'. She is very bonny and her eyes are pale blue.

Before I leave, I help move a massage chair that Bjørnar's son has brought for him, salvaged from a skip. We haul it out of the car and into the basement of the house. It is very heavy, made of black leather, and has a hand-held controller with multiple settings for the optimal relaxation of different muscle groups.

'It will be good for his back,' says Ingrid, kindly.

10
The Blue of Time
(Kulusuk, Greenland)

Late summer off the coast of Kulusuk Island, south-east Greenland, and a single iceberg sweats in the channel. The berg is vast, perhaps 100 feet from sea to summit, shaped like a mainsail with a rounded tip. It glistens white as wet wax. Its submerged bulk shows as a bottle-green aura.

Dark blue of the channel, sharp blue of the cloudless sky. Daytime moon above a shield-shaped mountain. On the far side of the channel, a glacier runs down to the water, six miles or so distant, the cliff of the calving face faintly visible.

It's low tide. On the foreshore of the village bay a man is leaning over something. He is straight-legged, bent at the waist. His sleeves are rolled up and his arms are red to the elbow. He wears a luminous yellow hi-vis jacket and hose-down clothes. The carcass of a porpoise lies slack across seaweedy rocks. He uses one hand to grip a flap of the black skin of the porpoise, then peels it back towards him, using the curved flensing knife in his other hand to cut away the meat as it comes. It looks like he is helping the porpoise out of a wetsuit.

A hundred or so wooden houses, each perched on an ice-smoothed table of gneiss. This is Kulusuk: more aviary than a village. The houses have brightly coloured outer panels of red, blue and yellow, with white dabs of anti-rust paint marking the nail-heads on the panels. Most are lashed down with steel cables for when the big winter storms come. The *piteraq* – the katabatic wind that rushes down off the ice cap – can reach hurricane force here, stripping the earth down to bare rock, leaving snowdrifts many feet high on the lee side of buildings, and shattering the shoreline sea ice.

There's no wind today. The air is warm. Unprecedentedly warm. The berg sweats. The man flenses the porpoise. Down at the breakwater, stout pale objects float a foot or so down, swaying slightly in the swell, tethered by rope to the lower rungs of the iron ladder that's bolted to the side of the breakwater. They're the bodies of ringed seals, heads and front flippers cut off, tied up by their tails. The bodies have been there a while. They glow faint green. Guts trail amid the kelp. It's been a poor month for the hunters of Kulusuk.

On the east side of the bay, in the lee of a crag, a slew of white wooden crosses drops down almost to the tideline. They're different sizes. Some have wonky crossbars. From a distance it looks like a snow patch or a tiny glacier, running off the steeper ground. It's a cemetery: one of the few sites in the village where enough topsoil has accumulated to bury a body.

The air is split by a high howl, and immediately thirty or forty other howls join it in chorus. The Kulusuk huskies are sitting and howling up at the sky, straight-backed, full-on wolf-howls. One is straining so hard the chain is taut as a bar, and the collar cuts at the howl, strangles it.

Four children and a husky pup are on a big trampoline, bouncing together, the children's feet stretching the net down almost to the rock on which the trampoline is set. The husky spreads out his legs, braces himself. When the howling starts, the pup howls too and then the children howl as well, bouncing and howling together.

The berg sweats, the man flenses the porpoise, the children and the dogs bounce and howl.

~

All through that hot summer of 2016, before I went to Greenland, ice around the world was yielding up long-held secrets. The cryosphere was melting, and as it melted things that would have better stayed buried were coming to the surface.

On the Yamal peninsula, between the Kara Sea and the Gulf of Ob, 4,500 square miles of permafrost thawed. Cemeteries and animal burial grounds turned to slush. Reindeer corpses that had died of anthrax seventy years earlier were exposed to the air. Twenty-three people were infected, their skin blackened with lesions. One, a child, died. Russian veterinarians travelled the region dressed in white anti-contamination suits, vaccinating reindeer and their herders. Russian troops burned infected corpses in high-temperature pyres. Russian agriculturalists said that nothing would ever grow in the region again. Russian epidemiologists predicted other releases from Arctic burial sites and shallow graves: smallpox from victims who had perished in the late 1800s, giant viruses that had been long-dormant in the frozen bodies of mammoths.

On the Siachen glacier in the Karakoram, where Indian and Pakistani troops have been fighting a forgotten war since 1984, the retreating ice was revealing spent shells, ice axes, bullets, abandoned uniforms, vehicle tyres, radio sets – and slaughtered human bodies.

In north-west Greenland, a buried Cold War US military base and the toxic waste it contained began to rise. Camp Century was excavated by the US army engineering corps in 1959. They tunnelled into the ice cap and created a hidden town: a two-mile network of passageways housing laboratories, a shop, a hospital, a cinema, a chapel, and accommodation for 200 soldiers, all powered by the world's first mobile nuclear generator. The base was abandoned in 1967. The departing soldiers took the reaction chamber of the nuclear generator with them. But they left the rest of the base's infrastructure intact under the ice, including the biological, chemical and radioactive waste it contained, assuming – as the Pentagon closure reports declared – that it would be 'preserved for eternity' by the perpetual snowfalls of northern Greenland. It is all interred there still: some 200,000 litres of diesel fuel and unknown amounts of radioactive coolant and other pollutants, including PCBs. But as global temperatures have risen, so snowmelt is forecast to exceed snow accumulation in the region of Camp Century. In a dynamic I have seen so often in the underland that it has become a master trope, troublesome history thought long since entombed is emerging again.

The heat in the Arctic that summer was record-breaking, and so was the melt. New lows were set for the extent of Arctic sea-ice coverage. In Nuuk, the Greenlandic capital, the temperature hit 24°C. Meteorologists in Denmark rechecked their measurements. No mistake. For the past decade, the ice cap had been losing mass at twice the rate of the previous century. That year it also began melting a month earlier than usual, and the flow rates on the meltwater rivers of the glaciers reached exceptional speeds. The glaciologists checked their models. No mistake.

The meltwater ran hard from April onwards, pooling as blue and green lakes up on the ice cap, flowing as rivers on the glaciers. The increased amounts of meltwater on the ice cap helped shift the albedo: more sunshine was being absorbed, increasing the temperature, resulting in more melt, and therefore more absorption – a classic feedback loop which winter would only pause.

The calving faces of Greenland's glaciers thundered. Icebergs sweated in Greenland's fjords. Polar scientists brought forwards their predictions of when the Arctic Ocean might be fully ice-free. The highest rates of ice loss were in the north-west and the south-east of the country, where I was heading.

Uneasy stories circulated about disappearances in the ice. A Russian businessman had flown in on the east coast, wearing a camel-skin coat and carrying a briefcase, and never flown out again. A Japanese hiker had vanished in the west of the country, been missing for weeks. Local people spoke half-jokingly of the *kisuwak*, the wild creature that roamed the ice and snatched unwary travellers – an animate version of the glacial crevasse or the silky-thin sea ice.

In that region, at this time of history, it felt as if there were many places where one might fall right through the world's surface.

~

'The year has been exceptional,' says Matt. 'The sea ice was gone from the fjords by June. The snowfall over the winter was minimal. No one's ever seen a year like it. Normally now the channel would be full of ice. A bear was seen swimming off Kulusuk two weeks ago. He must have been desperate. No one shot him.'

Matt has been in Kulusuk since he was nineteen. This is his sixteenth year. He and his partner Helen live in a blue-boarded house, just above the store and the school. They are both climbers, skiers and guides of formidable experience. They both carry themselves with the quiet competence of people whose abilities in wild country are exceptional, but who have no need to prove themselves unless circumstances demand it. Their commitment to the Greenlandic community they have joined is total, proved by the length of time

Matt has lived in the village and by the profound friendships he has formed there.

'Welcome to our home!' Matt says when we arrive. The house is light and airy inside, with pale wood floors and white walls. A large-scale map of the region is framed on one wall. The coastline is coralline in its complexity. We sit and drink tea together. As well as Matt and Helen, there are three of us, all good friends: me; Bill Carslake, a composer and conductor, gentle and funny of manner, who I have known for twenty years; and another Helen, Helen Mort, who I have known for only a year or two but already regard as one of the most talented people of my acquaintance. Helen M, as we come to call her in the mountains, to distinguish her from the other Helen – is a rock climber, a runner, and a writer of rare abilities. She is modest to a fault, gifted to an alarming degree and consistently subtle in her engagement with people and with landscapes. Together we have come to climb the peaks of Greenland's east coast, and to explore the underland of ice on this, the greatest glaciation outside Antarctica.

I go to the westerly window. It looks across the bay. A group of mothers and children are walking along the path by the sea. They are all wearing black head-nets cinched tight around their necks. They resemble a funeral procession, or a bee-keepers' outing.

'That's a new sight in Kulusuk,' says Matt, joining me at the window. 'Twenty years ago there were no mosquitoes; now, as things have warmed, mosquitoes and gnats have arrived. Some people here wear nets over their heads throughout the summer months.'

Kulusuk is one of a handful of small settlements on the east coast of Greenland – fingernail-holds on the edges of this great island. Fewer than 3,000 people live on around 1,600 miles of coastline. Like many of the smaller Greenlandic settlements, Kulusuk is a society ruptured by transition – a previously part-nomadic subsistence-hunting culture, into which modernity has intruded in the forms of stasis and alcohol.

Helen introduces me to Geo, a powerfully built Greenlander in his early sixties.

'Geo is my father,' says Matt, 'and I don't mean that sentimentally. He has become my father and I have become his son.'

When Geo smiles, which he does often, the crease-lines around his eyes run almost from ear to ear. Geo is a very good hunter, renowned for his boat-handling and dog-handling skills, and legendary for his toughness.

'Two winters ago, when a big storm blew in,' says Matt, 'the men were coming back from a hunting trip. The storm hit fast, and the snow was soon too thick for the dogs to pull the sleds. They had a high pass to get over to reach the village. People started to falter. It was a very serious situation. Geo went to the front of the team, put his head down, and broke trail for six hours. They got back safely.'

Geo lies Roman-style on the sofa in the main room, propped up on one arm, listening to the story being retold, smiling quietly. He, Matt and Helen communicate in a mixture of broken English and broken Greenlandic. The lack of a fluent shared language is no barrier to intimacy. They are physically at ease with one another. When they sit together, they often do so with an arm around each other's shoulders, or legs pressed together.

As a boy, Geo was taken to Denmark for a year, part of the ill-conceived 'Northern Danes' project of the 1960s, which sought to assimilate Greenlanders to the Danish way of life by forcing Greenlandic children to live with Danish families.

'Geo still shudders when you ask him about it,' says Helen.

He has visited England twice, as a guest of Matt and Helen – and on each occasion he has acquired a tattoo, one on each forearm. He rolls up his sleeves to show me: 'This one, Glasgow,' he says, pointing at a cross on his right forearm. 'This one, Kendal,' pointing at an anchor on his left.

'I took Geo out for a night on the town in Glasgow,' says Matt. 'We ended up in some pretty rough bars. Geo was a non-standard presence. In Filthy McNasty's I could see folk spotting Geo across the bar, thinking about coming over to take the piss – then having another look and thinking much the better of it. They had correctly assumed that Geo was hard beyond even the measure of Glasgow on a Friday night.'

Geo picks up the guitar that stands in the corner of the room, and sings a quiet, melancholy East Greenlandic song.

There is a knock on the door. It is Siggy, an Icelandic sailor with whom Matt once voyaged north up the coast. Siggy has a beautiful new-old boat, wooden-hulled, which he has sailed here from Reykjavik. He wears green moleskin trousers and speaks calmly.

'This is the year of no ice,' says Siggy. 'We can get anywhere, have been able to explore freely. We've been wearing T-shirts on deck.'

He shrugs.

'The weather should not be like this, but life has been made more easy for us sailors.'

I think of the Old English term *unweder* – 'unweather' – used to mean weather so extreme that it seems to have come from another climate or time altogether. Greenland is experiencing unweather.

Geo stops playing, lays down the guitar, and speaks matter-of-factly. 'In ten years, no snow, no ice, no hunting, no dogs.'

The sea ice is thinning to a degree that makes sailing easy for incomers, but hunting impossible for the native Greenlanders. The intricate stages of hardening through which sea ice annually cycles – frazil, grease, nilas, grey – are no longer being fulfilled in many places, for the temperature of the seawater is spiking above the key freeze-point of 28.6°F. When the men cannot travel safely over the sea ice, hunting becomes difficult. Seals haul out further offshore. Bears die of starvation rather than bullets. Inlets and fjords are dangerous to cross. Snowmobiles run the risk of plunging through thin ice, carrying their drivers with them. Hunting – one of the few aspects of traditional Greenlandic life that survived settlement – is under threat of erasure, this time by global temperature change.

Ice has a social life. Its changeability shapes the culture, language and stories of those who live near it. In Kulusuk, the consequences of recent changes are widely apparent. The inhabitants of this village are part of the precariat of a volatile, fast-warping planet. The melting of the ice, together with forced settlement and other factors, has had severe effects upon the mental and physical health of native Greenlanders, causing rates of depression, alcoholism, obesity and suicide to rise, especially in small communities. 'The loss of that landscape of ice,' writes Andrew Solomon, studying depression rates in Greenland, 'is not merely an environmental catastrophe, but also a cultural one.' The Inuktitut of Baffin Island in the Canadian Arctic have begun to use a word that refers at once to the changes in the weather, the changes in the ice, and the consequent changes in the people themselves. The word is *uggianaqtuq* – meaning 'to behave strangely, unpredictably'. Yet if any population knows what it is like to live with the unpredictability of ice it is surely the Inuit, who have been adapting to its shifts for millennia.

Later that day Helen introduces me to Frederick and Christina, two of the pillars of the Kulusuk community. Christina is Kulusuk born and bred, and she is the village's schoolteacher. Frederick is from West Greenland, but moved to Kulusuk with Christina years previously. They are both deeply cultured and self-aware;

disinclined to any kind of romanticism, with a strong sense of the fine margins of tolerance for life here, but also proud of the resilience proved by Kulusuk's continued existence.

'Climate change is felt in our lives here strongly,' says Frederick. 'New species have come here, old ones have gone. There is thunder and lightning sometimes in autumn. The sea ice used to be so deep always' – he gestures from the floor to the ceiling of the house, a distance of eight or nine feet – 'but each year it is thinner and this spring it was *this* thin' – he places his hands a forearm's length apart – 'and too much danger for the dog-sledding. It is harder to hunt. We can travel less far.'

He shrugs. 'It is a change to our spirit, as well as our lives.' Christina looks on, listens. She disappears into a side room, and emerges holding a gaudily painted wooden canoe, two feet or so long, in which stand in single file a zebra, a lion, a tiger and a giraffe.

'Our son made this at school,' says Christina. 'He called it Noah's Kayak, because it is saving the animals from the flood of global warming.'

There are no humans on board the kayak.

The melt is seen by some as an opportunity rather than a loss. Foreign investors have gathered as the ice has retreated, and access to Greenland's fabulous mineral wealth has become easier. 'There'll be a lot of billionaires made by what the melt reveals,' a geologist told me before I came to Greenland. 'Mining's coming to Greenland soon, and big style – in a country that's never before had anything much deeper than a quarry.'

The last few years have seen the granting of more than fifty mining licences in Greenland, allowing exploratory mining for gold, rubies, diamonds, nickel and copper, among other minerals. And on the southern tip of Greenland, close to a small town with high unemployment called Narsaq, lies one of the world's largest uranium deposits. Niels Bohr, the Nobel Prize-winning atomic physicist who worked on the Manhattan Project, visited Narsaq in 1957, shortly after the discovery of the deposit. A joint Chinese-Australian mining project now proposes to establish an open-pit mine behind Narsaq, in order to acquire not only uranium but also the rare earth minerals used in wind turbines, mobile phones, hybrid cars and lasers.

That evening in Kulusuk a lurid sunset brews above the village, lilac and orange backlighting a sawtooth ridge of peaks, with incandescent reefs of ribbed clouds. It is alpenglow of a kind – but of an incredible wattage.

'It's the ice cap that makes sunsets like this,' Matt explains. 'It's probably the biggest mirror in the world: hundreds of thousands of square miles of ice reflecting up the sun as it dips towards the horizon.'

We all walk together up a short switchback path to the top of the rock outcrop around which the village was built. I go to the western edge of the outcrop for a better view of the sunset in the fjord – and stop.

The little bay beneath me is the village's rubbish tip. Thousands of bin bags, a slew of plastic crates, cracked kayaks, melamine cupboards and white fridges have all been heaved over the cliff edge here to make the midden. It looks, in the dusk, like a tongue of ice flowing down towards the waterline: a glacier in advance, not retreat.

~

Ice has a memory. It remembers in detail and it remembers for a million years or more.

Ice remembers forest fires and rising seas. Ice remembers the chemical composition of the air around the start of the last Ice Age,

110,000 years ago. It remembers how many days of sunshine fell upon it in a summer 50,000 years ago. It remembers the temperature in the clouds at a moment of snowfall early in the Holocene. It remembers the explosions of Tambora in 1815, Laki in 1783, Mount St Helens in 1482 and Kumae in 1454. It remembers the smelting boom of the Romans, and it remembers the lethal quantities of lead that were present in petrol in the decades after the Second World War. It remembers and it tells – tells us that we live on a fickle planet, capable of swift shifts and rapid reversals.

Ice has a memory and the colour of this memory is blue.

High on the ice cap, snow falls and settles in soft layers known as firn. As the firn forms, air is trapped between snowflakes, and so too are dust and other particles. More snow falls, settling upon the existing layers of firn, starting to seal the air within them. More snow falls, and still more. The weight of snow begins to build up above the original layer, compressing it, changing the structure of the snow. The intricate geometries of the flakes begin to collapse. Under pressure, snow starts to sinter into ice. As ice crystals form, the trapped air gets squeezed together into tiny bubbles. This burial is a form of preservation. Each of those air bubbles is a museum, a silver reliquary in which is kept a record of the atmosphere at the time the snow first fell. Initially, the bubbles form as spheres. As the ice moves deeper down, and the pressure builds on it, those bubbles are squeezed into long rods or flattened discs or cursive loops.

The colour of deep ice is blue, a blue unlike any other in the world – the blue of time.

The blue of time is glimpsed in the depths of crevasses.

The blue of time is glimpsed at the calving faces of glaciers, where bergs of 100,000-year-old ice surge to the surface of fjords from far below the water level.

The blue of time is so beautiful that it pulls body and mind towards it.

Ice is a recording medium and a storage medium. It collects and keeps data for millennia. Unlike our hard disks and terrabyte blocks, which are quickly updated or become outdated, ice has been consistent in its technology over millions of years. Once you know how to read its archive, it is legible almost as far back – as far down – as the ice goes. Trapped air bubbles preserve details of atmospheric composition. The isotopic content of water molecules in the snow records temperature. Impurities in the snow – sulphuric acid, hydrogen peroxide – indicate past volcanic eruptions, pollution levels, biomass burning, or the extent of sea ice and its proximity. Hydrogen peroxide levels show how much sunlight fell upon the snow. To imagine ice as a 'medium' in this sense might also be to imagine it as a 'medium' in the supernatural sense: a presence permitting communication with the dead and the buried, across gulfs of deep time, through which one might hear distant messages from the Pleistocene.

Ice has an exceptional memory – but it also suffers from memory loss.

The weight on 2,000-year-old ice can reach half-a-ton per square inch. The air in this ice has been so compressed that cores brought up by deep drilling will fracture and snap as the air expands. This is why glaciers sound like shooting ranges. This is why if you were to drop a piece of very old blue ice in a glass of water or whisky, it might shatter the glass.

Deeper still – in ice aged between 8,000 and 12,000 years – the pressure becomes so great that air bubbles can no longer survive as vacancies within the structure of the ice. They vanish as visible forms, instead combining with the ice to form an ice-air mixture called clathrate. Clathrate is harder to read as a medium, and the messages it holds are fainter, more encrypted.

In mile-deep ice, individual layers can only just be made out as 'greyish ghostly bands . . . visible in the focused beam of a fibre-optic lamp'. And because ice flows – because it continues to flow even when under immense pressures – it distorts its record, its layers folding and sliding, such that sequence can be almost impossible to discern.

At the deepest points of the Greenland and Antarctic ice cap, where the ice is miles deep and hundreds of thousands of years old, the weight is so great that it depresses the rock beneath it into the Earth's crust. At that depth, the compressed ice acts like a blanket, trapping the geothermal heat emanating from the bedrock. That deepest ice absorbs some of that heat, and melts slowly into water. This is why there are freshwater lakes sunk miles below the Antarctic ice cap – 500 or more of these subglacial reservoirs, showing up as spectral dashed outlines on maps of the region, unexposed for millions of years, as alien as the ice-covered oceans thought to exist on Saturn's moon, Enceladus.

As a human mind might, late in life, struggle to remember its earliest moments – buried as they are beneath an accumulation of subsequent memories – so the oldest memory of ice is harder to retrieve, and more vulnerable to loss.

~

We load the boat chain-style on a rising tide, slipping on the kelpy rocks as we heave blue bear-proof barrels, weapons and packs along the line.

'Watch where you put things down,' says Helen. 'There's seal guts and cod heads and all sorts smeared on the rocks here.'

It takes half an hour to load and check. Then Geo guns the Yamaha 1200, spins the boat around from the dock, and we roar out across the channel, aiming for where the glacier called Apusiajik – 'the Little Ice' – meets the sea.

There is a high cry – haunting, falling away and then repeating, silver-gold in colour – that sets my neck tingling.

A red-throated diver, no, *three* red-throated divers, flying in formation northwards over the channel, the same direction as us. Big birds, heavy-set in the body but graceful in their lines, smooth in silhouette as if poured from water rather than made of feathers. I haven't heard the cry of a diver for a decade, since seeing one hunt on a loch in the shadow of Suilven in the far north-west of Scotland, and before that another decade earlier, on a forested lake in British Columbia.

'Truly a bird of the north,' says Matt.

We can hear the divers calling long after they are lost to sight.

Buck of the boat off the wave-chop. Salt spray, air cold and fast on the face. Sharp peaks rising in all directions. Fjords cutting away. A sense beginning to build in me of the scale of this landscape, beyond anything I have ever experienced or imagined: the vastness of the coastline, and always somewhere behind it to the west the ice cap itself, so huge it annihilates all features other than itself, all colours other than white and blue. I can feel a buzzing in my stomach, the surging excitement of a big journey starting. We will not see Kulusuk again for weeks.

The lower mountains are scabbed with snow. The exposed rock is golden, brown, red, white: warm marbled colours. It is some of the oldest surface rock in the world, and I know that it makes a torn-page match with the gneiss of the Outer Hebrides. Hundreds of millions of years ago, these two coastlines were united. A deep time kinship existed between this wildly unfamiliar region, and those Scottish islands in which I felt at home.

It is six miles across the channel from Kulusuk to Apusiajik, but it looks as if we could swim it. The glacier itself is five miles long, but it looks as if we could wander up it in a couple of hours with hands in pockets. We would die if we tried either.

The foreshortening illusion is powerful, born of the air's pristine clarity, and it is the first of countless misprisions of scale I will experience in Greenland. This is, I will learn, a landscape that plays tricks on the eye, dupes perception, induces forms of clarity that are in fact forms of delusion. Rock and ice walls reflect and redirect sound misleadingly: events occurring ahead of us seem to issue from behind. There are no usual units to which the eye has become accustomed: no buildings, no cars, no distant people. The terrain is built out of a few elements – rock, ice, water – that echo their own forms up and down the orders of magnitude.

Geo steers expertly, using one hand, past a group of black rock islands near the middle of the channel.

'There were orca around here a few days ago,' says Matt. 'And sei whales. We heard them before we saw them: the *hoo*ing of their blowholes.'

As we near Apusiajik, the water thickens with blue-white pebbles and boulders of ice that thunk on the hull. Geo steers an elegant course, but eventually the ice is too thick to avoid, and so he lowers speed and noses through it, *thud*, *thunk*, *thump*, *thud*, closing in on the snout of the glacier.

Apusiajik tumbles into the water. The tideline calving face is perhaps 2,500 feet in length, and pale blue at the points of freshest calving. Above the face the ice rolls over a drop, and a central bulge of rock is visible, splitting the roll-over, streaked black with lines of meltwater.

'That's new,' says Matt. 'A couple of years ago that wasn't there – it was pure ice.'

I will remember that island of fresh rock much later, when we take as our sleeping place another such ice-island, also recently revealed by the melt, far up a much bigger glacier.

Geo slows the boat, then pulls the engine back to an idle. We float along parallel to the face, keeping 1,500 feet or so distant, to give us time to get clear if a big calving happens. Geo points at the glacier, then turns back towards Kulusuk and a peninsula of bare rock that pushes out from the margin of the glacier, into the channel.

'Fifty years ago, when I was a boy,' he says, indicating the peninsula in the channel, 'ice was there.'

Then he points to an island yet further out in the channel.

'My father's time, ice was there.'

He points back to Kulusuk, then holds his hands to his ears, bunches his fingertips and flicks them open, miming an explosion.

'Before, in Kulusuk, we hear glacier boom! Now, no sound.'

In the course of Geo's life, the face of the Apusiajik glacier has retreated so far back and around that the noises of its calvings are no longer audible in the village. Melt has changed the soundscape of everyday life. The glacier is experienced as a silence.

~

We unload the boat chain-style on a falling tide, heaving the gear onto a rock-sand beach of white quartz and black mica. The tide has left little bergs stranded on the sand, along the line of the bay. They shine blue-silver in the late light. There is something exhausted about them. Other small bergs lap slowly towards land, or mill around in the offshore currents.

We hump the gear for about 900 feet, making four return trips, through a shallow boulder valley and onto a flattened plain of mossy topsoil and boulders, along which a stream runs on a gradual incline to the sea.

The plain is the path of a vanished glacier: moraines to seaward indicate the former extent of the glaciation. We are making camp in ghost ice.

I think of the accounts I have read of how small craft hugging the Greenland coastline will sometimes find their GPS navigation devices screaming alarm, warning of collision. The coordinates of the former extent of glaciers have been inputted into the mapping, but the retreat rate has been so fast that they are sailing into and *through* the digital phantom left behind by the ice.

The air around our tents is filled with white specks I cannot identify, which are not snow and are not dust, so that the atmosphere seems electrified, scintillating.

Two grey gulls fly overhead, cranking their wings against the rising easterly. A raven circles, croaking, then glides down to land on the erratic against which we have piled our gear. It folds its glossy wings, shakes itself down, and watches us with a cocked head, curiously.

We pitch the tents in a line, side to side, six feet separating each of them. Then we set to work on establishing a bear perimeter. Polar bears can smell a food source from up to twenty miles away. If you see a bear, you can be sure the bear has known of you for far longer, and has come to investigate. None of us wants to see a bear, for our sake and its sake. We have two weapons with us: massive-bore rifles that fire adapted shotgun shells containing single slugs rather than pellets. Each person carries flares at all times.

Around the camp we start to set a rectangular boundary of trip-wires which, if triggered, fire blank cartridges downwards into the earth, scaring away an inquisitive bear. We string the wires at a height of about two feet, so that they won't be snagged by any white foxes coming to scavenge.

It takes two hours to set camp to Matt's satisfaction. We sing as we work. Bill is a professional singer, blessed with a resonant bass voice. I warble happily. The sun lowers to the west. Two bergs move from left to right across the bay.

In a landscape as vast as the Arctic, the eye is surprised by details. Though the topsoil around the camp is only a few inches deep, it supports a diversity of mosses and plants. Club moss flourishes in the lee of boulders, and the rocks are painted with lichens: blotches of orange *Xanthoria parietina*, the intricate cartography of map lichens, and a crisp lettuce-like lichen I cannot name – acid green in colour, hard to the touch.

Everywhere are the emerald leaves of the tiny dwarf willow. I pick a leaf, half the size of my little fingernail, and hold it to the sun. It shines green, and I can see the delicate red vein-work that marks it. I know this *Salix* only from the Cairngorms, Britain's equivalent of the Arctic, where it grows sparsely on the highest parts of the plateau. Here it covers the ground, creeping sideways, its pitch-black branches a few millimetres thick at most.

We have pitched our tents *on top* of a forest, I realize. We are canopy dwellers.

I recall a joke I heard in Reykjavik. Question: 'How do you find your way out of a forest in Iceland?' Answer: 'Stand up.'

Now and then a muffled boom moves through the landscape,

arriving softly but forcefully as a push on the eardrums, a vibration in the flesh. It is the sound of calving ice, made by a slab of glacier crashing into the water from the Apusiajik face, round the mountain from us. *Sound is a blow delivered by air, through the ear, on the brain and the blood, and transmitted to the soul . . .*

Big bergs make slow journeys across the bay: a stricken U-boat, a cruise liner, the Scottie dog from a Monopoly set, white and clean, nodding its way over the course of the evening.

'Sun dogs!' calls Helen, pointing upwards with a smile. Glittering rainbow arcs stand at a convex to the curve of the sun itself.

Ice in the inlet, ice in the sky. Ice in the bay. Ice in the air above us. Sounds of ice from the glacier. We are sleeping where ice had once been.

That night the Northern Lights appear for the first time. A scarf of radar-green flutters in the sky. The mountains shoot jade search-lights into space.

We lie on our backs in the cold black air and watch the show, amazed into silence.

~

A week before leaving for Greenland, I go to the British Antarctic Survey on the outskirts of Cambridge to see a man called Robert Mulvaney. Mulvaney is an ice-core scientist, a palaeoclimatologist and glaciologist. He has spent his career studying the underland of ice: reading its memory for what it tells him about past climate and environment – and what it might foreknow about climate changes to come.

Mulvaney has worked for twenty drilling seasons in Antarctica and five in Greenland. In the field he grows a big beard and moustache; in the office he is clean-shaven. He shakes my hand hard, leads me briskly through the corridors of BAS, speaks quickly.

'I may seem like a relaxed person,' he says. 'I'm not. Not at all.'

He doesn't seem like a relaxed person. He seems like an impressive person who has spent most of a lifetime performing challenging tasks under conditions in which efficiency of effort is essential.

As a young man Mulvaney was a hard climber, and a hard caver too. I tell him about the Timavo, and about descending the Abyss of Trebiciano with Sergio puffing on his briarwood pipe.

'Ah, you were in the karst, then. I did some fairly far-out caving near there. Exploratory stuff in Yugoslavia, floating into wet systems on rafts, that kind of thing. I always preferred Yorkshire limestone, though. Drier.' He looks briefly nostalgic for life under the earth.

He takes me to his study, points me to a seat. 'I gave up serious caving and climbing when I'd lost too many friends to death and injury,' he says. 'So I became a sailor instead.'

Pinned on the noticeboard above his desk is a tattered pennant in Jamaican black, gold and green.

'I sewed that myself,' he says with unguarded pride, 'as we approached land at the end of my first Atlantic crossing.'

Next to it is a yellowed photograph of his wife and two daughters, waving at the camera out of the cockpit of a drastically tilted yacht, beached on wet mud. It's unclear if they are waving in greeting or distress.

'We'd run aground on a mud pinnacle somewhere off Essex,' Mulvaney says. 'If you haven't grounded when sailing the east coast, then you haven't sailed the east coast.'

Propped behind his computer is a handwritten postcard-sized sign in faded felt-tip, written in block capitals by a childish hand. It reads:

ROB MULVANEY GONE TO THE ANTARCTIC

Whereas Merlin and his fellow mycologists look down into the 'black box' of soil, Mulvaney and his fellow palaeoclimatologists look down into the 'white box' of ice. They use ice-penetrating phase-sensitive radar, which bounces off reflective horizons, to build up detailed images showing the internal layering and folding of deep ice. They use sonar: detonating explosions and mapping the echo returns. And they use core-drilling – the technique pioneered up in Camp Century by American scientists, while their military peers covertly tunnelled out the ice missile base.

Mulvaney has worked with core from its early days as a technology, and he has personally designed and engineered several of the standard drill types used in British climate science.

'For shallow drilling, down to twenty metres or so – that's about 200 years back in time – it's done by hand,' he says. 'Quick work. You stop off, set up, twist the drill down by hand. Any deeper than that and you have to move to electro-mechanical drilling: an engine-driven drill that's dropped down and then pulled up again by winch.'

He shows me a hand drill. It is a strikingly analogue tool. A 1.5-metre sleeve of metal, an internal drill bit with tool-steel teeth, a screw-form outer that guides the ice chippings up between the bit and the sleeve, and pop-out fins that prevent the torquing of the barrel when the drill is in action, but retract when it is being drawn back to the surface.

The drill is lowered, cuts its core, is retrieved, the core is disgorged, the drill is lowered again. Lower, bite, drill, raise, disgorge; lower, bite, drill, raise, disgorge. Repeat some 700 times to bore a kilometre of ice.

Ice-core science is industrial work, hard labour. Mulvaney once cored for ninety-two consecutive days, working up to fourteen hours a day in temperatures of -15°C. Ice-core scientists don't tend to be the kind of people who bring workplace lawsuits because the office air conditioning is set too low for comfort.

Ice-core science also tests patience. Once, Mulvaney tells me, he lost a drill 1,000 metres down. That was it. There was nothing to be done. It couldn't be fetched up.

'It took a year to set up the drill site, a year to drill to a kilometre, a second to lose the drill, and another year to relocate the drilling site.'

When core is brought up it is cut into standard 'bag' lengths, which are then wrapped, tagged and made ready for transfer out to the cold stores of laboratories around the world. Back in the labs, each core section is cut along its length into six parts, according to a standard profile. One of these parts is known as the Forever Archive, kept in case everything else is lost. The others are used for research.

In Greenland, Mulvaney was involved with a project known as NEEM, the North Greenland Eemian Ice Drilling Project. NEEM's aim was to drill and analyse core from the Eemian, the last interglacial period, which extended from around 130,000 to 115,000 years ago. The Eemian is of intense interest to scientists because it is thought to approximate to the climate processes and feedbacks that may be expected by the end of the twenty-first century. It has become, says Mulvaney, 'a hot spot for predictive research'. Fourteen nations were involved in the project.

At the NEEM research site in north-west Greenland, a 25-foot-deep drill-pit was chainsawed out of the ice and covered to create an 'ice cave'. Down in the ice cave, ambient temperatures were a balmy -20°C, and scientists were able to work twenty-four hours a day during the field season to extract and analyse the core. Over the course of two years they drilled more than two and a half kilometres to hit bedrock. The core they extracted was the first complete record of the Eemian.

What that core revealed was that intense surface melt of the Greenland ice cap had occurred during the warmth of the Eemian period. The meltwater had soaked into underlying snow and re-frozen, leaving tell-tale long-term signatures in the ice layers. Uncannily for the researchers, similar conditions repeated themselves during the coring work in the summer of 2012 – temperatures rose, rain fell, and the meltwater formed refrozen layers: Eemian echoes in the Anthropocene.

Mulvaney reaches behind his computer and picks up two small objects.

'Hold out your hand.'

He drops one of the objects into my palm. It is a small, heavy grey fang. I recognize it as the tooth of a coring drill bit. The cutting edge of the tooth is deformed, like a bullet after impact.

'That's one of the drill teeth that hit bedrock in Antarctica,' says Mulvaney proudly. 'Nine hundred and fifty metres down below Berkner Island.'

It looks good for nothing but spreading butter now.

'Is hitting bedrock an ice-core scientist's hallelujah moment?' I ask. 'Like a tycoon striking oil?'

'Oh yes, there's nothing better. Here, look at this, too.'

He hands me the other object, a small transparent plastic phial. I hold it up to the light. It contains a pinch of blond sand.

'These are the grains that came up in the last core before we hit bedrock on Berkner,' he said. 'This is the basal sediment. If you look at these under magnification, you'll see that they're rounded grains: they're aeolian – wind-blown quartz fragments, around 0.2 millimetres in diameter, smoothed and frosted.

'Show these to any geologist, and they'll tell you they were formed in desert-like conditions, and got rounded off by the wind.

So what we know from these is that, at some point, the land that now lies a kilometre below the ice was once a Sahara.'

'They're beautiful,' I say. 'Desert diamonds from the bottom of the world.'

'I can tell you're not a scientist,' he says.

Mulvaney takes me to the cold store. We open a heavy door and push through butcher's-shop hanging strips of heavy plastic.

The cold of the cold store is a killing cold, a knives-under-the-skin cold, a needles-in-the-eyes cold. It is so cold the ink in my pen freezes in under a minute. Mulvaney doesn't seem to notice. He wears a shirt with the sleeves rolled up. I am wearing three layers and wonder how long I can survive.

Mulvaney creaks the lid off a white polystyrene chest. It is filled with core sections in marked transparent bags. He rummages, then picks out a bag. Written on the side in black marker is '140,000 YA'.

'This one's from well before the last interglacial,' he says, giving it to me. I cradle it like a newborn baby, though it is very old, then I place it gently on a work surface, as far from an edge as possible.

He slides something from a plastic sleeve and passes it to me. It is a disc of ice a few millimetres thick that has been cut from the end of a section of core.

'That's young ice,' Mulvaney says. 'Baby ice. Maybe 10,000 years old, no more. Hold it up to the light.'

I lift it to the strip-light. It is instantly, witchily beautiful: silver and translucent, and seething within it like stars are scores of gleaming ice bubbles.

'Those are where the real gold is stored,' Mulvaney says. 'Each bubble is a museum.'

I remember Browne's use of the word 'conservatorie' in *Urne-Burialle*, to mean a space where something is conserved. Ice has long been one of our most brilliant 'conservatories': ice houses kept peaches and strawberries fresh long before the invention of fridges, chilled shipping containers move luxury perishables around the world, glaciers curate the bodies of the long dead, and in cryogenic facilities, billionaires with Lazarus delusions prepare the technology necessary to freeze their brains after death. In all of these scenarios, ice serves as a substance that slows change, and reaches far into both the future and the past.

'The search is now on for the oldest ice,' says Mulvaney. 'We want to drill to at least a million years, maybe even a million and a half, in Antarctica.

'It's a ten-year project at least,' he continues. 'First we need to locate the perfect drilling spot for ultra-deep drilling – and there's plenty of dispute over that. Strangely, the Japanese think it's near their territory, whereas the Russians think it's around Lake Vostock, where they have their base, and the British and Americans think it's around Dome C, where they work!'

He speaks with pride of the achievements of ice-core science.

'We helped get rid of lead in petrol. And we produced the CO_2/ temperature graphs that rang the bell on climate change. A few years ago, I thought my science was mostly coming to an end. What was there left to do now we'd called out global warming and cleaned up cars? Now I see a whole future opening up, in the search for the oldest ice. There's a climate puzzle that no one has been able to solve. Around one million years ago, the climate flips its periodicity from a 40,000-year frequency to a 100,000-year frequency. Why? No one knows. And if we can't explain *that* about the climate, how can we claim to know anything? If we can find and drill the oldest ice, well, we might just solve that puzzle. The secrets are in the depths.'

Before I leave I ask Mulvaney a last question, a version of the one

I asked Christopher, the dark-matter physicist, far below the earth at Boulby.

'Does working in spans of time as great as those you inhabit – 100,000 years, a million years – make the human present, our hours, our minutes, seem somehow brighter and more true, or does it crush them to irrelevance?'

He thinks for a few moments.

'Sometimes I hold a piece of rock and a piece of ice in my hand,' he says. 'Both have come from far under the surface, both carry messages from pre-human history. But in ten minutes' time the ice will have vanished, while the rock will still be here.'

Pause.

'This is why ice is exciting to me and rock is not. This is why I'm a glaciologist and not a geologist. Ice still thrills me with its durability and its perishability, even after all these years and all this core.'

~

Crunch and rasp of broken glass, the ice snapping at our feet. A hot, high Greenlandic sun, its light more white than yellow. Bergs in the bays, but a cloudless sky. We move in line, roped, sharp, *wired*.

From the bay that morning we follow a stream uphill from camp and enter a wide valley slung between peaks. There we come upon the shores of a shallow lake, unforeseen, its far shore set tight in the shadows of the easterly peaks. It appears frozen, but on approaching I realize that what seems to be ice is in fact alluvium: silt scoured from rock by the glaciers whose melt-streams feed the lake, giving it its burnish. Our arrival sends up a flock of seagulls, wings clapping the water as they take off.

We move along the lake's western shore, hopping from boulder to boulder, stepping on foot-hugging cushions of moss. The low-lying flora is vibrant: slews of pink fireweed, scarlet lichen beds, yellow willows.

An hour's work brings us to a low pass above the lake, and there the sound of our footfalls changes as we pass onto fine gravel, beached in a gorge between boulders. We rest. Matt unslings the weapon he always carries across his back, rolls his shoulders to ease them out. The clear cries of geese can be heard, growing in strength as they near, echoing off the mountain cirque to our east.

'That's a perfect fourth!' says Bill delightedly. He listens to landscapes like no one I have travelled with before, sees and hears them musically.

The geese pass high overhead, a dozen or so in a tight V. I take them to be pink-foots, and guess they are beginning their autumn migration south: probably to Iceland as their next stopover, and from there to England, where they might land honking on the fields around my parents' house in Cumbria.

'This valley is one of the great highways of this region,' says Matt. 'For creatures and also for people. It's the main dog-sled route from Kulusuk up to the northern fjords. From the village you come across the bay on the sea ice – if it's thick enough – make landfall not far from our camp, then up and over this low pass and down towards Igterajipima and on towards Sermiligaq. Geo and Helen and I have done it dozens of times. We ski it all the time, too, if we don't need the dogs. It's like a main road for us.'

I think of the aurora of the night before, the long green scarf that had shimmered down the length of the same valley. What was it that Barry Lopez called these old routes of movement and migration within the landscape? *Corridors of breath*. That was it – and the auroral light had seemed like a vivid, otherworldly breathing.

The gravel gorge is the route of a dry glacial stream, and it leads us directly to the snout of the glacier. This is the back of Apusiajik, the landward side, where it flows eastwards off the mountain that makes it. Where the tongue of the glacier dips to meet the rock, it is dirty with dust and debris. The tongue is hollowed where meltwater streams emerge from beneath it, leaving a brown carapace of hard ice arched above mouths of melt-tunnels that lead far back under the glacier.

We step up onto the carapace one after the other, stamping our feet, testing the ice for weakness. Each step booms, echoing through the underhang of the snout.

When you move onto a glacier, you enter *its* space. Sound changes, temperature drops, danger grows. The cold comes at you not in the form of fingers, probing, but as a cloud, an aura that surrounds you and settles in your core: *You're in my zone now.*

So much of an iceberg is below the water's surface; so much of a glacier is below the ice's surface. As a river flows calmly over gentle ground, so does a glacier. Where it moves over steeper ground – a 'roll-over' – or turns a corner, the ice disrupts and cracks. Crevasses are the glacier's equivalent of river rapids: expressions of turbulence in the flow.

Mountaineers speak of 'dry' zones and 'wet' zones on a glacier. In the wet parts of a glacier, the ice is covered by a layer of lying snow; in dry parts, there is no such covering. Wet zones are often easier to move on, but more perilous, as the dangers of crevasse and bergschrund are concealed, and it is hard to predict the weight-bearing qualities of the snow. When travelling on a wet glacier, the experience is one of near-continuous menace. A sense builds and stays of what lies beneath you: the great blue depths under the lying snow, the ever-present underland of ice. You are conscious of taking each step with care.

The lower reaches of the glacier that day are dry, and so we can see down into the ice's depths. There are little eye-shaped dolines, shimmering with cobalt meltwater. Fine fissures, only as wide as a finger or a palm or a forearm, narrow below us into blue. Crevasses yawn into chasms big enough to swallow a car or a house. Rounded pipes plunge vertically down, so straight and true it seems you might loose an arrow into one and hit bedrock.

Everywhere the underland of the glacier declares itself less as structure than as hue, a radiant blue brimming in every fissure or shaft. In Scandinavia, this blue light is sometimes known as the 'blood' of the glacier: an uncanny image for an uncanny phenomenon.

I stop to drink at a meltwater pool, dip my face to the ice, feel the blue blood-light soak into my eyes, my skull.

Our aim that day is a nameless peak, one of the summits whose upper corries breed the ice that gathers to become the Apusiajik glacier. The only map of the region, an unreliable 1:250,000, scarcely notices it. Its summit is a graceful curve of tawny rock rising from a glaciated cirque. It is very attractive indeed – and it is just one of the countless thousands of summits that rear from the ice and the fjords up and down this coast.

Far up the glacier we find a moulin, our first – and nothing to what we will find and descend many days later on the Knud Rasmussen glacier, far to the north of here. A moulin – the word is French for 'mill' – typically begins to form in a declivity on a glacier. Meltwater gathers in the declivity and, being slightly above freezing in temperature, warms the ice on which it pools. This increases the declivity, which in turn draws more water, which in turn begins to drill deeper as current and gravity come also to bear as boring forces. Under certain circumstances the meltwater will bore a hole into the glacier; grinding through the ice, sinking a shaft. Some moulins are skinny, only a few inches across. Some are hundreds of yards in diameter. Some reach only a few dozen feet into the ice before dispersing into side channels or sealing up completely. Some are up to a vertical mile in depth, and drop all the way to bedrock.

Moulins have become increasingly of interest to glaciologists and climate scientists for two reasons. Firstly because they are signs of rising surface melt-rates on glaciers and ice caps. And secondly because the deepest moulins duct water directly to the bed of the glacier. Because the meltwater is warmer than the ice, it transports thermal energy deep into the glaciers and melts more ice – so-called cryo-hydrologic warming. It is now also understood that the water can sometimes act as a lubricant, hastening the rate at which the ice slides over the rock beneath it, such that glaciers ride their own melting.

This quickened slide-speed can in turn quicken the rate at which glaciers calve into the sea, which in turn quickens the rate of sea-level rise. Across Greenland – as across Antarctica – glaciers are both shrinking and speeding up. The East Greenlandic glaciers presently have some of the fastest retreat rates and fastest flow rates of any on the planet. In warmer temperatures, meltwater lakes grow over days on the ice sheet, before abruptly draining through a self-created moulin in the course of a few hours.

A sub-science of speleo-glaciology has emerged, with scientists abseiling into moulins to retrieve information about temperature and flow rate, or sending data monitors into their depths. In north Greenland a NASA scientist named Alberto Behar launched a flotilla of yellow rubber ducks down a mile-long moulin to see if they would emerge at the tidal snout of the glacier: a low-tech way of mapping the interior of the ice, recalling the pine cones dropped into the karst rivers of Greece and Italy to fathom their courses.

The moulin we find that day is perhaps four feet wide, perfectly circular at the surface, and its blue shaft slides away at a diagonal into the depths of the ice. And it sings, the moulin *sings*, with a high, steady, neck-tingling cry. Air is moving within it and within the invisible system of melt-carved ice tunnels to which it connects, driven by water flow far down in the glacier's tunnel system.

Bill tilts his head towards the moulin, then looks up in wonder.

'That's an A, a D and a C sharp,' he says. 'It's the harmonic series of D!'

The moulin is a pipe of the vast aeolian organ of the glacier itself. I wish we could tune in, record its sounds, learn what it has to say.

'Sea ice is incredibly musical too,' says Helen. 'In the winter, it really hisses and whistles – around the tideline, especially, it seems somehow to hum.' I feel again the eerie sense of the ice as *alive*: the repertoire of its sounds, the variety of its forms, its colossal, shaping presence in this landscape.

As we approach the upper cirque of glacier, the ice becomes more contorted, the crevasses almost fully covered. We move over a soft white snowfield, aware we are walking above great depth. Everyone is vigilant, keeping the rope tight in case of sudden fall. I have again a sense of doors locking behind us; remember entering other intimidating labyrinths through which I have passed – the boulder ruckle in the Mendips, the catacombs in Paris, the descent of the Abyss of Trebiciano. Here, our footprints are our Ariadne's thread: the thin winding line that will show us the safe route out at the end of the day.

Matt has wondered if the bergschrund might be impassable, or if we might need to abseil into it and climb back up its far side: a committing and time-consuming move. But when we reach it, hot from the work of the ascent, there is a single viable crossing place: a pinch-point where the sides close to within a few feet of one another, the gap spanned by a snow bridge.

We cross one by one, treading softly, the climbers before and beyond on the rope standing braced to take a fall if the bridge collapses.

It comes to my turn. I intend to cross quickly, but for reasons I cannot explain I pause on the bridge. I look down into the depths of the bergschrund to the right, and feel a bloom of fear in my chest, like a drop of ink spreading in water. Below the snow bridge the sides of the bergschrund fall like a blue gorge, more than 150 feet deep, big enough to eat a truck and its trailer, its upper cliff overhanging, its true depth lost in shadow.

'Keep moving, Rob,' calls Helen from behind me, urgently. 'That's no place to stop.'

I realize I have stopped, have *been* stopped by the void and the glimpse into the depths that it gives or demands.

Half an hour later we step from the high ice and onto the lion-coloured rock of the summit ridge. We take off crampons, make a kit depot, rope up. Matt still has the rifle slung over his back.

'Surely you can leave that here and we can pick it up on the return?' I say to him. 'We're hardly going to meet bears up here?'

'Polar bears were encountered at 2,000 metres in 1913 on the first ascent of the highest peak in this region,' says Matt.

'Oh,' I say.

We move together up the ridge. There is no need to pitch it.

Lodged in the lichen of the summit boulders I find the pale quill of a raven feather and a single, implausible shell, bleached pure white.

We sit quietly together in the sunshine, on the warm rocks of that peak, and look out over the wildest land I have ever seen. Ridge upon ridge of rock spires, range upon range of summits extending as far south and north as the eye could see.

Fjord after fjord, inlet after inlet, island chains, peaks.

Blue ocean endless to the east, on which icebergs glint.

Shorelines pricked with gleams of white: thousands of beached bergs.

Green-water estuaries marbled with brown alluvial outwash, furling into flowerish patterns.

Over the valley, at the same altitude as us, lies a high circular corrie. In it sits a green lake of water, circular in form and cupped by seracs. It has the appearance of a font in a church, and its still surface catches the clouds and sunlight that move across it.

'Look behind you,' says Helen M, pointing.

There, away to the west, running laterally between the ridges of the highest peaks, is the ice cap itself.

It appears as a floating band of white, impossibly elevated, nacreous and faint. This is the 'Inner Ice' and it extends unbroken to the Arctic Ocean on the west side and the north, running for tens of thousands of square miles. Trillions of tons of ice, up to 11,000 feet thick, so great in their mass that they have warped the bedrock beneath them down into the Earth's crust by up to 1,180 feet below sea level. If melted at a stroke, the ice would reveal a vast concavity occupying the island's centre: flattened mountains, crushed valleys.

The Inner Ice looks off-worldly. I feel a longing to get up onto it, to traverse it, to be in that floating white for thirty days.

'Hey, there! Down in the bay, that dark shape in the water! Whale, I think.' Matt's eyes are incredibly sharp and so too is the air, the lensatic effect of its dustless clarity collapsing distance. We are two miles or more distant in space from the bay, but the whale can still be seen with the naked eye.

Except it isn't one whale, it's three. Three shadows in the green water of the bay, two large and one small, parents and a calf, feeding in the outflow where the glacial melt-river sweeps food into the sea. They move in the space between two big icebergs with turquoise underwater bulks.

We watch the whales through binoculars, breaching and disappearing, dark forms disclosing themselves then sinking back into invisibility.

A plume of gulls, a shake of silver, tracks their movements.

Far below us, half a day's travel away, we can also see the orange specks of our tents, and from this altitude we can clearly see the terminal and lateral moraines marking the former reach of the ice that had once spilled down the valley, which would have submerged our campsite under white.

'The Inuit don't come to the summits. Why would they?' says Matt. 'Every now and then, Geo will use the Inuit word for "beautiful", of a glacier or a place. But mostly this landscape is the venue of work, danger, and life for him. He loves the land too, though. I remember once he and I were in a boat near the calving face of a glacier, and he turned to me, nodded, smiled, and said, "I like to come hunting in this place in October."'

Bergs slip along the sea's horizon. The crumps of calvings reach us minutes after the events that have made them. A snow bunting flits among rocks to the north, startlingly fast.

We stay in that sunshine, on that marvellous summit, for an hour and an era. We don't talk much. Up there, language seems impossible, impertinent, sliding stupidly off this landscape. Its size makes metaphor and simile seem preposterous. It is *like* nowhere I have ever been. It shucks story, leaves the usual forms of meaning-making derelict.

Glint of ice cap, breach of whales, silt swirls in outflows, sapphire veins of a crevasse field.

A powerful dissonance overtakes my mind, whereby everything seems both distant and proximate at the same time. It feels as if I could lean from that summit and press a finger into the crevasses, tip a drop of water from the serac pool, nudge a berg along the skyline with my fingertip. I realize how configured my sense of distance has become from living so much on the Internet, where everything is in reach and nothing is within touch.

The immensity and the vibrancy of the ice are beyond anything I have encountered before. Seen in deep time – viewed even in the relatively shallow time since the last glaciation – the notion of human dominance over the planet seems greedy, delusory.

Up there on that summit, at that moment, gazing from the Inner Ice to the berg-filled sea, the idea of the Anthropocene feels at best a conceit, at worst a perilous vanity. I recall the Inuit word I first heard in northern Canada: *ilira*, meaning 'a sense of fear and awe', and also carrying an implication of the landscape's sentience with it. Yes. That is what I feel here. *Ilira*. It's comforting.

But then I think of the melt that is happening, that has happened, that *is hastening*. The cryosphere across the globe is troublingly on the move, as carbon dioxide levels rise and the planet warms. The roaring moulins, the sweating bergs, the collapsing permafrost yielding its grim contents; Geo describing how the sound of his village has changed as the glacier has retreated; the camp we have made in a ghost glacier; the dwindling sea ice; Mulvaney pulling up kilometre-deep core – delving down as a means of foretelling the climate future . . . And I think of Christina's son building his Noah's Kayak-Ark at school: the escape vessel for this newly melting world, with no room for humans on it.

Looking out from that summit, I no longer feel awed and exhilarated, but instead faintly sick. Sick at Greenland's scale – but also by our ability to encompass it. There is something obscene both to the ice and its meltings – to its vastness and vulnerability. The ice seems a 'thing' that is beyond our comprehension to know but within our capacity to destroy.

Three big bergs creep into view on the horizon: white sailing ships stealing up over the Earth's curve. The sun catches the upper edge of the first berg, sparks silver, then flares on its apex such that the berg appears to be aflame.

~

There is a passage in Aeschylus' *Agamemnon* known as the 'Mycenae Lookout' section. It concerns a watchman in a roof tower whose task is to look for the brazier fire on the horizon that will tell him that Troy has fallen, and to cry out if he sees it. At last, after many years of keeping watch, the watchman does see the fire flame on the horizon in the distance. But he finds he cannot cry the vital words. He is struck dumb, unable to articulate. In Aeschylus' memorable image, he feels as if 'βοῦς ἐπὶ γλώσσῃ μέγας βέβηκεν' – 'a great ox has stood on [my] tongue.' In Seamus Heaney's version, the watchmen feels his tongue to be 'deaden[ed] . . . like the dropped gangplank of a cattle truck'.

When I consider our attempts to speak the Anthropocene, I think of that watchman with the ox on his tongue, unable to cry out his warning, so that the danger draws ever closer. The idea of the Anthropocene repeatedly strikes us dumb. In the complexity of its structures and the range of its scales within time and space – from nanometric to the planetary, from picoseconds to aeons – the Anthropocene confronts us with huge challenges. How to interpret, or even refer to it? Its energies are interactive, its properties emergent and its structures are withdrawn. We find speaking of the Anthropocene, even speaking *in* the Anthropocene, difficult. It is, perhaps, best imagined as an epoch of loss – of species, places and people – for which we are seeking a language of grief and, even harder to find, a language of hope.

The cultural theorist Sianne Ngai suggests that, when shocked or grieving, we find ourselves able to speak of the experience only in 'thick speech'. When speaking thickly, Ngai says, we are challenged in our usual ability to 'interpret or respond'. A drastic slowdown and recursion of language occurs, a rhetorical enactment of fatigue and confusion. Tenses work against one another. There is a 'back-flowing', a loss of causal drive, a gathering of hesitancies and stutters. We speak an eddying speech, cloyed to the point of congealing.

Up there on the thinning ice, during those weeks in Greenland, I recognized this 'thick speech'. I would struggle often to stop language from sticking in my throat. The black-inked words in my notebooks seemed sluggish, tar-slow. Writing lost its point, clotted into purposelessness, there in an ice world that was both unhomely and untimely. Often it felt easier to say nothing; or rather, to observe but not to try to understand. I had an Anthropocene ox on my Holocene tongue.

~

We are descending the north-west ridge of the mountain, in the summit's cold shadow, when Helen M cries out.

'Look! Look up – shooting stars!'

How can there be shooting stars in broad daylight? I glance back at the summit and stop, amazed. The sun is silhouetting the peak, and the blue air above the top swarms with tiny silver points, swirling and darting with life-like energy and intent. There are hundreds of these glittering sprites, vanishing instantly when they pass into the shadow and out of the light. We all watch, mesmerized, for a minute or two. It is one of the most exquisite, eldritch sights I have ever seen in the mountains – these seething silver sparks, these scattering star-shards.

Later, we realize that it was probably willow snow, the white wisps of the dwarf willows shedding their seeds, which had been blown by the easterly wind and swept 2,000 feet up from the valley and over the summit, to where the hard Arctic sun backlit and silvered them, and the cold Arctic wind set them dancing.

We retrace our steps safely back down the glacier, unlocking in reverse the doors through which we passed on the way up: the bergschrund, the crevasse field, the roll-over . . . One by one at last we jump with thumps off the snout-ice and back onto the fine glacial gravel, which hushes under our feet.

Back out through the valley between the boulders, down to the shores of the lake, where we set the seagulls chattering up again in commotion.

The sun across the plain that evening at camp is low and white and bright, and it sets fire to the landscape. Cotton-grass heads glow like bulbs. The moss flames green. Each willow leaf, each pebble, each beached berg carries a flash of that late-day light.

The aurora that night comes as green fog-banks, rolling,

coalescing, ebbing. The first star shows over the glacier, then there are none, and then they come fast and faster.

We sit out together in silence again.

After an hour or so the aurora fades, burned out by moonrise. A full moon appears fast over the shoulder of the peak above our camp, as if lifting off the glacier we have climbed that day. We pass binoculars between us; viewed through lenses the moon is almost too bright for the eye. We can see crater rings, impact sites, low lunar seas and high lunar mountains. Its yellow light, borrowed from the sun, lends shadows to the rocks and the tents and to us. I feel an intense loneliness, made by the moonlight, that surprises me with its force.

A thunder of the glacier rouses me at two o'clock that night. I step out of the tent.

Sharp calls of sanderlings in the darkness. The moon still massive and yellow. Northern lights flickering as curtains of green above the ice cap, and a single streamer leading back up and over the summit of the peak we climbed.

The glacier roars again, incomprehensible, the reverberations taking twenty seconds to die away.

The next morning we wake to find the camp in a thick white mist, as if the ice has returned overnight and submerged us. Dew beads the tripwire lines. A raven circles above us, invisible, cawing.

Two days and two peaks later we break camp and leave for the Knud Rasmussen glacier, to seek a moulin that bores into its blue depths.

11

Meltwater

(*Knud Rasmussen Glacier, Greenland*)

We hear the moulin before we see it: a low rumble, rising in volume as we approach. It is set down in a shallow dip, a day's travel up the glacier, with three meltwater streams curling towards it, as the currents of foam had spiralled the Maelstrom in the Lofotens.

I circle around the moulin, keeping well back from its edge, to the point from which I can safely see furthest down into it. It is surely the most beautiful and frightening space into which I have ever looked. Its mouth is oval and around twelve feet across at its widest point. Its sides are of blue ice that are polished as glass, and scalloped in places. It drops vertically from the glacier's surface like the shaft of a well. Twenty feet down all light is lost and so is all sight. It seems that the moulin might bore through the glacier's depth until it hits bedrock hundreds of yards below. A torrent of meltwater pours into the void from its western lip.

We all feel some version of the moulin's draw that day. It acts on the landscape around it as a whirlpool acts upon the sea, such that everything seems to tend towards it. In its presence I feel a lean in my chest, an urge to step nearer to its edge, then nearer again. The moulin is certain and powerful, and it is a portal giving access to the blue underland of ice.

~

Seven days before we find the moulin, we reach the Knud Rasmussen glacier. It is a body of ice so great that it makes its own weather.

The glacier is invisible the afternoon we arrive, concealed by a bank of fog that runs the full span of the fjord, a mile or so wide but only a few hundred yards high. Above the fog is blue sky, below it is blue water, and behind it is blue ice. The cold mass of the unseen ice is condensing the moist air to create that hovering mist.

We cannot see the glacier but we can hear it. The Knud Rasmussen makes the Apusiajik glacier seem like an introvert. The first roar comes minutes after we have dropped our packs on the ledges of gneiss that will be our home until autumn. The noise comes without warning out of the fog-bank and shakes our bodies like bags of jelly.

'Boom!' says Helen. 'Welcome to the Knud Rasmussen. The ice is talking!'

High above us, faint rainbow patterns dapple the sky. These are the colours of the sun refracting in airborne ice crystals in the upper troposphere, four or five vertical miles above us.

Another explosion rolls out from behind the fog-bank.

We cannot see the glacier but we can feel it. It extends a chill around itself, dropping the air temperature by five degrees or more. The place we have chosen to camp is over a mile from the calving face, but even there we are within the glacier's aura. In the days we spend at the Knud Rasmussen, we become icy. We drink ice. We wash in ice. We sleep by and on the ice. Ice fills our ears and our dreams and our speech. Ice fills the water and the air and the rock. We enter the ice and the ice enters us.

~

The route to the Knud Rasmussen takes us far to the north of Apusiajik, and into a new order of remoteness and scale. We reach it through fjords like canyons, lined by slabbed walls of gneiss thousands of feet high, topped by spired peaks. A kind of rock unknown to me is visible in this region: crumbly, coarse-grained and the colour of chocolate, splitting the gneiss in broad veins up to 100 yards wide, and running for miles through peak and valley. You can follow the veins through the landscape, tracking them to the point where they disappear beneath the water of a fjord on one coast, then seeing them emerge again on the far side.

Even in this inhuman landscape, human conflict has left its mark. In a side valley, under a peak that rises to a fishtail-forked summit, we pass the remnants of an American Cold War base abandoned half a century earlier. The rusted skeleton of a hangar, its girders bent from repeated winter avalanches; a tractor with a snowplough fitted to its front, sunk into the shallow tundra; and thousands of oil drums, corroded to orange, stacked in clutches or standing in snaking lines. They give to the site the air of a hatchery, and remind me of the rusted fishing floats that gathered on the shores of Moskenes in the Lofotens. Everything artificial at the base has taken on the colours of the tundra – its dun hues of orange, brown and green. Lichen and moss flourish in the niches of relic infrastructure: Arctic camo.

Further down the same fjord, in a bay fed by a freshwater stream, is a monstrously beautiful berg. It gleams whitely in the sunlight, long and low-slung, rising never more than fifteen feet above the dark water that laps at it. Its upper ridge is elegantly curved, but what draws the eye are the deeply incised grooves that mark its flanks, running straight and parallel with one another, as if it has been methodically mauled. Each groove glows a slightly different shade of blue. Where the grooves shallow out, the ice is dimpled – the dips and rises glistening like flesh after injury.

The fjord forks into a Y, one arm cutting north-east and the other almost due north. As we cross the mouth of the northern fork, I see in the distance a glacier, the Karale, curving down to the tideline. To its west a smaller glacier is visible. It has retreated back above the water and ends in an arch of ice that I guess to be several hundred yards across. The arch shines with the blue light of old ice, and out of it comes a powerful melt-stream, rushing down to the sea.

'Geo and I once made it to the Karale from Kulusuk in two days by dog sled,' says Matt. 'We covered eighty kilometres each day, in desperate conditions. The weather was atrocious, and the sea ice was rotten. Many times Geo or I had to probe ahead of the sled with a harpoon, testing the ice's resilience so we didn't go through.'

We follow the north-eastern arm of the fjord. As we approach the fog-bank at its end, the water thickens with a mash of shrapnel ice and the odd berg sliding out with the tide. A mile short of the fog-bank, under a peak that rears from close to the water's edge, amid a boulder field of pale erratics, we find enough flat ground for our tents. There is a stream from a snowfield that offers fresh water. Upslope, there are tussocks of bilberry beginning to come into fruit. Directly over the fjord from us rises the sheer wall of a sharp peak, struck by a bolt of the chocolate rock.

We are within a few yards of the fjord edge and the rock is a continuous sloped curl of gneiss, running for hundreds of yards along the fjord shore, glittering with lines of quartzite and black mica.

Small blue bergs drift and click offshore.

'I'd like to die and be reborn as a boulder here,' I say. 'It's one of the most extraordinary places I've ever been.'

'It'll do us,' says Matt.

~

An hour short of dusk on the evening of our arrival, the fog-bank disperses to reveal the calving face of the Knud Rasmussen. The face runs the width of the fjord, curving out from the easterly shore to a sharp forwards point, then turning out of sight to the west.

The sea around the calving face is stained brown with silt, in contrast to the milky green of the outer water. The silt wells up from the melt-streams that are pouring out unseen below the fjord's surface. Birds gather on the silt blooms, feeding on their richness. They are the only scale-givers at this distance, and they are small as flies. Now and then birds near the face burst up, circle and mingle, then resettle on the water. Ten or twelve seconds later the noise of a minor calving reaches us.

The calving face gives a transect of the glacier's depths. Crevasses fissure downwards for hundreds of feet. There are rounded shafts: extensions of the moulin melt-systems. I can also see, even at this distance, the strata of the ice – its sedimentary formation. Whiter, wider bands thin out to blue and layerless ice far below.

The face is a Gothic city being pushed into the sea. Towers, belfries, chimneys, cathedrals, finials: all are going over the edge. Tunnels, crypts, cemeteries: all will be shattered into bergs. I think of the weight of the upper bodies pressing down in the cemetery of Saints-Innocents, until at last the dead crash through into the spaces around the curtilage of the burial site.

'That calving face is the terminus for ice that fell as snow far up on the Great Ice, tens of thousands of years ago,' says Helen.

Where the freshest calvings have happened, the ice is bluest. These marks of rupture seem not scar but revelation. This is the first sunlight that ice has seen for tens of thousands of years.

A ringed seal surfaces offshore, glances over at us, dips again and disappears in the milky green water. What must a calving event look like to a seal? I wonder. What must it *sound* like?

'There are certain glaciers that are held to be clearly malign around here,' says Matt. 'There's one that Kulusumi people just won't go near, because it has a reputation for hostility. If you *have* to cross near it, you don't speak, eat or even *look* at the glacier while making the crossing, because it calves so far below the waterline that it can kill you from beneath without warning. They call this *puitsoq*, "the ice that comes from below".'

In the lee of a boulder above the camp, I find a loose cache of thousands of individual dwarf willow leaves. They are brittle and black-brown, lying to a depth of three or four inches. They must have been accumulating there for years, gathered by the wind, frozen each winter and thawed again each summer. The vein lines are still visible on each leaf. I pick up a handful, rustle them through my fingers. They are weightless and sharp. In this dry air, with so little topsoil, deterioration rates of organic matter are decelerated. Time moves variously in this landscape, from the catastrophic suddenness of calving events to the patient process of leaf drift.

A berg glides past us, shaped like the eaves of a house. Seventeen gulls are perched on its ridge, all turned to windward.

~

Living by the Knud Rasmussen is like moving in next door to a thunderstorm. Each day we climb and explore further into the surrounding landscape. Each evening we return to our tents by the glacier. All day and all night the ice bellows, cries, echoes. There is no apparent connection between the air temperature and the activity of the calving face. Some of the loudest roars come in the dead of night, the coldest time, rousing us from our sleep with fears of polar bears.

'You think this is dynamic?' says Matt one morning. 'The Helheim glacier near Semersooq is now flowing into the sea at around *thirty-five* metres a day. That's one of the fastest glaciers in the world.'

The glacier is named after the underworld of the dead in Norse mythology: Helheim, 'the Realm of Hell', 'the Hidden Place', buried beneath the roots of the world-tree Yggdrasil. Our word 'hell', like the Icelandic word *helvíti*, comes from deep in language history: from the reconstructed Proto-Germanic noun **xaljo* or **haljo*, meaning 'underworld', 'concealed place', itself from a Proto-Indo-European root **kel-* or **kol-*, meaning at once 'to cover', 'to conceal' and 'to save'.

Around Greenland, some glaciers are retreating as they melt, while the flow rate of others is increasing, causing the upper ice to diminish. The softening ice cap is estimated to have lost around a trillion net tons in only four recent years. Lubricated by the moulins, many more tons of ice and meltwater are pouring into the fjords and the outer ocean, helping global sea levels to creep up, increment by increment.

One hot morning on a rest day, I lie on the slabs of gneiss where they run into the tideline, watching the ice through narrowed eyes, hoping I might see a calving event rather than only hear its aftereffects. But nothing moves that morning. I close my eyes and listen to the landscape – listen in a way I rarely do, letting each sound play and single itself out from the weave like a bright thread, trying to infer its source from its sound. I am trying to hear this landscape's undersong – the substrate sounds of a given place, the ambient murmur that goes often unheard or at least unlistened-for.

We cannot see behind ourselves, but we can hear behind ourselves. From all directions, sound flows in.

Glittering gull cries.

Crack of bergs beached on the tideline nearby, as the warmth of the sun pops ancient air bubbles.

Crockery clink of ice shards in the water, the lapping hush of slush-ice nudged by the incoming tide, the sloshing wallow of a bigger berg rolling as melt or current shifts its weight.

A waterfall on the far side of the fjord, crashing from a high cirque in a steady rush of sound, like corn being poured from a hopper.

Below it all, below even this undersong, a bedrock of something like white noise that I cannot granulate with my human ears – a distant hiss or hum that makes the finer sounds audible.

Bang! A gun-blast rips through the fragile weave and echoes back off the walls and water of the fjord. I jerk around. Matt is standing on a tideline rock. He fires each of the weapons twice in turn into the fjord to clear the barrels. *Bang! Bang!* His shoulder jerks back with each recoil. Water sprays up as if a big fish has breached. The reports are shockingly loud. The sound of each shot takes fifteen or twenty seconds to dissipate.

~

It happens that afternoon when we are all together, standing near the tents and talking inconsequentially, enjoying the lethargy of the rest day.

A shot-like *snap* begins it, whip-cracking across the fjord and the mountain walls.

'A hunter?' I say.

But it isn't a hunter, it is the glacier, and the sound of the crack marks the fall of a bus-sized block of ice from high on the calving face. We do not see it fall but we see it swill back up and bob.

Without that outrider of the main event, we might have missed what followed – an event that, as Helen puts it later, 'rarely occurs under witness'.

'There!' shouts Bill, but we are all already looking there, where the first block fell, for it seems that a white freight train is driving fast out of the calving face of the glacier, thundering laterally through space before toppling down towards the water, and then the white train is suddenly somehow pulling white wagons behind it from within the glacier, like an impossible magician's trick, and then the white wagons are followed by a cathedral – a blue cathedral of ice, complete with towers and buttresses, all of them joined together into a single unnatural sideways-collapsing edifice – and then a whole *city* of white and blue follows the cathedral as we shout and step backwards involuntarily at the force of the event, even though it is occurring a mile away from us, and we call out to each other in the silence before the roar reaches us, even though we are only a few yards from each other, and then all of the hundreds of thousands of tons of that ice-city collapse into the water of the fjord, creating an impact wave forty or fifty feet high.

And then something terrible happens, which is that out of the water where the city has fallen there up-surges, rising – or so it seems from where we are standing – right to the summit of the calving face itself, a black shining pyramid, sharp at its prow, thrusting and glistening, made of a substance that *has to be* ice but looks like no ice we have seen before, something that resembles what I imagine meteorite metal to be, something that has come from so deep down in time that it has lost all colour, and we are dancing and swearing and shouting, appalled and thrilled to have seen this repulsive, exquisite thing rise up that should never have surfaced, this star-dropped berg-surge that has taken three minutes and 100,000 years to conclude.

Twenty minutes later and the fjord is calm again.

The tide swills gently in rock pools. Lap of water on gneiss, pop of melting ice, sun glittering on the margins of the water, sedge-grass flicking in the wind.

The obscenity might never have occurred.

The berg has settled in the water as a sloping blue table, hundreds of square feet in area. Gulls land on this new territory in their dozens, shake out their wings, tuck one leg up into their breast feathers for warmth, hunker down.

I startle a single sanderling from a fold of bronze gneiss.

The next day at the tideline I find a small iceberg, rounded and dark blue, stranded in a rock pool. It is a relic of the dark star. I am just able to lift it. I carry it in both arms, cradling it, calling to the others. It numbs my hands and chest. It feels far heavier than it should. I stumble uphill towards the camp and place it on top of a boulder by the tents.

The sun shines through it. Air bubbles inside it show as silver: wormholes, right-angle bends, incredible zigzags and sharp layers.

That night an Arctic fox comes to our camp, a playful blue shadow.

The little berg takes two days to melt. It leaves a stain on the dark rock that won't vanish.

~

Ice, like oil, has long disobeyed our categories. It slips, slides, will not stay still. It confuses concepts, it confounds attempts to make it mean. In the 1860s, when glaciology was emerging as a science, the discourse of glaciers was riven by the dispute over whether ice should be classified as liquid, solid, or some other kind of colloid-like matter altogether.

It is unsurprising that ice should have proved so ungraspable to human habits of meaning-making, for ice is a shape-shifter and a state-shifter. It flies, it swims and it flows. It changes colour like a chameleon. Ice crystals at 30,000 feet set halos and parhelia shining around the sun and the moon. Ice falls as snow, as hail, as sleet; it crystallizes as feather and it gleams as mirror. Ice erases mountain ranges, but preserves air bubbles for millennia and is tender enough to bear a human body unriven for centuries. It is silent, and it creaks and thunders. It sharpens eyesight, and it breeds mirages.

We are now experiencing ice as a newly lively substance. For centuries, the polar regions were conventionally imagined as inert: the 'frozen wastes' of the north and south. Now, in the context of a warming planet, ice has become active again in our imaginations and landscapes. The 'frozen' poles are melting and the consequences of their melting are global. The Russian expression for 'permafrost', вечная мерзлота, translates as 'the eternally frozen ground' – the name looks increasingly inappropriate. Greenland, Antarctica and the Arctic are now front-line territories, in which the fate of ice will shape planetary futures.

A 'glacial pace' used to mean movement so slow as to be almost static. Today's glaciers, however, surge, retreat, vanish. The recession of Himalayan glaciers threatens the livelihoods and lives of more than a billion people in Asia, who depend on the water that is seasonally stored and released by these ice rivers. The West Antarctic Ice Sheet is breaking up, disassembling itself into bergs and sheets that drift unbiddably. Mapping cannot keep up with the shrinkage of the sea ice. Globe-makers can no longer confidently cap their globes with white. Ice has become *dirty*, in the sense of Mary Douglas's famous definition of dirt as 'matter out of place'.

In indigenous cultures that live adaptably in close contact with ice, it has always been an ambiguous entity, and the stories told about glaciers have often blurred boundaries between human and non-human activity. Glaciers appear in these stories as actors – aware and intentful, sometimes benign and sometimes malevolent. In Athapaskan and Tlingit oral traditions from south-western Alaska, for instance, as the anthropologist Julie Cruikshank documents, glaciers are both 'animate (endowed with life) and animating (giving life to) landscapes they inhabit'. In languages from this region, special verbs indicate the living power of what in English might be classified as passive landscape presences. These verbs recognize in the ice both its actions and, vitally, its powers to act. Linguistic anthropologists refer to the 'enlivening' influence of such verbs: their deep-level acknowledgement of a sentient environment which both listens and speaks recalls Robin Wall Kimmerer's wish for a 'grammar of animacy' that might acknowledge the autonomy of plant life.

Over the years of my time travelling on or near glaciers, I have read in translation dozens of the stories told about glaciers and ice across northern indigenous cultures. Many of them concern the dangerous underland of ice – a kingdom into which one might fatally plunge. A ubiquitous story tells, with regional variation, of a traveller 'falling through ice' (either through thin sea ice or into a crevasse) who is assumed to be dead, but then surfaces from this netherworld bearing tales of visions, hardship and survival. It is almost exactly the sequence of events and motifs that recurs in the most famous modern Western glacier-story of the underland, Joe Simpson's

Touching the Void. All of these stories relate to miraculous resurrections from depth. We had witnessed our own 'surfacing' on the day of the calving – except that it was ice itself, rather than any human agent, that had been to depth and then returned to the light.

In the days that follow the calving I reflected often on our response to it – the way our shouts turned from awe to something like horror as that shining black pyramid lurched up out of the water, sea streaming from it. My stomach lurched too as the ice came up: the sublime displaced by a more visceral response to this alien display. I have often sensed the indifference of matter in the mountains and found it exhilarating. But the black ice exhibited another order of withdrawnness, one so extreme as to induce nausea. Camus called this property of matter its 'denseness'. Confronted by matter in its raw forms, he wrote, 'strangeness creeps in':

> perceiving that the world is 'dense', sensing to what degree a stone is foreign and irreducible to us, with what intensity nature or a landscape can negate us. At the heart of all beauty lies something inhuman . . . the primitive hostility of the world rises up to face us across millennia . . . that denseness and that strangeness of the world is the absurd.

I recognized a version of that 'denseness' – that 'absurdity' – in Greenland to a degree that was new to me. Here was a region where matter drove language aside. Ice left language beached. The object refused its profile. Ice would not mean, nor would rock or light, and so this was a weird realm, in the old, strong sense of weird – a terrain that could not be communicated in human terms or forms. I thought back to Merlin, to the fungi and their buried kingdom of the grey – that shivering, slithering underland into which he had helped me look.

Greenland was a place where matter leaked through the usual screens. When the black-star calving had happened, the leak had turned into a torrent. Later I would find that torrent again, far down in the blue light of the moulin.

~

Big climbing days of glacier and peak come and go. The willow leaves turn from yellow to orange. One morning we leave our tents to find a first frost, starring the earth in its hollows.

We try the nameless mountain that rises behind the camp. From below, foreshortened, it seems a single slabbed face, thousands of feet high. It reveals itself to be more involved, though: abundant in features hidden from beneath. It has a glacial corrie at its heart. It has set-back shoulders that hold little lakes and permanent snowfields.

We climb it in seven pitches over seven hours, Helen M leading strongly, with scrambling and way-picking on mellower ground between the pitches. In the closed couloirs and on the slabs, I feel no nerves. On the ridges, fear squeezes my heart.

The peak's five-finned summit ridge is of clean golden rock, and from it we can see over and down into a huge horseshoe cirque to its east. The cirque is ringed by sharp mountains and has a collapsing serac face at its centre – a 600-foot wall spilling mint-blue ice rubble onto the snowfield beneath. A chilling wind lifts from the cirque, sapping my confidence.

Matt leads the final pitch: a chimney of cold, sound rock, up which we bridge and lay-back one by one. The serac face collapses three times while we climb, audible crumps echoing off the cirque walls. From the peak we can see far up the Knud Rasmussen. From here it seems less an ice river than an ice sea, flooding the peaks around it.

We reverse the summit pitch with cold hands. Lenticular clouds hover over the peaks of the Karale. A front is coming in. Late in the day the sun breaks through with slanted gold rays, backlighting big bergs in the fjord below us so they shine like opals.

That evening we sit together, exhausted, companionable. It is a time of cusps. Dusk, early September, just beneath the Arctic Circle, by a tideline glacier in East Greenland. The cusp of the day, the cusp of seasons, the cusp of the globe, the cusp of the land. The Arctic fox comes again to the camp, keeping to the shadows where it is blue-silver in colour.

We stay out late. The last light gathers on the water of the fjord, on the rims and edges of bergs, on quartz seams in the gneiss. Twilight specifies a landscape by means of such finely lit details, but it also disperses it. Relations between objects are loosened, such that shape-shifts occur. In the last minutes before full night falls, I experience a powerful hallucination, my tired eyes starting to see every pale boulder around our tent not as rock but as white bear, polar bear, crouched for the spring.

A big calving wakes me in the night. Minutes later, waves surge up the shoreline rocks.

The next morning, nine bergs – human in their size – have wandered into our bay overnight and become beached. They tick as they melt: nine ice-clocks.

~

We leave early the morning after next, carrying heavy packs: gear for several days away. We are heading inland, along the glacier, to establish an advance camp far up the Knud Rasmussen, and use that as a base to explore peaks and passes that lie further in.

We also want to search for a moulin wide enough to descend.

We will get onto the glacier via the moraine, pass through disrupted ice hard above the calving face, and then pick up flat ice in the glacier's centre, over which we can make good progress. That, at least, is the plan. Afterwards Matt will describe what we meet on the Knud Rasmussen as 'massively exploded terrain'. I will think of it as the Labyrinth. It makes the crevasse mazes of Apusiajik look like children's puzzles – and beyond it lies a Minotaur.

We follow the fjord shore to the calving face, cut up and over slopes of bilberry and willow until we meet the lateral moraine slope, a wall of rubble bulldozed to the valley side by the glacier's seawards drive.

Any boulder field on a steep gradient is a perilous place. I know someone who died in a boulder slope in the south-west USA, approaching a dihedral spire he planned to solo. He never even made it to the base of the route: ascending the boulder field, he triggered movement in a massive plate of rock, which slid into him at the waist, crushing his pelvis and trapping him fast.

So on a loaded moraine slope you walk like a cat. Your aim is to dislodge nothing, not even a grain of quartz. You move tenderly. You advance with soft tread, placing the balls of the feet first, not heel-jabbing with a stiff leg. You *never* pull on a rock with your hand; instead, you press down with your palm or fingertips so that any force you apply confirms the rock in its location. You *never* put your full foot-weight on a boulder without testing it first. You *never* move when someone is directly below you on the fall-line. You *never* put your foot or your arm into a gap between rocks, in case the one above drops down. Shins and forearms break easily in stone jaws.

We make it safely up the moraine face, four cats in a row – and me, a clumsy ox bringing up the rear. From the high point of its shoulder we can see up and along the glacier, and back down to the calving face. This close we have a sense of the face's scale. It is a sea cliff. The gulls look like pond-skaters on the silt gouts.

From there we pick cautiously down the far side of the moraine, desk-sized boulders rocking and rumbling underfoot as they take weight, and at last we step up one by one onto the black-glass fringe where the glacier meets the moraine, and from there we climb up onto the low billows of the Knud Rasmussen proper.

The night has left films of ice across standing water on the pools. This delicate ice tinkles when broken. The glacier is a frozen sea, as yet calm enough that we have no need of ropes or crampons.

Half a mile in and the sea becomes stormier. The billows of ice lift, become sharper in their contours, more hog's back than billow, then more shark's fin than hog's back. We rope up, axe up, crampon up. The consequences of a slip or trip are now severe. Our progress slows, as Matt probes to find a way through the crevasse maze. We speak less.

Crevasses open around us, a few feet deep only at first, soon dropping to twenty, thirty, fifty, countless feet deep. Colours change. The surface ice is whiter than at the snout. The crevasses glow a version of the unearthly blue we saw on Apusiajik. Here the blue is even more intense, more radiant, *older.*

Ice is blue because when a ray of light passes through it, it hits the crystal structure of ice and is deflected, bounces off into another crystal and is deflected again, bounces off into another, and another, and in this manner ricochets its way to the eye. Light passing through ice therefore travels much further than the straight-line distance to the eye. Along the way the red end of the spectrum is absorbed, and only the blue remains.

In glacial terrain of that seriousness, you move like light through ice. Time spins away and space misbehaves. You take an hour to travel half a mile in the desired direction. The straight-line distance to your destination is irrelevant, because the ice sends you on a bouncing and deflected course – a blue-line not a bee-line.

We are in the Labyrinth for four hours. At last Matt finds a way through and onto flatter ice, and we can unrope, eat, drink, stand in safety. I feel taut nerves slackening again. One of us cries briefly. We all feel hunted by this ice, haunted by it.

It is still hard going from there, uphill and inland, but the ice is calmer now and our progress is good. As we move, new tributary glaciers open up their vistas to either side. New peaks are glimpsed on the horizon. None has been climbed. They entice us. Our wish is to make a high bivouac that night, and from there strike out for a peak the next day: exploratory mountaineering, no maps to speak of, scant knowledge of the terrain ahead.

Hot sun now and the glacier's surface is thawing so fast we can see and hear it. Tiny plates of ice, formed in the hoar-frost forests that rise to the height of a centimetre each dawn, tilt and then blink out as they become water. The glacier hisses. It *crackles*. Sometimes a bank of slush ice collapses into a melt-stream, and the crystals rush down the channel like sizzling fat.

'Where does all this meltwater *go*?' I ask Matt.

'Down the moulins. We'll find them. You'll see.'

We do. Two smaller moulins first, a little bigger than the one we found on Apusiajik. And then the big one, a true maw gaping close to a lateral moraine band. Three meltwater streams curl towards it, braiding into a single current in the final yards then toppling into the drop.

We circle the moulin warily, as if approaching a wild creature. I put a rope on and Matt belays me to its brim. I lean a little out over the edge – and look straight down into deep blue, into the blood of the glacier. I feel my belly and my bones sucked towards the colour, step quickly back. *The void migrates to the surface . . .*

'This is the one,' says Matt. 'We can get down this one. We'll need to come back early though, very early, while the glacier is still frozen, before the melt-streams get running. But right now we need to find our bivouac site for tonight. I'd much rather be sleeping on rock than on ice.'

Where a tributary glacier sweeps down to pour into the Knud Rasmussen, a small rock island has been revealed. It is a recent artefact of the increased melt rates – an Anthropocene landmark not present on any existing maps, even on Google Earth – and it sticks out like a boulder in an ice-rapid where the tributary glacier tumbles 400 vertical feet to the Knud Rasmussen. We spot it from two miles away; wonder if it might give enough flat ground on which to camp.

Near dusk, we climb a slope of grey ice to reach it. Certainly, we are the first people ever to set foot on that new world, disclosed from the underland of ice. It is equivalent to perhaps half a tennis court in area.

'It's like walking on the moon', says Helen M in amazement. And it is. The rock is as the ice left it. A thick layer of grey stone dust coats everything. The bedrock has been smoothed by the passage of the ice, but its surface is scattered with loose round stones on which we stumble like drunks.

Big domes and bulges of ice rear immediately above the island, and it is from these that we gratefully catch meltwater in our bottles, slaking the thirst of the day's long work.

It takes half an hour to clear space for the tents, shovelling dust and moving stones. Bill, Helen M and I sing as we work. Bill's rich voice spills over the glacier as the sun sets, keeping our spirits up. Then we pitch the tents, and lash them down with rocks and cord in anticipation of the night's wind. Rock dust covers our hands and faces.

'Look, the mountains are on fire!' calls Helen, pointing.

They are, too: an intense light flows over the summits from the west, scalding the rock of the highest peaks so red that they seem to be running with lava.

~

The next dawn a low band of cloud stripes the land. We wake into silence, after a night of gusting wind. The air is calm. The glacier has been petrified by the overnight freeze.

We climb that day: a long ascent of a distant peak, the summit of which we fail to reach.

The morning after that we wake at five in half-light. We break camp on the rock island quickly, nervously. The air is calm. We crunch down the slope to meet the Knud Rasmussen, then pick up a line of moraine debris and follow it to the moulin.

Before we see it, noise tells us that even at that cold hour the moulin is churning, the mill is grinding. A stream of water tumbles steadily into it from its western lip.

'The sun's warming things up already,' says Helen. 'Every minute that passes, the flow will increase.'

We work fast. Matt manages the set-up. Two ropes, four belay points, each one double-pointed. Clear any rotten ice to reveal the hard ice, the sweet stuff that will hold a screw, then press the teeth of the ice-screw in until they bite, make sure it stands perpendicular to the surface, then with one hand steady the screw-barrel and with the other crank the handle. Any object foreign to ice will absorb heat and melt ice, so we must heap and pack brash ice around the screws and the karabiners.

It takes half an hour to rig things to Matt's satisfaction. The waterfall gains noticeably in power and noise. It is clear that, once inside the moulin, communication by voice will be almost impossible. We agree a simple sign-system: up, down, pause, and – forearms crossed to make an upheld X – *get me the fuck out of here*.

Tie on to descent rope, to haul rope, check and double-check knots. Stamp feet, pull hood hard over, run through final systems again. The moulin dropping away: a radiant blue sci-fi tube, ready to beam me down. Going over the edge I feel no fear, nor should I – just the familiar buzz in the scalp, the caul of bees.

The space of the moulin is immediately, intensely beautiful. The air has a blue aura and the ice surrounding me is sleek to the touch. I descend foot by foot, and the mouth of the moulin above me cinches its white oval tighter. Glancing down, I can see no base and the memory rises unbidden of flipping centimes into clear azure water from a boat in the Mediterranean as a child, watching as they spun silver through the depths, turning and flashing for thirty, forty, fifty seconds.

The deeper I go, the closer I come to the meltwater stream that is falling down the moulin, and then my crampons slip on the ice and I spin out from the face and into the torrent, which crashes down on my head with cold pummelling fists and the force of it punches me back out of the torrent, but from there I cannot catch the glassy sides of the moulin again and so I swing back further into the torrent, and there I am knocked out of it again, and so I begin to pendulum back and forth in and out of the torrent, and with each cold dousing I am losing strength, and I feel that I am trapped in a perpetual-motion machine that can run indefinitely even after I have ceased to function.

I glance up as I pendulum and can see Matt's face leaning out and looking down at me, mouthing words at me, but he is of the surface now and I am of the depth, and these are quite different places. He exists in that porthole of sky rimmed with white and gold light, but down here there is no colour or time other than blue. Up there Bill, Helen M and Helen are moving freely around on the glacier; down here there are only the glass of the ice, the torrent of the water and the obligations they enforce.

But this is too strange a site to leave unless I must, so I gesture to Matt that he should lower me, realizing that if I descend further I might be able to pull myself out of the flow, and so I drop deeper and, spinning around, I see that there is – sixty feet down into the glacier, a dozen centuries or so further on – a terrace of a kind, off which the water corkscrews deeper still in a twisting borehole too tight to admit me, but with a lateral blue side passage also leading away. I use the swing of the pendulum to catch the ice edge of the side entrance with my hand. I pull myself towards the passage and out of the flow, and see that below me is a fine spear-blade of ice, twelve feet or so long, that somehow grows *upwards* from the terrace, and I hook one of my feet around it and then perch on its point. At last I am secure, one hand gripping the passage edge, one foot on the spear-blade. I pause, catch my breath, glance at the porthole, thumb-up to Matt that I am fine. Braced there, I can study the space.

Twenty feet below me the meltwater current drills away and down into the glacier's underland, impossible for me to follow. The side passage leads off as a tunnel, though, and I can see a chamber filled with more blue at its end, and I want to follow the passage to that chamber. But I know that rope drag will occur soon after I begin to move sideways from the shaft, making progress difficult and meaning also that a slip in the side passage will bring me slamming back into the main shaft at speed. I wish I had ice-screws with me, so that I could set runners to manage the rope for the traverse of the tunnel. But I do not, and so there is no choice but to stay a while on that blade of ice in this otherworld, and then reluctantly, gratefully, to give the sign to Matt: *Get me the fuck out of here!*

He changes over the rig, and they haul me up and out, Helen, Helen M, Bill and Matt all running my weight on a Z-pulley prusik system, and I emerge from the moulin like a gopher from a burrow, head surfacing into the upper world which is full of laughter and *how-was-it*s and open mouths, and Helen is reaching forwards with a hand to pull me to safety, and the sun is streaming its gold on the silver of the ice, and I am blue to my bones for days afterwards from that deep time dive.

Later we send Bill down too, and from a depth of thirty feet he sings an aria from *Tosca*. The notes pour up through that great blue pipe-organ and fly joyfully out into the still air.

~

It is afternoon when we step off the Knud Rasmussen glacier for the last time, back near the fjord. The colours of the tundra leap to the eye, shocking in their brightness after the days on ice and rock. Sulphur blaze of grey-leaf willow on the turn, punk green of lichen, black mica shards in the rocks.

The leaves on the willows have reddened at their tips while we have been away.

Six ptarmigan churr among bilberries, their plumage on the turn to winter white. We are glad to see life that is not ice. They are unafraid of us. Bill reads them as a score, seeing their positions on the slope as six notes on a stave.

On reaching base camp, we drop our packs and bathe in the freezing water of the fjord, scrubbing days of dust and toil from our bodies among the icebergs, whooping and shouting.

That night there is the fieriest aurora display we have witnessed. We sit up in our sleeping bags to watch it. Green curtains bling and spangle inland, over the Knud Rasmussen, over the rock island, over the moulin. For the first time there are pink hues as well as green – the pink of willowherb. Search beams of green shoot up from summits to the west. The display is profuse, extravagant, spinning over thousands of miles of sky, *a busy working of nature wholly independent of the earth and seeming to go on in a strain of time not reckoned by our reckoning of days and years . . .*

'Have you noticed,' says Helen M, 'how the stars show in greater number through the aurora?'

She is right. I would have expected the Northern Lights to make the stars less visible, rather than more – the excess of light cancelling out the stars' glimmer. But instead it has the counter-intuitive effect of causing more stars to show, clusters of them, which vanish back into the blackness when the aurora flickers away. None of us can explain how the green light could be collaborative rather than competitive with the starlight.

That night I dream, clearly and for what seems like hours, that a fine blue moss has begun to grow under my skin, starting on my right forearm, then spreading up to my shoulder and across my chest. It is painless and luxurious.

~

Days afterwards, back in Kulusuk, on our final evening in the village, Helen, Matt and I go out kayaking in the bay with Nuka, one of the young men from the village. Nuka wears a black squared-off baseball cap, has a gold chain and a gold tooth. He is eighteen. He plays the guitar softly, with passion, like José González. He loves kayaking.

Cloud boils up and around Apusiajik. The late sun is bright and hard. A storm is coming. Gulls land in the water, sheet white in the storm-light. A single low-slung iceberg wanders the bay. Two men and a woman are hunched in the lee of a bay-side shack, drinking cans of Heineken.

We launch the kayaks from among boulders. We paddle out over cod heads and seal flukes, Nuka racing ahead with short fast strokes, then Matt accelerating after him, both men grinning with delight at being out on the water.

'This is where kayaking was *invented*!' shouts Matt.

He paddles directly at the small iceberg, hits it fast at its lowest point, and laughs as the kayak's front half ramps up onto the berg. Then he scooches himself off and splashes back into the water.

'Look!' Nuka is calling. He is holding up a dripping object, long and thin, with a wooden haft and a spear-point at one end.

'He's found a harpoon!' says Matt. Nuka takes aim at Matt, and throws the harpoon at his kayak. It drops safely short, Matt paddles over, grabs the floating harpoon and hurls it at me.

I have never played harpoon water-polo before, and I am not convinced that it is a traditional Greenlandic sport, but the rules seem clear enough: aim but don't maim.

We hurl the harpoon at one another, chase around the bay, paddle in bursts. Other boys from the village come out in their motor dinghies to buzz us, ripping their Evinrudes and cutting across our bows. To the north the Apusiajik glacier gleams its way down to the tideline. After a while we raft up and bob on the chop, looking back at the little village of Kulusuk perched on its bedrock, the white crosses of the shore-side cemetery showing clear in the sun.

When we return to shore, Nuka shows the harpoon proudly to Geo.

Geo shakes his head.

'This is not a harpoon,' he says to Nuka in Greenlandic.

He looks at us, takes it, grips it by its wooden haft like a walking stick with the point lowermost, and makes a downwards jabbing motion, stepping cautiously forwards as he does so, looking enquiringly ahead of himself as he presses the point against the earth.

It is not a harpoon, not a weapon at all, but rather a tool used to probe the depth of the sea ice ahead. It is a means of telling if it is safe to proceed – of testing the near future.

When I return to Britain, I learn that during the weeks we have been on the glaciers the Anthropocene Working Group of the Subcommission on Quarternary Stratigraphers has recommended the formal adoption of the Anthropocene as the current Earth epoch, with a start date of 1950 – coinciding with the dawn of the nuclear age.

12
The Hiding Place
(Olkiluoto, Finland)

Birches, birches, pines, birches, clearing, blue farmhouse. Low river valley, wooden bridge. Everything frozen: rivers, trees, turf, fields. Pink crag of granite, yellow ice-fall spilling from it. Boulders big as houses between the birches, among the pines. Black crow pulling red flesh off the white ribs of a dead fox. Jackdaw, jackdaw.

This is not a place for you.

The pirate radio station plays Blondie's 'Atomic'.

Snakes of spindrift race on the blacktop. Snow whirls in the headlights. Grey air that will not brighten. A boy riding a bike with sit-up-and-beg handlebars, his back very straight, fast past a blue letter-box on a white pole. Gneiss, silver-grey, quick with mica and ice.

This place is not a place of honour.

Over the bridge to the island. Salt marsh to either side of the bridge. Sea in shattered slabs of ice. Wind skittering stiff reeds, and starlings moving black above the reeds. The sea is frozen for half a mile offshore. Far out in the gulf, beyond sight, there are thirty-foot-high waves moving west through the half-light.

No highly esteemed deed is commemorated here.

Snowfall like static when the wind drops, like warp-speed when it blows. Double-layer chain-link fencing. Three huge structures showing through the blizzard, across the bay, towards the island's tip. Great grey outlines emerging and fading: dome, tower, slabbed walls. The sea has melted clear around them; the sea should not have done so. Two trucks crunch past on ice tyres.

Nothing valued is here. What is here is dangerous and repulsive to us.

The pirate radio station plays 'Disco Inferno' by the Tramps.

Snow flurries in the headlights. I have come to see a burial site and to bury something of my own. It will be dark when I reach the end of the world and it will be dark when I return to the surface.

Pay attention. We are serious. Sending this message was significant for us. Ours was considered an important culture.

We are going to tell you what lies underground, why you should not disturb this place, and what may happen if you do.

~

Deep in the bedrock of Olkiluoto Island in south-west Finland a tomb is under construction. The tomb is intended to outlast not only the people who designed it, but also the species that designed it. It is intended to maintain its integrity without future maintenance for 100,000 years, able to endure a future Ice Age. One hundred thousand years ago three major river systems flowed across the Sahara. One hundred thousand years ago anatomically modern humans were beginning their journey out of Africa. The oldest pyramid is around 4,600 years old; the oldest surviving church building is fewer than 2,000 years old.

This Finnish tomb has some of the most secure containment protocols ever devised: more secure than the crypts of the Pharaohs, more secure than any supermax prison. It is hoped that what is placed within this tomb will never leave it by means of any agency other than the geological.

The tomb is an experiment in post-human architecture, and its name is Onkalo, which in Finnish means 'cave' or 'hiding place'. What is to be hidden in Onkalo is high-level nuclear waste, perhaps the darkest matter we have ever made.

For as long as we have been producing nuclear waste we have been failing to decide how to dispose of it. Uranium was created in supernova explosions around 6.6 billion years ago, and is part of the space dust out of which the planet formed. It is as common in the Earth's crust as tin or as tungsten, and it is dispersed within the rocks on which we live. Slowly, expensively, miraculously, injuriously, we have learned how to convert uranium into power and into force. We know how to make electricity from uranium and we know how to make death from it, but we still do not know how best to dispose of it when its work for us is done. Over a quarter of a million tons of high-level nuclear waste in need of final storage is presently thought to exist globally, with around 12,000 tons being added to that figure annually.

Uranium is mined as ore in Canada, Russia, Australia, Kazakhstan and perhaps soon in the south of Greenland. The ore is crushed and milled; the uranium is leached out with acid, converted to a gas, enriched, consolidated and then processed into pellets. A single pellet of enriched uranium one centimetre in diameter and one centimetre long will typically release the same amount of energy as a ton of coal. Those pellets are sealed within gleaming fuel rods, usually made of zirconium alloy, which are bundled together in their thousands and then placed in the reactor core, where fission is initiated. Fission produces heat which is used to raise steam; the steam is ducted to turbines, turning their blades and producing electricity.

Once the fission process has slowed below a horizon of efficiency, the rods must be replaced. But they are still intensely hot and lethally radioactive. The unstable uranium oxide continues to emit alpha and beta particles, and gamma waves. If you were to stand next to an unshielded bundle of fuel rods fresh from the core, radioactivity would plunder your body, smashing cells and corrupting DNA. You would be likely to die within hours, vomiting and haemorrhaging.

So spent rods are slid out of the reactor by machine, kept always under water or another shielding liquid as they are moved, then typically stored in deep spent fuel pools for several years, before being sent for reprocessing or dry cask storage. Down in the fuel pools the water patiently absorbs the particle hail from the rods. Because this hail heats the water, it must be continuously circulated and cooled in order to prevent it boiling off and leaving the rods disastrously unshielded.

Even after decades in the pools, however, the rods are still hot, toxic and radioactive. The only way for them to become harmless to the biosphere is through long-term natural decay. For high-level waste this can take tens of thousands of years, during which time spent fuel must be kept secure: segregated from the air, from sun, from water and from life.

The best solution we have devised for securing such waste is burial. The tombs that we have constructed to receive these remains are known as geological repositories, and they are the Cloaca Maxima – the Great Sewer – of our species. Into low- and intermediate-level repositories go the lightly radioactive materials that are the by-products of nuclear power and weaponry: the items that will remain harmful only for scores of years – the clothes, the tools, the filter pads, the zips and the buttons. All are barrelled up and lowered into holes in silos that have been sunk below ground at storage sites around the world. Each new layer is packed in concrete, ready for its supercessor. The Waste Isolation Pilot Plant (WIPP) – the intermediate-level repository dug into the salt beds of New Mexico – is intended to receive 800,000 fifty-five-gallon soft-steel drums of military-origin transuranium waste, holding among other substances the radioactive shavings from US nuclear warhead manufacture. The WIPP drum chambers will in time form neat strata, standing as highly organized additions to the rock record – another taxon of Anthropocene future fossil.

The most dangerous waste, though – the toxic and radioactive spent fuel rods from reactors – requires even more secure burial: a special funeral and a special tomb. We have only ever attempted to construct a few such high-level waste repositories. Belgium has sunk a test site to research future deep repository possibilities, and has named the facility HADES. America's attempt at a high-level repository took place at an extinct super-volcano called Yucca Mountain in the Nevada desert, but construction was suspended after decades of controversy and protest, and the caverns tunnelled into the ignimbrite currently stand as empty halls. Among the reasons for the suspension of the project is Yucca Mountain's proximity to a 900-foot-wide earthquake zone, the Sundance Fault, which is itself undercrossed by a deeper fault called Ghost Dance. If Yucca Mountain were ever to be filled to capacity it would hold, writes John D'Agata, 'the radiological equivalent of two million individual nuclear detonations, about seven trillion doses of lethal radiation', enough to kill every human on Earth 350 times over.

By far the most advanced of all these deep storage facilities is Onkalo, the Hiding Place, set 1,500 feet down into 1.9-billion-year-old rock on the Bothnian coast of Finland. When the burial chambers of Onkalo are full with waste from the three power stations of Olkiluoto, they will hold 6,500 tons of spent uranium.

~

This is the way the world ends, this is the way the world ends, this is the way the world ends – not with a bang but a visitors' centre.

'Welcome to Olkiluoto Island,' says Pasi Tuohimaa. 'You made it!'

I have come to Onkalo the winter after the summer of great melt in Greenland, after the autumn of the moulin.

The reception area is clean and well funded. There are free-standing wardrobes, veneered on the outside with high-definition photographs of forest vistas. In the bathroom there is no piped music but there is piped birdsong. People piss to the calls of nuthatches, or perhaps they are tree-creepers.

Pasi takes me outside. A stepped boardwalk leads down from the back of the reception area to the sea marsh. Reeds brittle in the wind. The sea is frozen solid, yellow plates of ice piled among the bulrushes. Across the bay, passing in and out of visibility as the blizzard moves, are the outlines of three nuclear power stations. The third and most distant is mosque-like: a terracotta dome from which rises a minaret tower.

'The third is still under construction,' says Pasi. 'Not long now.'

The wind is very cold. We retreat to consider the scene from behind glass. The wide viewing windows have grey stickers of raptors on them to prevent bird-strike: generic falcon, generic hawk. The pressed timber frames of the windows present the scene of the bay beautifully. When the blizzard hides the power stations, we might be contemplating an early twentieth-century painting by Gallen-Kallela.

Pasi shows me around the permanent exhibition that explains how the nuclear power supply chain works from mine to consumer, and proves how radiation is a hazard only if incorrectly handled.

'People think nuclear waste is harmful for eternity,' says Pasi. 'It isn't! After 500 years, you could take spent uranium into your home.'

He opens his arms towards me. 'You could probably embrace it!'

He pauses, reconsiders.

'You would not want to keep it under your bed, but in your living room – no problem.'

He pauses again.

'You would not want to kiss it, but hugging is fine.'

He sounds like a father laying out the terms and conditions to his daughter's date.

'This is how we encapsulate the fuel rods for long-term storage,' he says, pointing to an eight-foot-long copper cylinder, a foot and a half in diameter. He raps it with his knuckles. It clunks.

'No fake – this is the real thing. Do you know how much copper trades for per kilogram? It is the best insulator: so inert.'

Inside the copper canister is a cast-iron canister which has been internally partitioned so that it resembles a noughts-and-crosses board, with gaps for the squares. Into these gaps will be slid the zirconium alloy fuel rods containing the spent uranium pellets. Each canister will weigh around twenty-five tons when complete; each canister will be nested in a bed of water-absorbing bentonite clay, inside a cored-out tube of gneiss, 1,500 feet down into the gneiss and granite bedrock.

I murmured the nesting order to myself, working outwards: *uranium, zirconium, iron, copper, bentonite, gneiss, granite* . . . I think back to the beginning of my journeys in the underland, and to the beginning of time, down in the dark-matter laboratory at Boulby Mine. At Boulby they encased xenon in lead in copper in iron in halite in hundreds of yards of rock in order to see back to the birth of the universe. At Onkalo they encased uranium in zirconium in iron in copper in bentonite in hundreds of yards of rock in order to keep the future safe from the present.

One of the exhibits in the display area has a life-size model of Albert Einstein sitting behind a desk, pen in hand, paper on desk.

'See who's here!' says Pasi, leading me to Einstein.

Einstein looks the worse for wear. His rubber face, which would be a poor likeness under the best of circumstances, has come unstuck from his neck. There is a gaping hole in his throat, through which I can see metal struts and hinges.

'Push the button,' urges Pasi, pointing at a red button on our side of the desk, designed to facilitate audience interaction with the exhibit.

I push it.

Einstein's upper body lurches towards us and stops with a jerk that dislodges the right-hand half of his grey moustache, which droops slowly forwards over his upper lip. A recorded voice that I do not take to be Einstein's begins to speak to us in Finnish.

Pasi frowns, then leans across the desk and tenderly presses Einstein's moustache back into place with his thumb.

~

The day before I go to Olkiluoto Island and down to the hiding place, I wait in the little nearby town of Rauma, reading the great folk epic of Finland, the *Kalevala*.

The *Kalevala* is a long poem of many voices and many stories which – like the *Iliad* and the *Odyssey* – grows out of diverse and deep-rooted traditions, from Baltic song to Russian storytelling. It existed chiefly as a mutable oral text for more than a thousand years, until in the nineteenth century the *Kalevala* was collected, edited and published by the Finnish scholar Elias Lönnrot, giving us the mostly fixed version we now have. Lönnrot's *Kalevala* is made up of many intertwining narratives that combine the mythical and the lyrical with the mundane and the logistical, and that together dramatize a northern people's engagement with a hard, beautiful landscape of forests, islands and lakes. In its layering of different ages of origin, the Finnish scholar Matti Kuusi compares the poem's own history of making with 'the numerous strata of a burial mound in which many generations . . . and their artefacts have been buried'.

The *Kalevala* is a haunting epic that has preoccupied me for some years, obsessed as it is with the power of word, incantation and story to change the world into which they are uttered. Its heroes are language masters and wonder-workers – and the greatest of them is called Väinämöinen, whose name translates memorably as 'Hero of the Slow-Moving River'.

In the room in which I read the *Kalevala* that day is a wall-sized photograph of Rauma, taken on a market day at some point in the late nineteenth century. The photograph has been blown up, so it is grainy. All the men have dressed for market day: they are wearing black suits and shoes and hats. They stand out clearly. All the women are wearing blinding white dresses and hats. The plate-camera's long exposure has drunk too deeply of the women's whiteness, though, so that they appear as ghostly, burned-out presences. I count the traces of eighty-seven of these overexposed women. They are leaning out of horse-drawn carts. They are clutching headscarves around their necks with one hand, while carrying shopping with another. Their dresses are ankle-length, and their hats are tall straw boaters with double bands. Here and there they have moved too fast and are blurred to the point of invisibility, lost in the blast.

I read the *Kalevala* for two hours in view of that photograph, and as I read I come to realize something so unsettling that the back of my neck prickles: despite its great age, the poem seems to possess foreknowledge of what is presently being undertaken on Olkiluoto Island.

Partway through the poem, Väinämöinen is given the task of descending to the underland. Hidden in the Finnish forests, he is told, is the entrance to a tunnel that leads to a cavern far underground. In that cavern are stored materials of huge energy: spells and enchantments which, when spoken, will release great power. To approach this subterranean space safely Väinämöinen must protect himself with shoes of copper and a shirt of iron, lest he be damaged by what it contains. Ilmarinen forges them for him. Clad in these insulating metals Väinämöinen approaches the tunnel mouth, which is disguised by aspens, alders, willows and spruce. He cuts down the trees to reveal the entrance. He enters the tunnel and finds himself in a deep 'grave', a 'demon . . . lair'. He has stepped, he realizes, into the throat of a buried giant called Vipunen whose body is the land itself.

Vipunen warns Väinämöinen not to bring to the surface what is buried in his caverns. He speaks of the 'grievous pain' of excavation. Why have you entered 'my guiltless heart, my blameless belly', Vipunen asks, 'to eat and to gnaw / to bite, to devour'? He warns Väinämöinen that he will end up visiting terrible violence upon humans if he continues on his course, that he will become 'a wind-borne disease / wind-borne, water driven / shared out by the gale / carried by chill air'. He threatens to imprison Väinämöinen by means of a containment spell so powerful that it is unlikely ever to be broken. It will take nine ram lambs born of a single ewe, together with nine bull oxen born of a single cow, together with nine stallions born of a single mare, pulling together to free him.

But Väinämöinen will not listen to Vipunen. He sings of his

conviction that the power buried underground should be returned to the surface:

Words shall not be hid
nor spells be buried;
might shall not sink underground
though the mighty go.

The *Kalevala* is fascinated by the underland; by the safe storage of dangerous materials and the safe retrieval of precious materials. At the poem's heart is a magical object or substance known as 'Sampo' or the 'Sammas'; constructed by the blacksmith Ilmarinen, another of the *Kalevala*'s supernatural heroes, and stored inside the 'copper slope' of a 'rocky hill', protected by a gate with ten locks. This enchanted artefact, most often figured as a mill or quern, brings power, wealth and fortune to whoever controls it. It is – in modern terms – a weapons system, a rich raw resource, a nation's organized industry, or a nuclear power station. The Sampo grinds out flour, it grinds out money – and it grinds out time. One of its given tasks is to grind out the age of the world, causing epochs to yield to one another in an immense cycle of precessions. *The world has changed too much . . . we are in the Anthropocene.*

~

We approach the entrance to the Hiding Place through flat, cleared land. The birches, pines and aspens have been felled and their stumps drilled out to make a square glade in the forest, close to the roadside. A doubled chain-link fence surrounds the site, to keep out moose, trespassers and terrorists. Snow settles on grey gravel. The blizzard has eased. In the yellow corrugated-steel central building a vending machine sells energy drinks with the brand name of Battery.

The landscape below which the Hiding Place is sunk has been flattened by the glacial ice that has rolled repeatedly over it in the past 2 million years. Erratic boulders big as buildings lie among trees where the last ice left them. The glaciers do not feel long gone, as if they will be back soon.

The mouth of the Hiding Place is a ramp blasted down into the gneiss. Lichen has already begun to colonize the exposed rock around the entrance: orange lipstick-kisses of *Xanthoria*. A shutter-gate locks off the ramp in case of accident. Now the gate is raised – and below it a tunnel angles down into darkness.

Shotcrete walls, unnaturally smooth. Green side-lights diminishing in size. Signs declare the speed limit at the end of the world to be 20 kmh. Utilities cables droop between brackets. A gurgle of water runs down a gutter. Air moves coldly up from below, stirring stone dust. *The earth is our tabernacle, a receptacle for all decompositions* . . . From the threshold the tunnel leads down and around in a steady crooked three-mile spiral before levelling out at the burial chambers themselves.

Seen in abstract, as if the rock that encases it does not exist, the Hiding Place has an elegant simplicity. There are three central shafts dropping vertically downwards from the surface: ventilation in, ventilation out and an elevator. Around these shafts the transport ramp turns in its helter-skelter, descending at last to a complex excavated space nearly 1,500 feet deep. Outwards from the central space extends a network of storage tunnels, into the floor of each line of which are bored the receptacle wells for the fuel rod canisters. When Onkalo is ready to receive its first deposition, there will be more than 200 storage tunnels, which together will hold the 3,250 canisters. In their form these tunnels resemble to me the chambers and galleries that boring beetles make under tree bark, creating space in which to lay their eggs and rear their larvae, before they kill the tree that feeds them.

Sometimes we bury materials in order that they may be preserved for the future. Sometimes we bury materials in order to preserve the future from them. Some kinds of burial aspire to repetition and re-inheritance (storage); others aspire to oblivion (disposal). At the Barbarastollen underground archive near Freiburg im Breisgau, a disused mine has been converted to a safe-house for German cultural heritage. More than 900 million images are stored there on microfilm in caskets, more than 1,300 feet below ground. The archive is designed to survive a nuclear war, and to preserve its contents for a minimum of 500 years. At Spitsbergen the Global Seed Vault freeze-stores an immense variety of seeds and plant matter, anticipating an epoch after catastrophe when the Earth's flora and biodiversity may need replenishing. Both of these vaults look forward to a time of future scarcity; both implicitly read the present as a time of plenty.

Onkalo, by contrast, is constructed with the desire that its contents never be retrieved. It is a place that confronts us with timescales that scorn our usual measures. Radiological time is not equivalent to eternity, but it does function across temporal spans of such breadth that our conventional modes of imagination and communication collapse in consideration of them. Decades and centuries feel pettily brief, language seems irrelevant compared to the deep time stone-space of Onkalo and what it will hold. The half-life of uranium-235 is 4.46 billion years: such chronology decentres the human, crushing the first person to an irrelevance.

But to think in radiological time is also, necessarily, to ask not what we will make of the future but what the future will make of us.

What legacies will we leave behind, not only for the generations that succeed us but also for the epochs and species that will come after ours? *Are we being good ancestors . . . ?*

The tunnel curls around and back. The air hums oddly. Unseen machines undertake obscure tasks. At a depth of 1,000 feet we enter a series of big side-chambers. In the first stands a yellow drilling engine, unmanned but with its eight halogen eyes glaring, its drill arms still drooling water. The keys are still in the ignition. The shotcrete chamber roof is slotted with silver and red bolt-plates. New drill holes in the roof weep onto us. The halogen casts hard shadows. I think of the lizard-machines in the drift-labyrinth at Boulby, waiting to be enveloped in their halite shrouds.

The bare walls of the chamber are covered in cave art: spray-paint markings in blue, red, apple green, nuclear yellow. The rock is adorned with numbers, pictograms, lines, arrows and other codes I cannot decipher, as remote in their meanings to me as the Bronze Age dancing figures of Refsvika.

~

The Greek word for 'sign', *sema*, is also the word for 'grave'. Around 1990 the research field of nuclear semiotics was born. As plans developed for the burial of radioactive waste, so the question emerged in America of how to warn future generations of the great and durable danger that lay at depth. It became important, the US Department of Energy decided, to devise a 'marker system' that could deter intrusion into a repository 'during the next 10,000 years'. The Environmental Protection Agency founded a 'Human Interference Task Force' charged with the imagining of such a system for the entombment sites under construction at Yucca Mountain and in the New

Mexico desert. Two separate panels were convened to consider the issue of the 'marker system', reporting to an overall Expert Judgment Panel. Among those invited to express interest in joining the panels were anthropologists, architects, archaeologists, historians, graphic artists, ethicists, librarians, sculptors and linguists, as well as geologists, astronomers and biologists.

The challenges faced by the panels were formidable. How to devise a warning system that could survive – both structurally and semantically – even catastrophic phases of planetary future. How to communicate with unknown beings-to-be across chasms of time to the effect that they must not intrude into these burial chambers, thus violating the waste's quarantine?

Several proposals developed by the panels involved forms of what is now known as hostile architecture, but which they referred to as 'passive institutional controls'. They suggested constructing above ground at the burial site a 'Landscape of Thorns' (fifty-foot-high concrete pillars with jutting spikes that impeded access and suggested 'danger to the body'), a 'Black Hole' (a mass of black granite or concrete that absorbed solar energy to become impassably hot) and 'Forbidding Blocks' (the bulks of which might intimidate a visitor into turning back).

The panel members realized, however, that such aggressive structures might act as enticements rather than cautions, suggesting 'Here be treasure' rather than 'Here be dragons'. Prince Charming hacked his way through the briars and thorns to wake Sleeping Beauty. Howard Carter excavated Tutankhamun's tomb despite the multiple obstructions placed in the way of access, and the warnings given in languages other than his own.

Other proposals from the panels involved versions of a transcendental signifier. Human faces could be carved into stone: pictograms or petroglyphs conveying horror. Munch's *The Scream* might be taken as a model, it was suggested, on the grounds that it could still somehow communicate terror to whatever being approached it in the distant future. Or a durable aeolian instrument might be constructed that tuned the far-future desert winds to a minor D, the note thought best to convey sadness.

The semiotician and linguist Thomas Sebeok argued on grounds of futility against the search for a transcendental signifier that could outlast all corruption and mutation. Such a sign did not exist, he said. Instead he proposed working towards what he called a long-term 'active communication system' that relayed the nature of the site using story, folklore and myth. Such a means of transmission – perpetuated by an elected 'atomic priesthood' – would be flexible, allowing retellings and adaptations to occur across generations. In this way what began as a simple set of warnings might be reconfigured as, say, a long poem or folk epic, made narratively new for each society in need of warning. Those ordained into the priesthood would have the responsibility of 'laying a trail of myths about the [burial sites] in order to keep people away'.

The Waste Isolation Pilot Project in New Mexico is currently due to be sealed in 2038. The plans for marking the site remain under development. Among those advising the project now are social scientists and writers of science fiction. Present plans for what Gregory Benford has called 'our society's largest conscious attempt to communicate across the abyss of deep time' include the following measures.

First the chambers and the access shafts will be backfilled. Then a thirty-foot-high berm of rock and tamped earth with a core of salt will be constructed, enclosing the above-ground footprint of the repository. Buried in the berm and the earth around it will be radar reflectors and magnets, discs made of ceramic, clay, glass and metal, engraved with warnings: 'Do Not Dig Or Drill'. The berm itself will be surrounded by an outer perimeter of 25-foot-high granite pillars, also bearing warning texts.

Set flat near the berm will be a map measuring 2,200 feet by 600 feet. The map will be slightly domed so that it sheds sand in the wind, and does not itself become buried. The continents will have granite edges, the oceans will be represented by caliche stone rubble, and marked on the map will be the world's significant radioactive burial sites. An obelisk will indicate the WIPP site: You Are Here.

This map at the Earth's end has echoes of Jorge Luis Borges's cautionary story 'On Exactitude in Science', which imagines a world in which the art of cartography aspires to such representative perfection that the Cartographers of the Empire construct 'a map of the Empire whose size was that of the Empire'. But of course this one-to-one scale map proves both unusable and overwhelming. The 'following Generations', perceiving the danger of such a map, leave it to erode. 'In the Deserts of the West,' ends Borges's story, 'still today, there are Tattered Ruins of that Map, inhabited by Animals and Beggars.'

Close to the WIPP map what is called a 'Hot Cell' will be constructed: a reinforced concrete structure extending some sixty feet above the earth and thirty feet down into it. 'Hot' because it will house small samples of the interred waste, in order to demonstrate the radioactivity of what is buried far beneath.

Within the curtilage of the berm an information chamber will be built of granite and reinforced concrete, designed to last a minimum of 10,000 years. The chamber will carry stone slabs into which will be inscribed more maps, timelines, and scientific details of the waste and its risks, written in all current official UN languages, and in Navajo.

Buried directly below the information chamber will be a 'Storage Room'. This room will have four small entrances, each secured by a sliding stone door. In the room will be messages of warning cut into stone and simply phrased:

We are going to tell you what lies underground, why you should not disturb this place, and what may happen if you do.

This site was known as the WIPP (Waste Isolation Pilot Plant Site) when it was closed in 2038 AD.

The waste was generated during the manufacture of nuclear weapons, also called atomic bombs.

We believe that we have an obligation to protect future generations from the hazards that we have created.

This message is a warning about danger.

We urge you to keep the room intact and buried.

That configuration of berm, map, Hot Cell, information chamber and buried Storage Room – all set atop the casks of pulsing radioactive molecules entombed deep in the Permian strata – seems to me our purest Anthropocene architecture yet, and the greatest grave that we have so far sunk into the underland. Those repeated incantations – pitched somewhere between confession and caution – seem to me our most perfected Anthropocene text, our blackest mass.

But I know also that even those words will decay over the course of deep time – blasted from the stone by desert wind, eaten from it by atmospheric moisture, or lost in translation. For language has its half-life too, its decay chain. The written history of humanity is only around 5,000 years old, when cuneiform first emerged. Our language systems are dynamic, our inscription systems vulnerable to destruction or distortion. Most ink is perishable in direct sunlight, fading within months towards invisibility. Even if lettering is inscribed in durable substances, there is no guarantee that it will be legible to future audiences. Today perhaps a thousand people in the world can understand cuneiform.

Those in charge of the burial chambers at Onkalo are largely unconcerned about how to communicate warnings to future generations. They know that, at their latitude, the forest will soon begin to grow over abandoned land, concealing the above-ground presence of the site. They know too that once the forest has grown it will not be long, in terms of Earth time, until the glaciers return to this region. They know that the passage of the ice will smooth out all signs of what has been done here, placing the whole terrain under erasure.

~

We reach the lowest point of Onkalo. An arched side tunnel leads off the terminal chamber. The tunnel's floor is flat and screeded. Sunk into that floor are two cored-out cylindrical spaces. These are burial holes awaiting their bodies. Each hole is eight feet deep and five feet in circumference, protected by a circular yellow guard-rail.

At the tunnel's mouth sit a grey melamine table and a brown plastic chair. Until the lethal canisters arrive this is a workplace, and as in all workplaces there are forms that require filling in and legs that need resting.

A series of brown plastic panels are bolted to the side of the tunnel, and on them an unknown finger has sketched pictures in the stone dust that clings to the plastic. There are three panels. On the left-hand panel the finger has drawn a landscape with a storm, a tree, a house. On the centre panel, a rabbit sitting on a cloud. On the right-hand panel is a human face with a crinkled smile.

The belly of Onkalo is not the deepest place that I have been during the years of underland travel, but it seems at this point the darkest. I have a strong sense of the weight of time above and around us, bearing down on veins and tissue.

Far above us, waves crash eastwards through the Gulf of Bothnia, the sea shifts under its cracked jacket of ice, a multinational workforce prepares a turbine housing to receive the largest blades ever fitted in a nuclear power station, the sun swings over a shattered Syria, atmospheric CO_2 increases its parts per million and the Knud Rasmussen glacier hastens its calving into the fjord.

It all feels very distant, the busyness of another planet.

'There was a joke among the designers and engineers at Onkalo during the early years of its construction,' says Pasi abruptly, rapping the stone with his knuckles, 'that as they began drilling and blasting, the first thing they would uncover would be a copper canister, containing spent fuel rods . . .'

I think with a jolt of the *Kalevala*, with the powerful Sampo grinding out its epochal changes, with its embedded warnings from centuries ago about the dangers of disinterral from below ground, about the need for copper to insulate from harm, and about the dreadful disease that will ravage air, water and all life if it is brought in untimely fashion to the surface.

I think of Sebeok's 'atomic priesthood' charged with conveying warnings across generations in the form of folklore and myth. I think of the last line of the poem pinned on the tin sheet above the sinkhole into which people had been clubbed and pushed and bayoneted up in the Slovenian beech woods. *A curse be upon anyone who might attempt to erase this record* . . . I have a swift, chilling sense of the *Kalevala* as part of a messaging system, the warnings of which we have not heeded or even heard.

The stillness of the stone around is now crushing. I remember being in the bedding plane in the Mendips with Sean, the pressure exerted by the unmoving black stone. Other memories arrive, from further back, rising unbidden in my mind. I am with my father, using the claw of a hammer to prise up the floorboard of the house in which I grew up, in order to lodge a time capsule in a jam jar there. What did we put in the jar? A little die-cast aeroplane, a bomber? Yes. A letter to an unknown future recipient. Rice to absorb moisture and prevent the perishing of paper and ink. A Polaroid photograph of me and my brother. Is that right? At this distance, details have decayed. I can only clearly recall the fact of placing the jar – fat jar, thin mouth, brass lid – and nailing down the floorboard above it. Gone. Safe. A message to the future.

Time begins to fission into overlapping shadow-times. Burial thoughts from the underland crowd in. Neil Moss, his body still there in the shaft in the Peak District, entombed in concrete to prevent future harm to others. The Mesolithic bodies in the Mendips, chrysalized by calcite, *almost converted into stone* . . . My father's wish that his ashes be scattered to the winds in three places, so that in this way there is no grave to which we will be tied after his death, and the medium of his remembrance will be an atmosphere, a skein of associations.

I sit down, tired, on the brown plastic chair at the end of the world. Pasi is still in the side tunnel, talking to a worker. I imagine walking down and around a corner in the main tunnel, where it turns out of sight of Pasi. In the right-hand wall of the tunnel are three boreholes, each about the diameter of my shoulder. I imagine reaching as far as I can into the middle borehole, and I imagine that when I retrieve my arm a weight has been lifted from me and a promise has been kept.

Once the canisters of waste have been deposited in Onkalo and all the reception cylinders are replete, the spiralling access ramp will be backfilled, the ventilation shafts will be backfilled, the lift shaft will be backfilled and at last the mouth of the tunnel entrance will be backfilled – 2 million tons of bedrock and bentonite, sealing those canisters in place, keeping the future safe from the present.

Then I see that on another of the plastic panels bolted to the wall of the terminal chamber there is a handprint in the dust: spread fingers, the pad of the thumb pressed clear. It is the print of a right hand, left there at some point for the keeping of balance, for the taking of rest – or just for the making of a mark.

I think of the black and red hand-prints left on the cave walls at Chauvet, of the red figures of the dancers with their outstretched arms, of the spray-can hand stencil on the catacomb wall in Paris, of Helen reaching a hand down to haul me out of the moulin. I think of the many people I have encountered in and through the underland who have been committed to shared human work rather than retreat and isolation. Many of them have been mappers, really, of networks of mutual relation, endeavouring to stitch their thinking into unfamiliar scales of time and space, seeking not the scattered jewels of personal epiphany but rather to enlarge the possible means by which people might move and think together across landscapes, in responsible knowledge of deep past, deep future and the inhuman earth.

Suddenly, surprisingly, there is something hopeful – no, something *moving* – about this mundanely functional space I have reached. The melamine desk and the moulded chair. The plastic panels with their doodled art. Pasi's passion for Onkalo. The copper canisters, the visitors' centre, Einstein's drooping moustache. Here a vast problem is being solved, gradually and practically, by a community of people to the best of their abilities. Here the hard labour of collective decision-taking and world-making is being carried out, imperfectly but necessarily, and with a care that extends not only for a decade or a generation but far forwards into a post-human future.

Maybe this is among the best things we can try to do, I think, as the Sampo grinds through the world's epochs: to be good ancestors. I remember a paragraph I have copied out into a notebook, from a book called *After Nature*:

> People are best able to change their ways when they find two things at once in nature: something to fear, a threat they must avoid, and also something to love, a quality . . . which they can do their best to honour. Either impulse can stay the human hand, but the first stops it just short of being burnt or broken. The second keeps the hand poised, extended in greeting or in an offer of peace. This gesture is the beginning of collaboration, among people but beyond us, in building our next home.

~

When we return to the surface the wind has eased but the snowfall has strengthened. Dusk is coming. All sight is through fading grey light. Mid-afternoon and already the day is over.

Back over the bridge from the island. Salt marsh at either side of the bridge. The sea in shattered pieces. A blue letter-box on a white pole. Boulders big as houses among the pines, between the birches.

My headlights making tunnels in the dusk ahead. Birches, pines, birches, birches. Everything frozen.

On the way back to Rauma a yellow dashboard warning light pings on. The back right tyre is losing pressure. I can feel the car's grip on the icy road starting to loosen. I pull over, crunch to a halt, get out. The tyre is almost flat. Deep forest runs to right and left of the road. The car detects the air temperature to be -12°C. I am getting cold fast. I don't have enough warm clothes with me. I look in the boot. There is a spare tyre but no jack. This is not a good situation. I do not know what to do.

Five minutes later I see the headlights of an approaching vehicle, the first to pass. I stand by my car and raise a hand into the air, asking for help, not expecting to receive it. But the car pulls over, and a man gets out. I explain the situation, my helplessness, that I was driving back from the island when it happened. He says he is a worker from Olkiluoto on his way home after finishing a shift.

'I'm sorry. You must be tired,' I say. 'Thank you for stopping.'

'It is just no problem,' he says.

He has a jack. Ten minutes later he has changed the tyre and stowed the flat in the boot. He cleans the oil and grease from his fingers with a cloth. Then he puts out a hand, I shake it in gratitude, and we drive off one after the other into the darkness.

13
Surfacing

The way out of the underland is where nine springs flow clear from the bedrock.

Months after Onkalo, when the year has warmed, I take my youngest son to the chalk uplands a mile or so from our house. He is four years old and I am forty-one. We cycle most of the way, then I lay the bike in the grass and he and I walk hand-in-hand the few hundred yards to a half-acre copse of beech and ash called Nine Wells Wood. Nine Wells lies close to the railway line, close to the hospital, and like many small woods its extent seems much greater once entered than it appears from the outside.

The hour or so that he and I spend together in the wood is happy and calm. There I am able to focus on him, to walk at his pace, to think what it is like to see the world as a four-year-old might see it. The sun is high and strong, and light streams through the canopy, falling in splinters around us.

We make our way to the end of the wood where the springs rise. The springs have organized themselves in a circle around a hollow in the chalk, filling a pool that is perhaps a foot deep and six feet across. The water in the pool is so clear as to be invisible, save for the root-like reflection of the branches above that it carries.

The sides of the hollow are slippery, so I hold on to the trunk of an elder with one hand and grip his arm with the other, and in this manner he and I are able to slither down to the edge of the pool and crouch there.

He is amazed by the fact of the springs. He cannot comprehend that water should issue from the earth like this, that the stone should *flow* in this way.

We count the springs off, one by one. They declare themselves only by the ripples they stir on the surface.

'The water is black,' he says, and this puzzles me until I realize that because the water is so clear he is seeing straight through it to the pool bed, which is dark with fallen leaves and twigs.

To prove the water's existence I dip a hand and drink. The water – straight from the chalk – tastes different from any other I know; somehow *round* in the mouth. And cold. Stone-cold. I hold a cupped hand of the water up for him and he drinks from it too, tentatively at first, then greedily, gripping my wrist, enjoying the water's coolness on that warm day.

Of the nine springs, he likes best the one with the strongest flow. I like the smallest, though, the one on the inaccessible far side of the pool, just below water level. There the chalk is whitest and the spring shows itself only as the faintest of ripples, and as a triangular rift in the chalk giving onto inky blackness.

Sitting there on the ground by the springs, him sitting on me, I let my mind run against the flow of the water, following its path back into the rift in the chalk, and down through the interstices of the rock. I think of what has been excavated and interred here over thousands of years of human presence – Neolithic causewayed enclosures, Bronze Age burial barrows, a sunken Iron Age ring fort, a medieval cemetery, a Second World War anti-tank trench, a buried Cold War observation post a few hundred yards distant, down into which a designated observer was to retreat in the event of a nuclear strike, with no room for his wife or children, who were to be abandoned by order of the government.

I hug my son. A young woman appears on the path above the pool, looks down into the spring-hollow and smiles on seeing us. She is walking her collie. The dog darts around, barking. We talk for a little, low to high, about the springs, the wood, the weather. On her calf she has a circular tattoo of a map that shows the Arctic Circle from Canada all the way round to Greenland, as if seen from a vantage point somewhere above the North Pole.

Lumps of white chalk lie among the ivy, glowing in the day-dusk of the wood. Dragonflies hunt the spring stream where it flows away from us. Beneath and around us, invisibly, the fungal network connects tree to tree.

The young woman walks on, calling for her dog, which has disappeared. My son and I talk quietly about nothing much. We feel small in the universe, and together.

Later, as we are leaving, he runs on ahead down a tunnel of briar and blackthorn. The tunnel is at first in shadow, but as I watch him run he passes into a place where the sunshine falls so brightly that he is burned up by it, lost to my sight, and suddenly the knowledge that he will die strikes me and every leaf falls from the trees around us and the air greys to ash and colour is utterly lost – and then life and hue pour back into the world as quickly as they were drained from it, and the leaves flicker greenly on the trees again.

I run to catch up with him, calling loudly, and he turns to face me at the edge of the wood. As I kneel down on the earth he raises a hand in the air, fingers spread wide. I reach my hand towards his and meet it palm to palm, finger to finger, his skin strange as stone against mine.

NOTES

List of abbreviations used in the notes

ALDP : *Arts of Living on a Damaged Planet*, ed. Anna Tsing, Heather Swanson, Elaine Gan and Nils Bubandt (Minneapolis: University of Minnesota Press, 2017)

ANP : Richard Bradley, *An Archaeology of Natural Places* (London: Routledge, 2006)

TAP : Walter Benjamin, *The Arcades Project*, trans. Howard Eiland and Kevin McLaughlin (London: Harvard University Press, 1999)

TK : *The Kalevala*, trans. Keith Bosley (Oxford: Oxford University Press, 2008)

Epigraphs

Page

v *'Is it dark down there . . . under-land of Null?'*: Helen Adam, 'Down There in the Dark', in *A Helen Adam Reader*, ed. Kristin Prevallet (Orono, ME: National Poetry Foundation, 2007), p. 34.

v *'The void migrates to the surface . . .'*: *Advances in Geophysics*, ed. Lars Nielsen, vol. 57 (Cambridge, MA: Academic Press, 2016), p. 99.

Chapter 1: Descending

Pages

12 *'deep subterranean fact'*: Elaine Scarry, *The Body in Pain: The Making and Unmaking of the World* (Oxford: Oxford University Press, 1985), p. 3.

12 *'the awful darkness inside the world'*: Cormac McCarthy, *Blood Meridian* (1985; New York: Vintage, 1992), p. 117.

12 *'They lay full length . . . he could not move'*: Alan Garner, *The Weirdstone of Brisingamen* (1960; London: HarperCollins, 2014), pp. 177–8.

13 *'flat tradition . . . resolutely flat perspectives'*: Stephen Graham, *Vertical: The City from Satellites to Bunkers* (London: Verso, 2016), pp. 4–7.

13 *'Force yourself to see more flatly'*: Georges Perec, *Species of Spaces and Other Pieces*, trans. John Sturrock (1974; Harmondsworth: Penguin, 1997), p. 51.

14 *Anthrax spores are being released from reindeer corpses*: on this and other forms of Arctic surfacing see Sophia Roosth's fine essay 'Virus, Coal, and Seed: Subcutaneous Life in the Polar North', *Los Angeles Review of Books*, 21 December 2016 <https://lareviewofbooks.org/article/virus-coal-seed-subcutaneous-life-polar-north/>.

14 *'doorway to the underworld'*: Melissa Hogenboom, 'In Siberia There is a Huge Crater and It is Getting Bigger', BBC, 24 February 2017.

14 'Wenn du mich siehst, dann weine': see R. Brázdil, P. Dobrovolny et al., 'Droughts in the Czech Lands, 1090–2012 AD', *Climate of the Past* 9 (August 2013), 1985–2002.

14–15 *'The problem is not that things become buried . . . dark force of "sleeping giants"'*: Þóra Pétursdóttir, 'Drift', in *Multispecies Archaeology*, ed. Suzanne E. Pilaar Birch (London: Routledge, 2018), pp. 85–102, p. 98; see also Þóra Pétursdóttir, 'Climate Change? Archaeology and Anthropocene', *Archaeological Dialogues* 24:2 (2017), 182–93; *'sleeping giants'* is quoted from Graham Harman, *Immaterialism* (Cambridge: Polity Press, 2016), p. 7.

15 *'Deep time' is the chronology of the underland*: the coining of the phrase 'deep time' is usually attributed to John McPhee in *Basin and Range* (New York: FSG, 1981); John Playfair wrote of 'the abyss of time' as he examined the Siccar Point unconformity with James Hutton in June 1788.

16 *'netherworld . . . I saw them'*: 'Gilgamesh, Endiku and the Nether World', Version A, in J. A. Black, G. Cunningham, E. Fluckiger-Hawker, E. Robson and G. Zólyomi, *The Electronic Text Corpus of Sumerian Literature* (Oxford: 1998–) <http://etcsl.orinst.ox.ac.uk/section1/tr1814.htm>.

17 *'People were making journeys into the darkness'*: Alistair Pike, quoted in Emma Marris, 'Neanderthal Artists Made Oldest-Known Cave Paintings', *Nature*, 22 February 2018.

18 *'The descent beckons / as the ascent beckoned'*: William Carlos Williams, 'The Descent', in *The Collected Poems of William Carlos Williams, Volume II 1939–1962*, ed. Christopher MacGowan (New York: New Directions, 1988), p. 245.

18 *'the feet of the dead . . . touch those of the living, who stand upright'*: Richard Bradley, drawing on the work of Tim Ingold, *ANP*, p. 12; see Tim Ingold, *The Appropriation of Nature* (Manchester: Manchester University Press, 1986), p. 246.

19 *the first of the objects . . . help me see in the dark*: the whalebone owl and the demon casket were made and given to me on the Isle of Harris in the Outer Hebrides by the sculptor Steve Dilworth, about whose extraordinary life and work more can be read in the chapter entitled 'Gneiss' in my book *The Old Ways: A Journey on Foot* (London: Hamish Hamilton, 2012). Images of his sculptures and practice can be seen at <http://www.gallery-pangolin.com/artists/steve-dilworth>.

Chapter 2: Burial

Pages

25 *'surprised with the appearance . . . converted into stone'*: *Bristol Mercury & Universal Advertiser*, 16 January 1797. This source among others is quoted in full in A. Boycott and L. J. Wilson, 'Contemporary Accounts of the Discovery of Aveline's Hole, Burrington Combe, North Somerset', *Proceedings of the University of Bristol Spelaeological Society* 25:1 (2010), 11–25. I draw here also on R. J. Schulting, '". . . Pursuing a Rabbit in Burrington Combe": New Research on the Early Mesolithic Burial Cave of Aveline's Hole', *Proceedings of the University of Bristol Spelaeological Society* 23:3 (2005), 171–265.

28 *'is hollow . . . some huge subterranean sea'*: Arthur Conan Doyle, 'The Terror of Blue John Gap', in Arthur Conan Doyle, *Tales of Terror and Mystery* (1902; Cornwall: House of Stratus, 2009), p. 58.

29 *'I do not trust space an inch'*: Tim Robinson, *My Time in Space* (Dublin: Lilliput, 2001), p. 114.

30 *'To be human means above all to bury'*: Robert Pogue Harrison, *The Dominion of the Dead* (Chicago: University of Chicago Press, 2003), p. xi. See also Rebecca Altman's fine essay 'On What We Bury', *ISLE* 21:1 (Winter 2014), 85–95.

30–31 *In a cave system called Rising Star . . . some 300,000 years ago*: see John Hawks et al., 'New Fossil Remains of Homo Naledi from the Lesedi Chamber, South Africa', *eLife* 6 (2017).

31 *'between fourty and fifty Urnes . . . nether part of the Earth'*: Thomas Browne, *Religio Medici and Urne-Buriall*, ed. Stephen Greenblatt and Ramie Targoff (1658; New York: NYRB Classics, 2012), pp. 103, 114–15, 112.

32–3 *Twelve thousand years ago in a limestone cave . . . inside her chamber*: see Leore Grossman et al., 'A 12,000-Year-Old Shaman Burial from the Southern Levant (Israel)', *PNAS* 105:46 (2008), 17665–9.

40 *The most notorious story in British caving history . . . known as Moss Chamber*: I draw in this description on several sources, principally: James Lovelock, *Life and*

Death Underground (London: G. Bell and Sons, 1963), pp. 11–27; Dave Webb and Judy Whiteside, 'Fight for Life: The Neil Moss Story' <www.mountain.rescue.org.uk/assets/files/The Oracle/history and people/NeilMossStory.pdf>; and *Fight for Life: The Neil Moss Story*, dir. Dave Webb (2006).

44 *'For the first time in millennia . . . the vast majority of people a few generations ago'*: Harrison, *The Dominion of the Dead*, p. 31.

Chapter 3: Dark Matter

Pages

55 *shielded from the surface by 3,000 feet of halite, gypsum . . . clay and topsoil*: on the strata sequence at Boulby, see 'Lithological Log of Cleveland Potash Ltd', Borehole Staithes No. 20, drilled September–December 1968 to a depth of *c.*3500 feet (BGS ID borehole 620319, BGS Reference NZ71NE14).

58 *'the revelation of a new order . . . and darkness as well'*: Kent Meyers, 'Chasing Dark Matter in America's Deepest Gold Mine', *Harper's Magazine* (May 2015), 27–37: 28.

58 *'As if . . . you could infer the meadow'*: Rebecca Elson, 'Explaining Dark Matter', in *A Responsibility to Awe* (Manchester: Carcanet, 2001), p. 71.

75 *'I suddenly thought . . . it seems to have stuck'*: Paul Crutzen, quoted in Howard Falcon-Lang, 'Anthropocene: Have Humans Created a New Geological Age?', BBC, 11 May 2011 <http://www.bbc.co.uk/news/mobile/science-environment-13335683>.

75 *'mankind* [sic] *. . . millions of years to come'*: Paul Crutzen and Eugene Stoermer, 'The Anthropocene', *International Geosphere-Biosphere Newsletter* 41 (May 2000) <https://www.mpic.de/mitarbeiter/auszeichnungen-crutzen/the-anthropocene.html>.

75 *As the Pleistocene was defined by the action of ice . . . at a global scale*: for several years now I have taught a graduate course at Cambridge called 'Cultures of the Anthropocene'. The literature of and on the idea of the Anthropocene is vast, various, disputatious and growing. Some of the texts I find most interesting are detailed in the bibliography and are drawn on in this brief discussion of the concept and its implications for deep time, politics and ethics.

76 *'stratigraphically optimal'*: Anthropocene Working Group of the Subcommission on Quaternary Stratigraphy, 'When Did the Anthropocene Begin? A Mid-Twentieth-Century Limit is Stratigraphically Optimal', *Quaternary International* 383 (2015), 204–7.

77 *'Are we being good ancestors?'*: Jonas Salk, 'Are We Being Good Ancestors?', *World Affairs* 1:2 (1992), 16–18.

78 *'palaeontology of the present'*: W. J. T. Mitchell, *What Do Pictures Want? The Lives and Loves of Images* (Chicago: University of Chicago Press, 2005), p. 325.

79 *A trace fossil is the sign . . . absence serves as sign*: see also Ilana Halperin, 'Autobiographical Trace Fossils', in *Making the Geologic Now: Responses to Material Conditions of Contemporary Life*, ed. Elizabeth Ellsworth and Jamie Kruse (New York: Punctum, 2013), pp. 154–8.

81 *'At night, according to . . . beneath the earth'*: Bede, *The Reckoning of Time*, trans. Faith Wallis (725; Liverpool: Liverpool University Press, 1999), p. 97.

82 *Occasionally the miners hacked their ways into geodes . . . down there in the crust*: on Pennine mining cultures, see Peter Davidson's glittering chapter, 'Spar Boxes: Northern England', in his *Distance and Memory* (Manchester: Carcanet, 2013), pp. 42–58.

Chapter 4: The Understorey

Pages

89 *'underground social network . . . fungal species'*: Suzanne Simard, 'Notes from a Forest Scientist', afterword to Peter Wohlleben, *The Hidden Life of Trees*, trans. Jane Billinghurst (Vancouver/Berkeley: Greystone Press, 2016), p. 247.

90 *'forged their duality . . . making a forest'*: Simard, in Wohlleben, *Hidden Life of Trees*, p. 249.

90–91 *'co-operative system . . . forest wisdom . . . mothers'*: Suzanne Simard, 'Exploring How and Why Trees "Talk" to Each Other', *Yale Environment 360*, 1 September 2016 <https://e360.yale.edu/features/exploring_how_and_why_trees_talk_to_each_other>.

91 *'the wood wide web'*: see Suzanne Simard et al., 'Net Transfer of Carbon between Ectomycorrhizal Tree Species in the Field', *Nature* 388:6642 (1997), 579–82.

91 *'The wood wide web . . . languages of the forest network'*: Simard, in Wohlleben, *Hidden Life of Trees*, p. 249.

96 *'plants are physiologically separate . . . functioning of ecosystems'*: see E. I. Newman, 'Mycorrhizal Links between Plants: Their Functioning and Ecological Significance', *Advances in Ecological Research* 18 (1988), 243–70: 244.

98 *'a busy social space . . . cross-species world underground'*: Anna Tsing and Rosetta S. Elkin, 'The Politics of the Rhizosphere', *Harvard Design Magazine* 45 (Spring/

Summer 2018) <http://www.harvarddesignmagazine.org/issues/45/the-politics-of-the-rhizosphere>

98 *'Next time you walk through a forest . . . lies under your feet'*: Anna Tsing, 'Arts of Inclusion, or How to Love a Mushroom', *Manoa* 22:2 (2010), 191–203: 191.

99 *'we had roots that grew . . . one tree and not two'*: Louis De Bernières, *Captain Corelli's Mandolin* (Reading: Secker and Warburg, 1996), p. 281.

100 *and of the hyphae that are weaving . . . a version of love's work*: Ginny Battson has also written – beautifully – on mycelia and/as love, in a short online essay, 'Mycelium of the Forest Floor. And Love', 12 October 2015 <https://seasonalight.wordpress.com/2015/10/12/mycelium-of-the-forest-floor-and-love/>.

101 If only your mind were a slightly greener thing . . . drown you in meaning: Richard Powers, *The Overstory* (New York: W. W. Norton, 2018), p. 4.

102 *Fungi were among the first organisms . . . changing conditions of the Anthropocene*: for more on the cultural and political histories of fungi, and how they entangle with our own, see Anna Tsing, *The Mushroom at the End of the World: On the Possibility of Life in Capitalist Ruins* (Princeton: Princeton University Press, 2017). I have also drawn in this discussion on Karen Barad, 'No Small Matter: Mushroom Clouds, Ecologies of Nothingness, and Strange Topologies of Spacetimemattering', in *ALDP*, pp. G103–G120.

102–3 *Scientists working in Chernobyl after the disaster . . . processing it in some way*: see N. N. Zhdanova et al., 'Ionizing Radiation Attracts Soil Fungi', *Mycological Research* 108:9 (2004), 1089–96; and E. Dadachova and A. Casadevall, 'Ionizing Radiation: How Fungi Cope, Adapt, and Exploit with the Help of Melanin', *Current Opinion in Microbiology* 11:6 (2008), 525–31.

103 *'Learning to see mosses is more like listening than looking'*: Robin Wall Kimmerer, *Gathering Moss: A Natural and Cultural History of Mosses* (Corvallis: Oregon State University Press, 2003), p. 11.

103 *'mosses . . . the limits of ordinary perception'*: Kimmerer, *Gathering Moss*, p. 10.

104 *'holobionts'*: Lynn Margulis, 'Symbiogenesis and Symbionticism', in *Symbiosis as a Source of Evolutionary Innovation: Speciation and Morphogenesis*, ed. Lynn Margulis (Boston: MIT Press, 1991), pp. 1–14: p. 3.

104 *'consisting of trillions of bacteria, viruses and fungi . . . sharing a common life'*: Glenn Albrecht, 'Exiting the Anthropocene and Entering the Symbiocene', *PYSCHOTERRATICA*, 17 December 2015 <https://glennaalbrecht.com/2015/12/17/exiting-the-anthropocene-and-entering-the-symbiocene/>.

104 *'To dwellers in a wood . . . voice as well as its feature'*: Thomas Hardy, *Under the Greenwood Tree* (1872; London: Penguin, 2012), p. 3.

104 *'live in a world that watches . . . sensate, personified. They feel'*: Richard Nelson, *Make Prayers to the Raven: A Koyukon View of the Northern Forest* (Chicago: University of Chicago Press, 1986), p. 14.

105 *'the word for world is forest'*: Ursula K. Le Guin, *The Word for World is Forest* (1972; London: Orion Books, 2015).

111 *'all its technical vocabulary . . . no words to hold this mystery'*: Robin Wall Kimmerer, *Braiding Sweetgrass: Indigenous Wisdom, Scientific Knowledge, and the Teachings of Plants* (Minneapolis: Milkweed, 2013), p. 49.

111 *'fluent botany . . . gift of seeing'*: Kimmerer, *Braiding Sweetgrass*, pp. 48–9.

112 *'A bay is a noun . . . well[ing] up all around us'*: Kimmerer, *Braiding Sweetgrass*, p. 55.

112 *'grammar of animacy'*: Robin Wall Kimmerer, 'Speaking of Nature', *Orion Magazine*, 14 June 2017, *passim*.

112 *mammal language*: J. H. Prynne, 'On the Poetry of Peter Larkin', *No Prizes* 2 (2013), 43–5: 43.

113 *'geotraumatics'*: Cybernetic Culture Research Unit, 'Barker Speaks', in *CCRU: Writings 1997–2003* (Falmouth: Time Spiral Press, 2015), p. 155.

113 *'planetary dysphoria'*: Emily Apter, 'Planetary Dysphoria', *Third Text* 27:1 (2017), 131–40.

113 *'apex-guilt'*: aliciaescott, 'Field Study #007, The Extinction Event', *Bureau of Linguistical Reality*, 1 September 2015 <https://bureauoflinguisticalreality.com/2015/09/01/field-study-007-the-extinction-event/>.

113 *'species loneliness'*: Kimmerer, *Braiding Sweetgrass*, p. 208.

113 *'by human intelligence . . . the wood wide web'*: Albrecht, 'Exiting the Anthropocene and Entering the Symbiocene'.

Second Chamber

Pages

121 *'if we need to go into caves in a nuclear war . . . a lot of food*: British Pathé, 'Caveman 105 Days Below', *YouTube*, 13 April 2014 <https://www.youtube.com/watch?v=YSdBBv5LY84>.

122 *'I can only think clearly in the dark . . . darkness in Europe'*: Ludwig Wittgenstein, quoted in Tim Robinson, *Connemara: The Last Pool of Darkness* (London and Dublin: Penguin, 2009), p. 1. In the same book, Robinson tells the story of artist Dorothy Cross's habit of diving down to feed the conger eels at the bottom of the harbour.

Chapter 5: Invisible Cities

Pages

133 *'convolutes'*: 'Translators' Foreword', in *TAP*, p. xiv.

133 *'collective dream'*: *TAP*, p. 152.

133 *'It is more arduous to honour . . . memory of the nameless'*: these words, from Benjamin's preparatory notes to 'Theses on the Philosophy of History', are etched into glass at Dani Karavan's memorial to Benjamin at Portbou.

135 *'subterranean city . . . upper world'*: *TAP*, pp. 85–98.

135 *'Our waking existence . . . lose ourselves in the dark corridors'*: *TAP*, p. 84.

135 *'key'*, *'underworld'*: *TAP*, p. 403, p. 84.

135 *'make some sign to the world one is leaving'*: *TAP*, p. 88.

135 *'hatchway[s] leading from the surface to the depths'*: *TAP*, p. 98.

135 *'guard the threshold'*: *TAP*, p. 214.

135 *'protect and mark the transitions'*: *TAP*, p. 88.

136 *'lightning-scored, whistle-resounding darkness . . . entered and traversed'*: *TAP*, pp. 84–5.

139 *'Paris has another Paris under herself . . . its arteries and its circulation'*: Victor Hugo, *The Essential Victor Hugo*, trans. E. H. and A. M. Blackmore (1862; Oxford: Oxford University Press, 2004), p. 395.

140 *So started one of the most remarkable episodes of Paris's history*: the years of the disinterral of Paris's cemeteries are vividly discussed in Graham Robb, *Parisians: An Adventure History of Paris* (London: Picador, 2010); and Andrew Hussey, *Paris: The Secret History* (London: Penguin, 2007), among other sources.

142 *'Temporary Autonomous Zone'*: Hakim Bey, *T.A.Z.: The Temporary Autonomous Zone, Ontological Anarchy, Poetic Terrorism* (Brooklyn: Autonomedia, 2003).

143 *An unofficial 'university' of the catacombs was established*: see, for a fascinatingly detailed account of one aspect of the multiple encryptions of recent cataphile culture, Sean Michaels, 'Unlocking the Mystery of Paris' Most Secret Underground Society', Gizmodo, 21 April 2011 <https://gizmodo.com/5794199/unlocking-the-mystery-of-paris-most-secret-underground-society-combined>.

147 I found a Hollow place . . . It was quite soft: see Samuel Taylor Coleridge, *The Notebooks of Samuel Taylor Coleridge*, ed. Kathleen Coburn, vol. 1 (London: Routledge and Kegan Paul, 1957), entry 949.

148 *'an identical copy of their city . . . who is alive and who is dead'*: Italo Calvino, *Invisible Cities*, trans. William Weaver (1972; London: Vintage, 1997), pp. 98–9.

149 *'there is a layer of urban stratigraphy . . . unearthed below ground'*: Wayne Chambliss, personal communication, May 2018.

149 *'infrastructure that supports urban life . . . above the surface of the earth'*: Pierre Bélanger, 'Altitudes of Urbanisation', *Tunnelling and Underground Space Technology* 55 (January 2016), 5–7: 5.

149 *'Complex subterranean spaces . . . above and below ground'*: Graham, *Vertical*, p. 5.

151 *'The cold in these underground corridors . . . exchanged addresses'*: *TAP*, p. 89.

152 The city of the dead antedates . . . every living city: Lewis Mumford, *The City in History: Its Origins, Its Transformations, and Its Prospects* (New York: Harcourt & Brace, 1961), p. 7.

154 *'most photographed barn in America'*: Don DeLillo, *White Noise* (London: Penguin, 1986), p. 128.

155 *'feeding the rat'*: Al Alvarez, *Feeding the Rat: A Climber's Life on the Edge* (London: Bloomsbury, 2013).

155 *'recod[es] people's normalised relationships to city space'*: Bradley Garrett, *Explore Everything: Place-Hacking the City* (London: Verso, 2014), p. 6.

156 *I was especially struck by the manic systematicity of much explorer practice*: see, for more on the connection-delirium of contemporary infrastructure-mappers, Shannon Mattern's dazzling essay 'Cloud and Field', *Places Journal* (August 2016) <https://placesjournal.org/article/cloud-and-field>.

156–7 *'London deserted . . . what a place to explore!'*: Edward Thomas, 'Chalk Pits', in *Selected Poems and Prose* (1981; London: Penguin, 2012), pp. 77–8.

170–71 *I wonder at what will remain of our cities . . . trace impressions of its presence*: I draw here on, among other sources, Jan Zalasiewicz's work on cities and the rock record, including an interview with him by Andrew Luck-Baker for 'Leaving our Mark: What Will Be Left of Our Cities', 1 November 2012 <https://www.bbc.co.uk/news/science-environment-20154030>.

Chapter 6: Starless Rivers

Pages

177 *Starless rivers run through classical culture, and they are the rivers of the dead*: see for a detailed examination of geology and mythology in this context, Julie Baleriaux, 'Diving Underground: Giving Meaning to Subterranean Rivers', in *Valuing*

Landscape in Classical Antiquity, ed. Jeremy McInerney and Ineke Sluiter (Leiden: Brill, 2016), pp. 103–21; and Salomon Kroonenberg, *Why Hell Stinks of Sulfur: Mythology and Geology of the Underworld* (London: Reaktion, 2013).

178 'Flectere si nequeo superos . . . *the River of Hell*': Virgil, *The Aeneid*, trans. Peter Davidson (personal communication).

178 '*vanishing lakes*': see Johann von Valvasor, 'An Extract of a Letter Written to the Royal Society out of Carniola, by Mr John Weichard Valvasor, R. Soc. S. Being a Full and Accurate Description of the Wonderful Lake of Zirknitz in that Country', in *Philosophical Transactions, Giving Some Accompt of the Present Undertakings, Studies, and Labours, of the Ingenious in Many Considerable Parts of the World*, ed. Henry Oldenburg and Francis Roper, vol. 16 (London: Printed for T.N. by John Martyn, 1687). I draw in this chapter also on the defining work of Trevor Shaw, *Foreign Travellers in the Slovene Karst: 1486–1900* (Ljubljana, Založba ZRC, 2008); and Trevor Shaw and Alenka Čuk, *Slovene Caves & Karst Pictured 1545–1914* (Ljubljana: Založba ZRC, 2012).

181 '*limitless tempest*': Rainer Maria Rilke, letter to Lou Andreas-Salomé, 11 February 1922, in *Rainer Maria Rilke, Lou Andreas-Salome: Briefwechsel* (Zurich: M. Niehans, 1952), p. 464 (translation mine).

181 '*Ancient tangled deeps . . . never to be sought*': Rainer Maria Rilke, 'Sonnet 17', in *Sonnets to Orpheus*, trans. Martyn Crucefix (London: Enitharmon Press, 2012), p. 47.

184 '*We are the bees of the invisible . . . the great golden hive of the invisible*': Rainer Maria Rilke, '106. To Witold von Hulewicz, Postmark: Sierre, 13.11.25', in Rilke, *Selected Letters 1902–1926*, trans. R. F. C. Hull (London: Quartet Encounters, 1988), p. 394.

186 '*The Timavo River flows from the mountains . . . springs beside the sea*': Posidonius, *Posidonius*, ed. Ludwig Edelstein and I. G. Kidd, trans. I. G. Kidd (Cambridge: Cambridge University Press, 1988), p. 46.

186–7 *Systematic exploration of the river's hidden extent . . . dive into the ink*: I draw here in part on an excellent series of four articles in Italian tracing the course and history of the Reka/Timavo by Pietro Spirito that appeared in *Il Piccolo* between 2 and 23 August 2014, gathered under the title 'Alla scoperta del Timavo'.

187 '*The Timavo is a dream . . . metre by metre*': Marco Restiano, quoted in Pietro Spirito, 'Nei cantieri sottoterra da anni si dà la caccia al fiume che non c'è', *Il Piccolo*, 23 August 2014 (translation mine).

192 '*When you're in the cave . . . unknown land that people didn't know existed*': Hazel Barton, 'This Woman is Exploring Deep Caves to Find Ancient Antibiotic

Resistance', interview with Shayla Love, *Vice*, 20 April 2018 <https://www.vice.com/en_id/article/j5an54/hazel-barton-is-exploring-deep-caves-to-find-ancient-antibiotic-resistance-v25n1>.

193 *'A peak can exercise the same irresistible power of attraction as an abyss'*: Théophile Gautier, trans. Claire Elaine Engel, originally in *Les Vacances du Lundi* (1869; Paris: G. Charpentier et E. Fasquelle, 1907), p. 13.

193 *'a passion for depth . . . man had been before'*: Lovelock, *Life and Death Underground*, p. 66.

194 'Au revoir, papa': Jacques Attout, *Men of Pierre Saint-Martin* (London: Werner Laurie, 1956), p. 96.

194 *'The show has hardly begun'*: Attout, *Men of Pierre Saint-Martin*, p. 102.

195 *'Never again shall I celebrate . . . vast and luminous'*: Attout, *Men of Pierre Saint-Martin*, pp. 38–9.

196 *'Because it's there'*: George Mallory, quoted in 'Climbing Mount Everest is Work for Supermen', *New York Times*, 18 March 1923.

196 *In* The Darkness Beckons, *Martyn Farr tells the story of . . . but also its destruction*: Martyn Farr, *The Darkness Beckons* (1980; Sheffield: Vertebrate Press, 2017); see also 'Dead Man's Handshake: The Linking of Kingsdale Master Cave and Keld Head, 1975–9', in Chris Bonington, *Quest for Adventure* (London: Hodder and Stoughton, 1990).

198–9 *For years I could only understand . . . Budapest's underwater maze*: I draw in this discussion of cave diving on Farr, *The Darkness Beckons*; and Antti Apunen, *Divers of the Dark: Exploring Budapest's Underground Caves*, trans. Marju Galitsos (Helsinki: Tammi, 2015).

198 *'I have had such beautiful moments . . . just total serenity'*: Don Shirley, quoted in Sebastian Berger, 'Ghosts of the Abyss: The Story of Don Shirley and Dave Shaw', *Telegraph*, 6 March 2008.

199 *'I have perceived non-existence . . . the oceanic secret'*: Natalia Molchanova, 'The Depth', trans. Victor Hilkevich <http://molchanova.ru/en/verse/depth>.

207 *'Conquistadors of the useless'*: Lionel Terray, *Conquistadors of the Useless: From the Alps to Annapurna*, trans. Geoffrey Sutton (1963; Sheffield: Bâton Wicks, 2000).

Chapter 7: Hollow Land

Pages

222–6 *Between 1941 and 1945 the limestone of southern central Europe . . . continues to wound the present*: I draw in these pages chiefly on Pamela Ballinger's outstanding

History in Exile: Memory and Identity at the Borders of the Balkans (Princeton: Princeton University Press, 2002); also John Earle, *The Price of Patriotism* (London: Book Guild, 2005); Pavel Stranj, *The Submerged Community*, trans. Mark Brady (Trieste: Editoriale Stampa, 1992); Jan Morris's wonderful *Trieste and the Meaning of Nowhere* (London: Faber and Faber, 2001); also Maja Haderlap, *Angel of Oblivion*, trans. Tess Lewis (New York: Archipelago, 2016); and the generously shared knowledge of Lucian Comoy, John Stubbs and Stephen Watts, among others.

225 '*the terrain of memory*': Ballinger, *History in Exile*, p. 15.

225 '*autochthonous . . . rights*': Ballinger, *History in Exile*, p. 252.

226 'lieux de mémoire': Pierre Nora and Charles-Robert Ageron, *Les Lieux de Mémoire*, 3 vols. (Paris: Éditions Gallimard, 1993).

229 The shadow past . . . rain through karst: Anne Michaels, *Fugitive Pieces* (London: Bloomsbury, 1997), p. 17.

229 I think there is no innocent landscape, that doesn't exist: Anselm Kiefer, in interview with Jim Cuno, 'Interviewing Anselm Kiefer', 13 December 2017 <http://blogs.getty.edu/iris/audio-interviewing-anselm-kiefer/>.

229 *Kiefer longs for . . . the earth's own stigmata*: I draw here on conversations about Kiefer, place-guilt and absolution with Kryštof Vosatka. The discussion of 'occulting landscapes' in this chapter was also developed in response to 'Project Cleansweep', photographer Dara McGrath's documentation of the sites of displaced violence in Britain, and in conversation with Rob Newton.

229 '*paralysing horror . . . evident even in that remote place*': W. G. Sebald, *The Rings of Saturn*, trans. Michael Hulse (1995; London: Vintage, 2002), p. 3.

230 The violent event persists . . . is blinding: E. Valentine Daniel, 'Crushed Glass, or, Is There a Counterpoint to Culture?', in *Culture/Contexture: Explorations in Anthropology and Literary Studies*, ed. E. Valentine Daniel and Jeffrey M. Peck (Berkeley: University of California Press, 1996), p. 370.

235 '*A mountain has an inside*': Nan Shepherd, *The Living Mountain* (1977; Edinburgh: Canongate, 2011), p. 16.

238 *These Alps became weaponized peaks . . . the caves of the slopes and valleys*: I draw here and elsewhere on Mark Thompson, *The White War: Life and Death on the Italian Front* (New York: Basic Books, 2009); and John Schindler, *Isonzo* (London: Praeger, 2001).

238 '*elastic geography . . . seeks to challenge, transform or appropriate*': Eyal Weizman, *Hollow Land: Israel's Architecture of Occupation* (London: Verso, 2007), pp. 6–7.

239 '*a complex architectural construction . . . attempts to partition it*': Weizman, *Hollow Land*, p. 15.

239 '*laboratory of the extreme*': Weizman, *Hollow Land*, p. 9.

241 Find beauty, be still: W. H. Murray, *Mountaineering in Scotland and Undiscovered Scotland* (London: Diadem Books, 1979), p. 4.

Third Chamber

Pages

246 You will find on the right . . . Do not even draw nigh this spring: R. Janko, 'Forgetfulness in the Golden Tablets of Memory', *Classical Quarterly* 34:1 (1984), 89–100: 96. More on the *Totenpässe* can be found in Fritz Graf and Sarah Iles Johnston's *Ritual Texts for the Afterlife: Orpheus and the Bacchic Gold Tablets* (London: Routledge, 2007).

246 '*coruscations*': J. M. Peebles, *The Practical of Spiritualism. Biographical Sketch of Abraham James. Historic Description of his Oil-Well Discoveries in Pleasantville, P.A., through Spirit Direction* (Chicago: Horton and Leonard Printers, 1868), p. 77.

247 *Early this millennium, on the sweltering north coast of Java . . . ancient poisonous sludge*: see for more details on the geology and interpretations of the 'mud volcano', Nils Bubandt, 'Haunted Geologies: Spirits, Stones, and the Necropolitics of the Anthropocene', in *ALDP*, G121–G142.

248 but they are no longer . . . in the same order: Kate Brown, 'Marie Curie's Fingerprint: Nuclear Spelunking in the Chernobyl Zone', in *ALDP*, G33–G50: G34. I am grateful to Kate Brown for allowing me to draw on her remarkable research for this scene.

Chapter 8: Red Dancers

Pages

264 '*rite of passage . . . mental ordeals*': Hein Bjerck, 'On the Outer Fringe of the Human World: Phenomenological Perspectives on Anthropomorphic Cave Paintings in Norway', in *Caves in Context: The Cultural Significance of Caves and Rockshelters in Europe*, ed. Knut Andreas Bergsvik and Robin Skeates (Oxford: Oxbow Books, 2012), p. 60. See also Anders Hesjedal, 'The Hunters' Rock Art in Northern Norway: Problems of Chronology and Interpretation', *Norwegian Archaeological Review* 27:1 (1994), 1–28.

264 '*ritual actions . . . the outer fringe of the human world*': Bjerck, 'On the Outer Fringe', p. 55.

265 *'land meets the sea . . . come closest together'*: *ANP*, pp. 13 and 29.

265–6 *'hel-shoes . . . the path from the grave to the world beyond'*: *ANP*, p. 145.

266 *Terje Norsted and Bjerck both propose*: see Terje Norsted, 'The Cave Paintings of Norway', *Adoranten* (2013), pp. 5–24.

270 *'thin places'*: the phrase is attributed to George MacLeod, founder of the Iona Community.

273 Time isn't deep . . . more as drift: Þóra Pétursdóttir, in conversation with me, Oslo, April 2017.

276 *'a shooting star'*: Bjerck, 'On the Outer Fringe', p. 49.

276 *'cavescape'*: Bjerck, 'On the Outer Fringe', p. 58.

279 *'Art is born like a foal that can walk straight away . . . they arrive together'*: John Berger, 'Past Present', *Guardian*, 12 October 2002.

280 *'flashed onto a mammoth . . . And a frieze of other animals thirty feet long'*: Jean-Marie Chauvet, quoted by John Berger and Simon McBurney in *The Vertical Line: Can You Hear Me, in the Darkness?*, Artangel Arts (Strand Tube Station, 1999). <https://www.artangel.org.uk/the-vertical-line/can-you-hear-me-in-darkness/>.

281 *'in an enormous present . . . everything that surrounds us'*: Simon McBurney, 'Herzog's Cave of Forgotten Dreams: The Real Art Underground', *Guardian*, 17 March 2011.

281–2 *'It was as if time had been abolished . . . the painters were here too'*: Jean-Marie Chauvet, quoted by Jean Clottes in *World Rock Art* (Michigan: Getty Conservation Institute, 2002), p. 44; these lines also appear in *Cave of Forgotten Dreams* (2010), dir. Werner Herzog.

283 *'became known just as everything visible . . . potential of the universe to be otherwise'*: Kathryn Yusoff, 'Geologic Subjects: Nonhuman Origins, Geomorphic Aesthetics, and the Art of Becoming *In*human', *cultural geographies* 22:3 (2015), 383–407: 391.

283 *'I am simply struck . . . the notion of our death appears to us'*: Georges Bataille, *The Cradle of Humanity: Prehistoric Art and Culture*, ed. and trans. Stuart Kendall and Michelle Kendall (New York: Zone Books, 2005), p. 85. Quoted by Yusoff in 'Geologic Subjects', 392.

Chapter 9: The Edge

Pages

297 *a battle for the soul of Norway*: see Richard Milne, 'Oil and the Battle for Norway's Soul', *Financial Times*, 27 July 2017; and also *Atlantic* (2016), dir. Risteard O'Domhnaill and featuring Bjørnar Nicolaisen.

297 *'natural resources should be managed . . . safeguarded for future generations'*: the Constitution of Norway, as laid down on 17 May 1814 by the Constituent Assembly at Eidsvoll and subsequently amended, most recently in May 2018 <https://www.stortinget.no/globalassets/pdf/english/constitution english.pdf>.

306 *'a sheer unobstructed precipice of black shining rock'*: Edgar Allan Poe, 'A Descent into the Maelstrom', in *The Fall of the House of Usher and Other Writings*, ed. David Galloway (1841; London: Penguin, 2003), p. 177.

306–7 *'wilderness of surge . . . the abyss of the whirl'*: Poe, 'A Descent into the Maelstrom', pp. 178–82.

308 *'I became possessed . . . ghastly radiance they shot forth'*: Poe, 'A Descent into the Maelstrom', pp. 188–9.

308–9 *In 1818 an American army officer . . . potential for resources and habitation*: see Duane A. Griffin, 'Hollow and Habitable within: Symmes' Theory of Earth's Internal Structure and Polar Geography', *Physical Geography* 25:5 (2004), 382–97.

310 *'oceans of oil'*: Jamie L. Jones, 'Oil: Viscous Time in the Anthropocene', *Los Angeles Review of Books*, 22 March 2016 <https://lareviewofbooks.org/article/oil-viscous-time-in-the-anthropocene>.

310 *'We need new acreage . . . step up our exploration activities'*: Mayliss Hauknes, Statoil spokesperson, quoted in 'Statoil Seeking New Acreage', Rigzone, 1 October 2016 <https://www.rigzone.com/news/oil_gas/a/16859/statoil_seeking_new_acreage/>.

310 *'underexplored Cretaceous basins'*: 'Ceduna Sub-Basin', Karoon Gas Australia Ltd <http://www.karoongas.com.au/projects/ceduna-sub-basin>.

311 *'destructive currents of the kind found in the Maelstrom'*: Bjørn Gjevig, quoted in Malcolm W. Browne, 'Deadly Maelstrom's Secrets Unveiled', *New York Times*, 2 September 1997.

312 *We have now drilled some 30 million miles . . . hunt for resources*: see Reza Negarastani's extraordinary theory-fiction, *Cyclonopaedia: Complicity with Anonymous Materials* (Melbourne: re.press, 2008).

317 *'solastalgia . . . existential distress caused by environmental change'*: Glenn Albrecht, 'Solastalgia, a New Concept in Human Health and Identity', *Philosophy Activism Nature* 3 (2005), 41–4: 43.

317 *'Worldwide, there is an increase in ecosystem distress syndromes . . . human distress syndromes'*: Glenn Albrecht et al., 'Solastalgia: The Distress Caused by Environmental Change', *Australian Psychiatry* 15:1 (2007), 95–7: 95.

319 *'monstrous transformer'*: Graeme Macdonald, '"Monstrous Transformer": Petro-fiction and World Literature', *Journal of Postcolonial Writing* 53 (2017), 289–302.

320 *photographs I have seen recently of hermit crabs . . . Avon night cream*: see also D. K. A. Barnes, 'Remote Islands Reveal Rapid Rise of Southern Hemisphere Sea Debris', *Scientific World Journal* 5 (2005), 915–21.

320 *'empire of things'*: Frank Trentmann, *Empire of Things: How We Became a World of Consumers, from the Fifteenth Century to the Twenty-First* (New York: HarperCollins, 2016).

320 *'a swelling topography of scrapped modernity . . . confronting us with its pestering presence'*: Þóra Pétursdóttir and Bjørnar Olsen, 'Unruly Heritage: An Archaeology of the Anthropocene' (Tromsø: UiT The Arctic University of Norway, 2017), p. 2 <https://www.sv.uio.no/sai/forskning/grupper/Temporalitet%20-%20materialitet/lesegruppe/olsen-unruly-heritage.pdf>.

320 *'What we excrete comes back to consume us'*: Don DeLillo, *Underworld* (New York: Scribner, 1997), p. 791.

320 *'hyperobjects'*: see Timothy Morton, *Hyperobjects: Philosophy and Ecology after the End of the World* (Minneapolis: University of Minnesota Press, 2013).

320 *'viscous'*: Morton, *Hyperobjects*, p. 27.

320 *'plastiglomerate'*: see Patricia L. Corcoran et al., 'An Anthropogenic Marker Horizon in the Future Rock Record', *GSA Today* 24.6 (June 2014), 4–8.

321 *'New People'*: John Wyndham, *The Chrysalids* (1955; London: Penguin, 2018), p. 158.

Chapter 10: The Blue of Time

Pages

329 *On the Yamal peninsula, between the Kara Sea . . . frozen bodies of mammoths*: see Noah Sneider's fine essay 'Cursed Fields', *Harper's Magazine* (April 2018), 40–51.

329 *On the Siachen glacier in the Karakoram . . . slaughtered human bodies*: see Rob Nixon, quoting Arundhati Roy, in 'The Swiftness of Glaciers: Language in a Time of Climate Change', *Aeon Magazine*, 19 March 2018 <https://aeon.co/ideas/the-swiftness-of-glaciers-language-in-a-time-of-climate-change>.

330 *'preserved for eternity'*: L. K. Clark et al., 'Sanitary Waste Disposal for Navy Camps in Polar Regions', *Journal of the Water Pollution Control Federation* 34:12 (1962), 1229.

331 *In that region, at this time of history . . . through the world's surface:* see for more on climate change and 'untimeliness', Cymene Howe, '"Timely": Theorizing the Contemporary', 21 January 2016 <https://culanth.org/fieldsights/800-timely>.

335 *Ice has a social life*: see Cymene Howe's ongoing project *Melt: The Social Life of Ice at the Top of the World*, which examines cryo-human interrelations and the implications of climate-induced geohydrological change in the Arctic and beyond.

335 '*The loss of that landscape of ice . . . also a cultural one*': Andrew Solomon, *Far and Away: How Travel Can Change the World* (London: Scribner, 2016), p. 259.

335 uggianaqtuq: see S. Gearheard, 'When the Weather is Uggianaqtuq: Inuit Observations of Environmental Changes, Version 1' (Boulder, Colorado: NSIDC – National Snow and Ice Data Center, 2004) <http://nsidc.org/data/NSIDC-0650>.

339–40 *The weight on 2,000-year-old ice . . . sequence can be almost impossible to discern*: I draw in this discussion on, among other sources, Richard B. Alley, *The Two-Mile Time Machine* (Princeton: Princeton University Press, 2000), pp. 41–58.

340 '*greyish ghostly bands . . . focused beam of a fibre-optic lamp*': Alley, *The Two-Mile Time Machine*, p. 50.

346 Sound is a blow delivered by air . . . transmitted to the soul: Plato, *Timaeus and Critias*, trans. Robin Waterfield (Oxford: Oxford University Press, 2008), p. 65.

355 Corridors of breath: Barry Lopez, *Arctic Dreams: Imagination and Desire in a Northern Landscape* (1986; New York: Bantam, 1987), p. 152.

363 *Sick at Greenland's scale . . . our ability to encompass it*: Elizabeth Kolbert experienced the identical response of nausea when reporting from west Greenland in the same weeks that I was in the east of the country. 'Again, I was hit, and vaguely sickened, by Greenland's inhuman scale,' she writes in her fine essay 'Greenland is Melting', *New Yorker*, 24 October 2016 <https://www.newyorker.com/magazine/2016/10/24/greenland-is-melting>.

363 '*deaden[ed] . . . gangplank of a cattle truck*': Seamus Heaney, 'Mycenae Lookout', in *The Spirit Level* (London: Faber and Faber, 1996), p. 29.

364 '*thick speech*': Sianne Ngai, *Ugly Feelings* (Cambridge, MA: Harvard University Press, 2005), p. 252.

364 '*interpret or respond*': Ngai, *Ugly Feelings*, p. 250.

364 '*back-flowing*': Ngai, *Ugly Feelings*, p. 249.

Chapter 11: Meltwater

Pages

380 '*matter out of place*': Mary Douglas, *Purity and Danger: An Analysis of Concepts of Purity and Taboo* (1966; London: Routledge, 2002), p. 44.

380 *'animate (endowed with life) . . . landscapes they inhabit'*: Julie Cruikshank, *Do Glaciers Listen? Local Knowledge, Colonial Encounters, and Social Imagination* (Vancouver: University of British Columbia Press, 2005), p. 3.

380 *'grammar of animacy'*: Kimmerer, 'Speaking of Nature'.

381 *'denseness . . . that strangeness of the world is the absurd'*: Albert Camus, 'Absurd Walls', in *The Myth of Sisyphus*, trans. Justin O'Brien (London: Hamish Hamilton, 1973), p. 19.

392 a busy working of nature . . . reckoning of days and years: Gerard Manley Hopkins, 'Sept. 24 1870', in *The Journals and Papers of Gerard Manley Hopkins*, ed. Humphry House and Graham Storey (Oxford: Oxford University Press, 1959), p. 200.

Chapter 12: The Hiding Place

Pages

398 *Deep in the bedrock of Olkiluoto Island . . .*: I have been writing about 'deep time' since my first book, *Mountains of the Mind* (London: Granta, 2003). In respect of radiological as well as geological time, I draw in this chapter and elsewhere on, among other sources, John McPhee, *Annals of the Former World* (New York: FSG, 1998); Stephen Jay Gould, *Time's Arrow, Time's Cycle* (Cambridge, MA: Harvard University Press, 1987); Andy Weir, 'Deep Decay: Into Diachronic Polychromatic Material Fictions', PARSE 4 (2017) <http://parsejournal.com/article/deep-decay-into-diachronic-polychromatic-material-fictions/>; Vincent Ialenti, 'Adjudicating Deep Time: Revisiting the United States' High-Level Nuclear Waste Repository Project at Yucca Mountain', *Science & Technology Studies* 27:2 (2014), 27–48, and 'Death and Succession among Finland's Nuclear Waste Experts', *Physics Today* 70:10 (2017), 48–53. After travelling to Onkalo and completing a first draft of this chapter, I watched Michael Madsen's documentary *Into Eternity* (2010), which also examines the WIPP site-marking plans, and – in a brilliant final scene – visually collapses the 2011 excavations at Onkalo with an imagined far-future disinterral of the chambers.

401 *'the radiological equivalent of . . . seven trillion doses of lethal radiation'*: John D'Agata, *About a Mountain* (New York: W. W. Norton, 2011), p. 35.

405 *'the numerous strata of a burial mound . . . artefacts have been buried'*: Matti Kuusi quoted in Keith Bosley, 'Introduction', *TK*, p. xxi. Bosley's introduction and translation are both excellent, and I draw especially on the introduction in this paragraph contextualizing the *Kalevala*.

406 *'grave . . . demon lair'*: *TK*, p. 202.

406 *'grievous pain'*: *TK*, p. 206.

406 *'my guiltless heart . . . to bite, to devour'*: *TK*, p. 205.

406 *'a wind-borne disease . . . carried by chill air'*: *TK*, p. 208.

407 *'Words shall not be hid . . . though the mighty go'*: *TK*, p. 213.

407 *'copper slope . . . rocky hill'*: *TK*, p. 548.

408 The earth is our tabernacle, a receptacle for all decompositions . . .: Michael Serres, *Statues: The Second Book of Foundations*, trans. Randolph Burks (London: Bloomsbury, 2015), p. 17.

410 *The Greek word for 'sign'*, sema, *is also the word for 'grave'*: see Harrison, *The Dominion of the Dead*, p. 20.

410 *'marker system . . . during the next 10,000 years'*: Kathleen M. Trauth et al., 'Expert Judgment on Markers to Deter Inadvertent Intrusion into the Waste Isolation Pilot Plant', *Sandia National Laboratories*, SAND92–1382. UC–721 (1993) <https://prod.sandia.gov/techlib-noauth/access-control.cgi/1992/921382.pdf>, pp. 1–8.

410 *'Human Interference Task Force'*: Thomas Sebeok, 'Communication Measures to Bridge Ten Millennia (Technical Report)', *Research Centre for Language and Semiotic Studies, for Office of Nuclear Waste Isolation*, BMI/ONWI-532 (1984), p. iii.

411 *'passive institutional controls'*: Trauth et al., 'Expert Judgment on Markers', pp. 1–12.

411 *'Landscape of Thorns'*: Trauth et al., 'Expert Judgment on Markers', pp. F-61–F-62.

411 *'danger to the body'*: Trauth et al., 'Expert Judgment on Markers', p. F-42.

411 *'Black Hole'*: Trauth et al., 'Expert Judgment on Markers', pp. F-70–F-71.

411 *'Forbidding Blocks'*: Trauth et al., 'Expert Judgment on Markers', pp. F-74–F-75.

412 *'active communication system'*: D'Agata, *About a Mountain*, p. 93.

412 *'atomic priesthood'*: Sebeok, 'Communication Measures to Bridge Ten Millennia', p. 24.

412 *'laying a trail of myths . . . keep people away'*: D'Agata, *About a Mountain*, p. 93.

412 *'our society's largest conscious attempt . . . the abyss of deep time'*: Gregory Benford, *Deep Time: How Humanity Communicates across Millennia* (New York: Avon Books, 1999), p. 85.

413 *The map will be slightly domed*: see for details and diagram, Trauth et al., 'Expert Judgment on Markers', p. F-76.

413 *'a map of the Empire . . . inhabited by Animals and Beggars'*: Jorge Luis Borges, 'On Exactitude in Science', in Borges, *Jorge Luis Borges: Collected Fictions*, trans. Andrew Hurley (London: Penguin, 1998), p. 325.

413 ‘*Hot Cell*’: Trauth et al., ‘Expert Judgment on Markers’, pp. 3–7.

414 ‘We are going to tell you what lies underground . . . keep the room intact and buried’: see Trauth et al., ‘Expert Judgment on Markers’, Appendix F.

419 ‘*People are best able to change . . . in building our next home*’: Jedediah Purdy, *After Nature: A Politics for the Anthropocene* (Cambridge, MA: Harvard University Press, 2015), p. 288.

SELECT BIBLIOGRAPHY

'Place is always moving, like a sleeping cat,' observes Toshiya Tsunoda, beautifully. Sometimes you have to stay still to see its subtle shifts, the dream-shudders of its skin. Much of the research and thinking for *Underland* happened not underground but in libraries and through books. This bibliography details some of the many texts I consulted over the years, and which helped me try to find both language and form for subjects that – by definition of their involvement with the underland – often resisted easy containment or expression. I have asterisked those texts that were especially interesting or influential to me, or to which I am particularly indebted for information. Asserted facts, suggested details and thought-splinters on the part of *Underland*'s narrator may be tested with reference to the works cited here and in the notes. I am deeply grateful to the many explorers, artists, writers and scholars who have descended into the darkness before me.

~

Adam, Helen, *A Helen Adam Reader*, ed. Kristin Prevallet (Orono, Maine: National Poetry Foundation, 2007)

Adorno, Theodor, and Horkheimer, Max, *Dialectic of Enlightenment*, trans. John Cumming (1944; London: Verso, 1997)

Albrecht, Glenn, 'Solastalgia, a New Concept in Human Health and Identity', *Philosophy Activism Nature* 3 (2005)

——, 'Exiting the Anthropocene and Entering the Symbiocene', *PSYCHOTERRATICA*, 17 December 2015 <https://glennaalbrecht.com/2015/12/17/exiting-the-anthropocene-and-entering-the-symbiocene/>

*——, et al., 'Solastalgia: The Distress Caused by Environmental Change', *Australian Psychiatry* 15:1 (2007)

aliciaescott, 'Field Study #007, The Extinction Event', *Bureau of Linguistical Reality*, 1 September 2015 <https://bureauoflinguisticalreality.com/2015/09/01/field-study-007-the-extinction-event/>

*Alley, Richard B., *The Two-Mile Time Machine* (Princeton: Princeton University Press, 2000)

Altman, Rebecca, 'On What We Bury', *ISLE* 21:1 (Winter 2014)

Alvarez, Al, *Feeding the Rat: A Climber's Life on the Edge* (London: Bloomsbury, 2013)

Anon., 'Russia's Melting Ice Could Release More Threats to Humanity', *National*, 11 August 2016 <https://www.thenational.ae/world/russia-s-melting-ice-could-release-more-threats-to-humanity-1.159511>

Anthropocene Working Group of the Subcommission on Quaternary Stratigraphy, 'When Did the Anthropocene Begin? A Mid-Twentieth-Century Limit is Stratigraphically Optimal', *Quaternary International* 383 (2015)

Apter, Emily, 'Planetary Dysphoria', *Third Text* 27:1 (2017)

*Apunen, Antti, *Divers of the Dark: Exploring Budapest's Underground Caves*, trans. Marju Galitsos (Helsinki: Tammi, 2015)

Art Map, 'Beneath the Ground: From Kafka to Kippenberger' <https://artmap.com/k20/exhibition/beneath-the-ground-from-kafka-to-kippenberger-2014>

Attout, Jacques, *Men of Pierre Saint-Martin* (London: Werner Laurie, 1956)

~

*Ballinger, Pamela, *History in Exile: Memory and Identity at the Borders of the Balkans* (Princeton: Princeton University Press, 2002)

Barnes, D. K. A., 'Remote Islands Reveal Rapid Rise of Southern Hemisphere Sea Debris', *Scientific World Journal* 5 (2005)

Barton, Hazel, 'This Woman is Exploring Deep Caves to Find Ancient Antibiotic Resistance', interview with Shayla Love, *Vice*, 20 April 2018 <https://www.vice.com/en_id/article/j5an54/hazel-barton-is-exploring-deep-caves-to-find-ancient-antibiotic-resistance-v25n1>

Bataille, Georges, *The Cradle of Humanity: Prehistoric Art and Culture*, ed. and trans. John S. Kendall and Leslie M. Kendall (New York: Zone Books, 2005)

Battson, Ginny, 'Mycelium of the Forest Floor. And Love', 12 October 2015 <https://seasonalight.wordpress.com/2015/10/12/mycelium-of-the-forest-floor-and-love/>

Bede, *The Reckoning of Time*, trans. Faith Wallis (725; Liverpool: Liverpool University Press, 1999)

Bélanger, Pierre, 'Altitudes of Urbanisation', *Tunnelling and Underground Space Technology* 55 (2016)

*Benford, Gregory, *Deep Time: How Humanity Communicates across Millennia* (New York: Avon Books, 1999)

*Benjamin, Walter, *The Arcades Project*, trans. Howard Eiland and Kevin McLaughlin (London: Harvard University Press, 1999)

Bennett, Jane, *The Enchantment of Modern Life: Attachments, Crossings, and Ethics* (Princeton: Princeton University Press, 2011)

Berger, John, 'Past Present', *Guardian*, 12 October 2002

—— and McBurney, Simon, *The Vertical Line: Can You Hear Me, in the Darkness?*, Artangel Arts (Strand Tube Station, 1999) <https://www.artangel.org.uk/the-vertical-line/can-you-hear-me-in-darkness/>

Berger, Sebastian, 'Ghosts of the Abyss: The Story of Don Shirley and Dave Shaw', *Telegraph*, 6 March 2008

Bergsvik, Knut Andreas, and Skeates, Robin, *Caves in Context: The Cultural Significance of Caves and Rockshelters in Europe* (Oxford: Oxbow Books, 2012)

Bernstein, J. M., 'Re-Enchanting Nature', *Journal of the British Society for Phenomenology* 31:3 (2000)

Bey, Hakim, *T.A.Z.: The Temporary Autonomous Zone, Ontological Anarchy, Poetic Terrorism* (Brooklyn: Autonomedia, 2003)

Black, J. A., Cunningham, G., Fluckiger-Hawker, E., Robson, E., and Zólyomi, G., *The Electronic Text Corpus of Sumerian Literature* (Oxford: 1998–) <http://etcsl.orinst.ox.ac.uk/section1/tr1814.htm>

Blum, Hester, 'Speaking Substance: Ice', *Los Angeles Review of Books*, 21 March 2016 <https://lareviewofbooks.org/article/speaking-substances-ice/>

Bögli, Alfred, and Franke, Herbert W., *Luminous Darkness: The Wonderful World of Caves* (Chicago: Rand McNally, 1966)

Bonington, Chris, *Quest for Adventure* (London: Hodder and Stoughton, 1990)

Bonnefoy, Yves, *The Arrière-Pays*, trans. Stephen Romer (London: Seagull Books, 2012)

Borges, Jorge Luis, *Jorge Luis Borges: Collected Fictions*, trans. Andrew Hurley (London: Penguin, 1998)

Borodale, Sean, *Bee Journal* (London: Cape, 2012)

*——, *Asylum* (London: Cape, 2018)

Boycott, A., and Wilson, L. J., 'Contemporary Accounts of the Discovery of Aveline's Hole, Burrington Combe, North Somerset', *Proceedings of the University of Bristol Spelaeological Society* 25:1 (2010)

*Bradley, Richard, *An Archaeology of Natural Places* (London: Routledge, 2006)

Braje, Todd, et al., 'Evaluating the Anthropocene: Is There Something Useful about a Geological Epoch of Humans?', *Antiquity* 90 (2016)

Brázdil, R., Dobrovolny, P., et al., 'Droughts in the Czech Lands, 1090–2012 AD', *Climate of the Past* 9 (August 2013)

British Pathé, 'Caveman 105 Days Below', *YouTube*, 13 April 2014 <https://www.youtube.com/watch?v=YSdBBv5LY84>

Browne, Malcolm W., 'Deadly Maelstrom's Secrets Unveiled', *New York Times*, 2 September 1997

*Browne, Thomas, *Religio Medici and Urne-Buriall*, ed. Stephen Greenblatt and Ramie Targoff (1658; New York: NYRB Classics, 2012)

Byrne, Denis, *Surface Collection: Archaeological Travels in Southeast Asia* (Plymouth: AltaMira Press, 2007)

~

Cadoux, Jean, et al., *One Thousand Metres Down: A Journey to the Starless River*, trans. R. L. G. Irving (London: Allen and Unwin, 1957)

*Calvino, Italo, *Invisible Cities*, trans. William Weaver (1972; London: Vintage, 1997)

Camus, Albert, *The Myth of Sisyphus*, trans. Justin O'Brien (London: Hamish Hamilton, 1973)

Carroll, Lewis, *Alice's Adventures in Wonderland, and Through the Looking-Glass and What Alice Found There; with ninety-two illustrations by John Tenniel* (1865; London: Macmillan and Co, 1902)

Casselman, Anne, 'Strange but True: The Largest Organism on Earth is a Fungus', *Scientific American*, 4 October 2007 <https://www.scientificamerican.com/article/strange-but-true-largest-organism-is-fungus/>

Casteret, Norbert, *The Descent of Pierre Saint-Martin*, trans. John Warrington (London: Dent, 1955)

'Ceduna Sub-Basin', Karoon Gas Australia Ltd <http://www.karoongas.com.au/projects/ceduna-sub-basin>

Chakrabarthy, Dipesh, 'The Climate of History: Four Theses', *Critical Inquiry* 35:2 (2009)

Cilek, Václav, 'Bees of the Invisible: Awakening of a Place (part 2)', trans. Teresa Stehlikova, *Cinesthetic Feasts*, 5 July 2015 <https://cinestheticfeasts.wordpress.com/2013/07/05/genius-loci-cilek-p-2/>

——, *To Breathe with Birds: A Book of Landscapes*, trans. Evan W. Mellander (Philadelphia: University of Pennsylvania Press, 2015)

Clark, L. K., et al., 'Sanitary Waste Disposal for Navy Camps in Polar Regions', *Journal of Water Pollution Control Federation* 34:12 (1962)

Clark, Timothy, *Ecocriticism on the Edge: The Anthropocene as a Threshold Concept* (London: Bloomsbury, 2015)

'Climbing Mount Everest is Work for Supermen', *New York Times*, 18 March 1923

Clottes, Jean, *World Rock Art* (Michigan: Getty Conservation Institute, 2002)

*Cohen, Jeffrey Jerome, *Stone: An Ecology of the Inhuman* (Minneapolis: University of Minnesota Press, 2015)

Coleridge, Samuel Taylor, *The Notebooks of Samuel Taylor Coleridge*, ed. Kathleen Coburn, vol. 1 (London: Routledge and Kegan Paul, 1957)

Constitution of Norway, as laid down on 17 May 1814 by the Constituent Assembly at Eidsvoll and subsequently amended, most recently in May 2018 <https://www.stortinget.no/globalassets/pdf/english/constitutionenglish.pdf>

Cook, Jill, *Ice Age Art: Arrival of the Modern Mind* (London: The British Museum Press, 2013)

Corcoran, Patricia L., et al., 'An Anthropogenic Marker Horizon in the Future Rock Record', *GSA Today* 24:6 (June 2014)

*Cruikshank, Julie, *Do Glaciers Listen? Local Knowledge, Colonial Encounters, and Social Imagination* (Vancouver: University of British Columbia Press, 2005)

Crutzen, Paul, and Stoermer, Eugene, 'The Anthropocene', *International Geosphere-Biosphere Newsletter* 41 (2000) <https://www.mpic.de/mitarbeiter/auszeichnungen-crutzen/the-anthropocene.html>

Cybernetic Culture Research Unit, *CCRU: Writings 1997–2003* (Falmouth: Time Spiral Press, 2015)

~

Dadachova, E., and Casadevall, A., 'Ionizing Radiation: How Fungi Cope, Adapt, and Exploit with the Help of Melanin', *Current Opinion in Microbiology* 11:6 (2008)

D'Agata, John, *About a Mountain* (New York: W. W. Norton, 2011)

Daniel, E. Valentine, and Peck, Jeffrey M. (eds.), *Culture/Contexture: Explorations in Anthropology and Literary Studies* (Berkeley: University of California Press, 1996)

Davies, Jeremy, *The Birth of the Anthropocene* (Berkeley: University of California Press, 2016)

Dawdy, Shannon Lee, *Patina: A Profane Archaeology* (Chicago: University of Chicago Press, 2016)

De Bernières, Louis, *Captain Corelli's Mandolin* (Reading: Secker and Warburg, 1996)

Debord, Guy, *Theory of the Dérive* (1956; London: Atlantic Books, 1997)

Dee, Tim, 'Naming Names', Caught by the River, 25 June 2014 <https://www.caughtbytheriver.net/2014/06/naming-names-tim-dee-robert-macfarlane/>

Deleuze, Gilles, *The Fold: Leibniz and the Baroque*, trans. Tom Conley (London: Continuum, 2006)

——, and Guattari, Felix, *Nomadology: The War Machine*, trans. Brian Massumi (New York: Semiotext(e), 1986)

DeLillo, Don, *White Noise* (London: Penguin, 1986)

*——, *Underworld* (New York: Scribner, 1997)

Douglas, Mary, *Purity and Danger: An Analysis of Concepts of Purity and Taboo* (1966; London: Routledge, 2002)

Doyle, Arthur Conan, *Tales of Terror and Mystery* (1902; Cornwall: House of Stratus, 2009)

Dufresne, David (dir.), *Fort McMoney* (i-doc) (2013)

~

Earle, John, *The Price of Patriotism* (London: Book Guild, 2005)

Edgeworth, Matt, et al., 'Diachronous Beginnings of the Anthropocene: The Lower Bounding Surface of Anthropogenic Deposits', *Anthropocene Review* 2:1 (2015)

Ehrlich, Gretel, *This Cold Heaven: Seven Seasons in Greenland* (New York: Pantheon Books, 2001)

Ellsworth, Elizabeth, and Kruse, Jamie (eds.), *Making the Geologic Now: Responses to Material Conditions of Contemporary Life* (New York: Punctum, 2013)

Elson, Rebecca, *A Responsibility to Awe* (Manchester: Carcanet, 2001)

Engel, Claire Elaine, *Mountaineering in the Alps: An Historical Survey* (1950; London: George Allen and Unwin, 1971)

~

Falcon-Lang, Howard, 'Anthropocene: Have Humans Created a New Geological Age?', BBC, 11 May 2011 <http://www.bbc.co.uk/news/mobile/science-environment-13335683>

Farr, Martyn, *Darkworld: The Secrets of Llangattock Mountain* (Llandysul: Gomer Press, 1997)

——, *The Darkness Beckons* (1980; Sheffield: Vertebrate Press, 2017)

Farrier, David, '"Like a Stone": Ecology, Enargeia, and Ethical Time in Alice Oswald's Memorial', *Environmental Humanities* 4 (2014)

——, 'Reading Edward Thomas in the Anthropocene', *Green Letters* 18:2 (2014)

Finer, Jem, 'Score for a Hole in the Ground' <http://www.scoreforaholeintheground.org/>

Fittko, Lisa, *Escape through the Pyrenees* (Evanston, IL: Northwestern University Press, 1991)

Franke, Herbert W., *Wilderness under the Earth*, trans. Mervyn Savill (London: Lutterworth Press, 1958)

Freud, Sigmund, *The Interpretation of Dreams*, ed. and trans. James Strachey (1899; London: George Allen and Unwin, 1954)

Frost, Robert, *Mountain Interval* (New York: H. Holt and Company, 1916)

~

Gardam, Jane, *The Hollow Land* (London: Julia MacRae Books, 1990)

Garner, Alan, *The Weirdstone of Brisingamen* (1960; London: HarperCollins Children's Books, 2014)

Garrett, Bradley, *Explore Everything: Place-Hacking the City* (London: Verso, 2014)

——, et al., *Subterranean London: Cracking the Capital* (London: Prestel, 2015)

*——, et al. (eds.), *Global Undergrounds: Exploring Cities Within* (London: Reaktion Books, 2016)

Gautier, Théophile, *Les Vacances du Lundi* (1869; Paris: G. Charpentier et E. Fasquelle, 1907)

Gearheard, S., 'When the Weather is Uggianaqtuq: Inuit Observations of Environmental Changes, Version 1' (Boulder, Colorado: NSIDC – National Snow and Ice Data Center, 2004) <http://nsidc.org/data/NSIDC-0650>

Ghosh, Amitav, 'Petrofiction', *New Republic*, 2 March 1992

Gibbard, P. L., and Walker, M. J. C., 'The Term "Anthropocene" in the Context of Formal Geological Classifications', *Geological Society of London, Special Publications* (2013)

Gould, Stephen Jay, *Time's Arrow, Time's Cycle* (Cambridge, MA: Harvard University Press, 1987)

Graf, Fritz, and Johnston, Sarah Iles, *Ritual Texts for the Afterlife: Orpheus and the Bacchic Gold Tablets* (London: Routledge, 2007)

*Graham, Stephen, *Vertical: The City from Satellites to Bunkers* (London: Verso, 2016)

Griffin, Duane A., 'Hollow and Habitable within: Symmes' Theory of Earth's Internal Structure and Polar Geography', *Physical Geography* 25:5 (2004)

Grossman, Leore, et al., 'A 12,000-Year-Old Shaman Burial from the Southern Levant (Israel)', *PNAS* 105:46 (2008)

Grusin, Richard (ed.), *The Nonhuman Turn* (London: University of Minnesota Press, 2015)

~

Haderlap, Maja, *Angel of Oblivion*, trans. Tess Lewis (New York: Archipelago, 2016)

Haraway, Donna, 'Anthropocene, Capitalocene, Plantationocene, Chthulucene: Making Kin', *Environmental Humanities* 6 (2015)

——, *Staying with the Trouble: Making Kin in the Chthulucene* (Durham, N. C.: Duke University Press, 2016)

Hardy, Thomas, *Under the Greenwood Tree* (1872; London: Penguin, 2012)

Harman, Graham, *Immaterialism* (Cambridge: Polity Press, 2016)

*Harrison, Robert Pogue, *The Dominion of the Dead* (Chicago: University of Chicago Press, 2003)

Hawks, John, et al., 'New Fossil Remains of Homo Naledi from the Lesedi Chamber, South Africa', *eLife* 6 (2017)

Heaney, Seamus, *The Spirit Level* (London: Faber and Faber, 1996)

Herzog, Werner (dir.), *Cave of Forgotten Dreams* (2010)

Hesjedal, Anders, 'The Hunters' Rock Art in Northern Norway: Problems of Chronology and Interpretation', *Norwegian Archaeological Review* 27:1 (1994)

Hoffmann, D. L. et al., 'U-Th Dating of Carbonate Crusts Reveals Neandertal Origin of Iberian Cave Art', *Science* 359:6378 (February 2018)

Hogenboom, Melissa, 'In Siberia There is a Huge Crater and It is Getting Bigger', BBC, 24 February 2017 <http://www.bbc.com/earth/story/20170223-in-siberia-there-is-a-huge-crater-and-it-is-getting-bigger>

Hopkins, Gerard Manley, *The Journals and Papers of Gerard Manley Hopkins*, ed. Humphry House and Graham Storey (Oxford: Oxford University Press, 1959)

Household, Geoffrey, *The Courtesy of Death* (London: Michael Joseph, 1967)

*——, *Rogue Male* (1939; London: Chatto and Windus, 2002)

Howe, Cymene, '"Timely": Theorizing the Contemporary', *Cultural Anthropology* <https://culanth.org/fieldsights/800-timely>

Hugo, Victor, *The Essential Victor Hugo*, trans. E. H. and A. M. Blackmore (1862; Oxford: Oxford University Press, 2004)

Hussey, Andrew, *Paris: The Secret History* (London: Penguin, 2007)
Hutton, Noah (dir.), *Deep Time* (2015)

~

Ialenti, Vincent, 'Adjudicating Deep Time: Revisiting the United States' High-Level Nuclear Waste Repository Project at Yucca Mountain', *Science & Technology Studies* 27:2 (2014)
——, 'Death and Succession among Finland's Nuclear Waste Experts', *Physics Today* 70:10 (2017)
Ingold, Tim, *The Appropriation of Nature* (Manchester: Manchester University Press, 1986)
*International Commission on Stratigraphy, 'International Chronostratigraphic Chart' (v2016/04) <http://www.stratigraphy.org/ICSchart/ChronostratChart2016-04.pdf>

~

Janko, R., 'Forgetfulness in the Golden Tablets of Memory', *Classical Quarterly* 34:1 (1984)
Jones, Jamie L., 'Oil: Viscous Time in the Anthropocene', *Los Angeles Review of Books*, 22 March 2016 <https://lareviewofbooks.org/article/oil-viscous-time-in-the-anthropocene>

~

Kafka, Franz, *The Complete Stories*, trans. Willa and Edwin Muir (New York: Schocken, 1971)
——, *Metamorphosis and Other Stories*, trans. Willa and Edwin Muir (Aylesbury: Penguin, 1977)
**The Kalevala*, trans. Keith Bosley (Oxford: Oxford University Press, 2008)
*Kimmerer, Robin Wall, *Gathering Moss: A Natural and Cultural History of Mosses* (Corvallis: Oregon State University Press, 2003)
*——, *Braiding Sweetgrass: Indigenous Wisdom, Scientific Knowledge, and the Teachings of Plants* (Minneapolis: Milkweed, 2013)
——, 'Learning the Grammar of Animacy', *Anthropology of Consciousness* 28:2 (2017)

——, 'Speaking of Nature', *Orion Magazine*, 14 June 2017.

Kircher, Athanasius, *Mundus Subterraneus, in XII Libros Digestus* (Amsterdam, 1678)

*Klingan, Katrin, et al., *Textures of the Anthropocene: Grain, Vapor, Ray*, 3 vols. (Cambridge, MA: MIT Press, 2015)

*Kolbert, Elizabeth, *The Sixth Extinction: An Unnatural History* (New York: Henry Holt, 2014)

——, 'Greenland is Melting', *New Yorker*, 24 October 2016 <https://www.newyorker.com/magazine/2016/10/24/greenland-is-melting>

Kpomassie, Tété-Michel, *An African in Greenland* (London: Secker and Warburg, 1983)

Kroonenberg, Salomon, *Why Hell Stinks of Sulfur: Mythology and Geology of the Underworld* (London: Reaktion, 2013)

~

Larkin, Philip, *The Whitsun Weddings* (London: Faber and Faber, 1964)

Latour, Bruno, 'Agency at the Time of the Anthropocene', *New Literary History* 45:1 (2014)

Le Guin, Ursula K., *The Word for World is Forest* (1972; London: Orion Books, 2015)

'Lithological Log of Cleveland Potash Ltd', Borehole Staithes No. 20, drilled September–December 1968 to a depth of *c.*3500 feet (BGS ID borehole 620319, BGS Reference NZ71NE14)

*Lopez, Barry, *Arctic Dreams: Imagination and Desire in a Northern Landscape* (1986; New York: Bantam, 1987)

Lovelock, James, *Life and Death Underground* (London: G. Bell and Sons, 1963)

Lowenstein, Tom, 'Excavation and Contemplation: Peter Riley's Distant Points', in *The Gig: The Poetry of Peter Riley* 4/5 (2000)

Luciano, Dana, 'Speaking Substances: Rock', *Los Angeles Review of Books*, 12 April 2016 <https://lareviewofbooks.org/article/speaking-substances-rock/>

Luther, Kem, *Boundary Layer: Exploring the Genius Between Worlds* (Corvallis: Oregon State University Press, 2016)

~

Macaulay, Thomas Babington, *Ranke's History of the Popes* (London: Longman, Brown, Green, and Longmans, 1851)

McBurney, Simon, 'Herzog's Cave of Forgotten Dreams: The Real Art Underground', *Guardian*, 17 March 2011

McCarthy, Cormac, *Blood Meridian* (1985; New York: Vintage, 1992)
McCarthy, Tom, *Satin Island* (London: Cape, 2014)
Macdonald, Graeme, 'Oil and World Literature', *American Book Review* 33:3 (2012)
——, '"Monstrous Transformer": Petrofiction and World Literature', *Journal of Postcolonial Writing* 53 (2017)
McGrath, Dara, 'Project Cleansweep' <http://daramcgrath.com/Project_Cleansweep_Cover_Page.html>
McInerney, Jeremy, and Sluiter, Ineke (eds.), *Valuing Landscape in Classical Antiquity* (Leiden: Brill, 2016)
Maclean, FitzRoy, *Eastern Approaches* (London: Jonathan Cape, 1949)
MacNeice, Louis, *Collected Poems* (London: Faber and Faber, 2007)
McPhee, John, *Basin and Range* (New York: FSG, 1981)
*——, *Annals of the Former World* (New York: FSG, 1998)
Madsen, Michael (dir.), *Into Eternity* (2010)
*Manaugh, Geoff, *The BLDG BLOG Book: Architectural Conjecture, Urban Speculation, Landscape Futures* (San Francisco: Chronicle, 2009)
Margulis, Lynn (ed.), *Symbiosis as a Source of Evolutionary Innovation: Speciation and Morphogenesis* (Boston: MIT Press, 1991)
Marris, Emma, 'Neanderthal Artists Made Oldest-Known Cave Paintings', *Nature*, 22 February 2018
Mattern, Shannon, 'Cloud and Field', *Places Journal* (August 2016) <https://places journal.org/article/cloud-and-field>
Meyers, Kent, 'Chasing Dark Matter in America's Deepest Gold Mine', *Harper's Magazine* (May 2015)
Michaels, Anne, *Fugitive Pieces* (London: Bloomsbury, 1997)
Michaels, Sean, 'Unlocking the Mystery of Paris' Most Secret Underground Society', Gizmodo, 21 April 2011 <https://gizmodo.com/5794199/unlocking-the-mystery-of-paris-most-secret-underground-society-combined>
Miéville, China, *The City and the City* (London: Pan Books, 2009)
——, *Three Moments of an Explosion* (London: Macmillan, 2015)
Milne, Richard, 'Oil and the Battle for Norway's Soul', *Financial Times*, 27 July 2017
Mitchell, W. J. T., *What Do Pictures Want? The Lives and Loves of Images* (Chicago: University of Chicago Press, 2005)
Molchanova, Natalia, 'The Depth', trans. Victor Hilkevich <http://molchanova.ru/en/verse/depth>
Moore, Jason W., *Capitalism in the Web of Life* (London: Verso, 2015)
Morris, Jan, *Trieste and the Meaning of Nowhere* (London: Faber and Faber, 2001)

Mortimer, John Robert, *Forty Years' Researches in British and Saxon Burial Mounds of East Yorkshire. Including Romano-British discoveries, and a description of the ancient entrenchments on a section of the Yorkshire Wolds . . . With over 1000 illustrations from drawings by Agnes Mortimer* (London: A. Brown and Sons, 1905)

Morton, Timothy, *Hyperobjects: Philosophy and Ecology after the End of the World* (Minneapolis: University of Minnesota Press, 2013)

——, 'Poisoned Ground: Art and Philosophy in the Time of Hyper-Objects', *Symploke* 21:1–2 (2013)

Muecke, Stephen, 'Global Warming and Other Hyperobjects', *Los Angeles Review of Books*, 20 February 2014 <https://lareviewofbooks.org/article/hyperobjects>

Mumford, Lewis, *The City in History: Its Origins, Its Transformations, and Its Prospects* (New York: Harcourt & Brace, 1961)

Murray, W. H., *Mountaineering in Scotland and Undiscovered Scotland* (London: Diadem Books, 1979)

~

Negarastani, Reza, *Cyclonopedia: Complicity with Anonymous Materials* (Melbourne: re.press, 2008)

Nelson, Richard, *Make Prayers to the Raven: A Koyukon View of the Northern Forest* (Chicago: University of Chicago Press, 1986)

Nelson, Victoria, *The Secret Life of Puppets* (Cambridge, MA: Harvard University Press, 2001)

Newman, E. I., 'Mycorrhizal Links between Plants: Their Functioning and Ecological Significance', *Advances in Ecological Research* 18 (1988)

*Ngai, Sianne, *Ugly Feelings* (Cambridge, MA: Harvard University Press, 2005)

Nielsen, Lars (ed.), *Advances in Geophysics*, vol. 57 (Cambridge, MA: Academic Press, 2016)

Nixon, Rob, 'The Swiftness of Glaciers: Language in a Time of Climate Change', *Aeon Magazine*, 19 March 2018 < https://aeon.co/ideas/the-swiftness-of-glaciers-language-in-a-time-of-climate-change>

Nora, Pierre and Ageron, Charles-Robert, *Les Lieux de Mémoire*, 3 vols. (Paris: Editions Gallimard, 1993)

Norsted, Terje, 'The Cave Paintings of Norway', *Adoranten* (2013)

~

O'Domhnaill, Risteard (dir.), *Atlantic* (2016)

O'Neill, Joseph, *Land under England* (London: New English Library, 1978)

Oldenburg, Henry, and Roper, Francis, *Philosophical Transactions, Giving Some Accompt of the Present Undertakings, Studies, and Labours, of the Ingenious in Many Considerable Parts of the World*, vol. 16 (London: Printed for T.N. by John Martyn, 1687)

Olsen, Bjørnar, *In Defense of Things: Archaeology and the Ontology of Objects* (Plymouth: AltaMira Press, 2017)

~

Peebles, J. M., *The Practical of Spiritualism. Biographical Sketch of Abraham James. Historic Description of His Oil-Well Discoveries in Pleasantville, P.A., through Spirit Direction* (Chicago: Horton and Leonard Printers, 1868)

Perec, Georges, *Species of Spaces and Other Pieces*, trans. John Sturrock (1974; Harmondsworth: Penguin, 1997)

Pétursdóttir, Þóra, 'Climate Change? Archaeology and Anthropocene', *Archaeological Dialogues* 24:2 (2017)

*——, 'Drift', in *Multispecies Archaeology*, ed. Suzanne E. Pilaar Birch, (London: Routledge, 2018)

——, and Olsen, Bjørnar, 'Unruly Heritage: An Archaeology of the Anthropocene', (Tromsø: UiT The Arctic University of Norway, 2017) <https://www.sv.uio.no/sai/forskning/grupper/Temporalitet%20-%20materialitet/lesegruppe/olsen-unruly-heritage.pdf>

Plato, *Timaeus and Critias*, trans. Robin Waterfield (Oxford: Oxford University Press, 2008)

Playfair, John, 'Biographical Account of the Late Dr James Hutton, F.R.S. Edin.', *Transactions of the Royal Society of Edinburgh* 5 (1805)

Poe, Edgar Allan, *The Fall of the House of Usher and Other Writings*, ed. David Galloway (London: Penguin, 2003)

Posidonius, *Posidonius*, ed. Ludwig Edelstein and I. G. Kidd (Cambridge: Cambridge University Press, 1988)

Postlethwaite, John, *Mines and Mining in the (English) Lake District* (Whitehaven: W. H. Moss and Sons, 1913)

Powers, Richard, *The Overstory* (New York: W. W. Norton, 2018)

Prynne, J. H., *The White Stones* (Lincoln: Grosseteste, 1969)

——, 'On the Poetry of Peter Larkin', *No Prizes* 2 (2013)

*Purdy, Jedediah, *After Nature: A Politics for the Anthropocene* (Cambridge, MA: Harvard University Press, 2015)

~

Rigzone, 'Statoil Seeking New Acreage', 1 October 2016 <https://www.rigzone.com/news/oil_gas/a/16859/statoil_seeking_new_acreage/>

Riley, Peter, *The Derbyshire Poems* (Exeter: Shearsman Books, 2012)

Rilke, Rainer Maria, *Rainer Maria Rilke, Lou Andreas-Salome: Briefwechsel* (Zurich: M. Niehans, 1952)

——, *Selected Letters 1902–1926*, trans. R. F. C. Hull (London: Quartet Encounters, 1988)

——, *Sonnets to Orpheus*, trans. Martyn Crucefix (London: Enitharmon Press, 2012)

Robb, Graham, *Parisians: An Adventure History of Paris* (London: Picador, 2010)

*Robinson, Tim, *My Time in Space* (Dublin: Lilliput, 2001)

——, *Connemara: The Last Pool of Darkness* (London: Penguin, 2009)

Roosth, Sophia, 'Virus, Coal, and Seed: Subcutaneous Life in the Polar North', *Los Angeles Review of Books*, 21 December 2016 <https://lareviewofbooks.org/article/virus-coal-seed-subcutaneous-life-polar-north/>

~

Salk, Jonas, 'Are We Being Good Ancestors?', *World Affairs* 1:2 (1992)

Sanderson, John, *The Travels of John Sanderson in the Levant, 1584–1602: With His Autobiography and Selections from His Correspondence*, ed. William Foster (Abingdon: Routledge, 2016)

*Savoy, Lauret, *Trace: Memory, History, Race and the American Landscape* (Berkeley: Counterpoint, 2015)

Scarry, Elaine, *The Body in Pain: The Making and Unmaking of the World* (Oxford: Oxford University Press, 1985)

Scheurmann, Ingrid, and Scheurmann, Konrad, *For Walter Benjamin*, 3 vols. (Bonn: AsKI e.v. and Inter Nationes, 1994)

Schindler, John, *Isonzo* (London: Praeger, 2001)

Schuller, Kyla, 'Speaking Substances: Bodies', *Los Angeles Review of Books*, 23 March 2013 <https://lareviewofbooks.org/article/bodies/>

Schulting, R. J., '"... Pursuing a Rabbit in Burrington Combe": New Research on the Early Mesolithic Burial Cave of Aveline's Hole', *Proceedings of the University of Bristol Spelaeological Society* 23:3 (2005)

Seaborn, Adam, *Symzonia: A Voyage of Discovery* (New York: J. Seymour, 1820)

Sebald, W. G., *The Rings of Saturn*, trans. Michael Hulse (1995; London: Vintage, 2002)

*Sebeok, Thomas, 'Communication Measures to Bridge Ten Millennia (Technical Report)', *Research Centre for Language and Semiotic Studies, for Office of Nuclear Waste Isolation*, BMI/ONWI-532 (1984)

Serres, Michael, *Statues: The Second Book of Foundations*, trans. Randolph Burks (London: Bloomsbury, 2015)

Shaw, Trevor, *Foreign Travellers in the Slovene Karst: 1486–1900* (Ljubljana: Založba ZRC, 2008)

——, and Čuk, Alenka, *Slovene Caves & Karst Pictured 1545–1914* (Ljubljana: Založba ZRC, 2012)

Shellenberger, Michael, and Nordhaus, Ted (eds.), *Love Your Monsters: Postenvironmentalism and the Anthropocene* (Oakland: The Breakthrough Institute, 2011)

Shepherd, Nan, *The Living Mountain* (1977; Edinburgh: Canongate, 2011)

Simard, Suzanne (interview with Diane Toomey), 'Exploring How and Why Trees "Talk" to Each Other', *Yale Environment 360*, 1 September 2016 <https://e360.yale.edu/features/exploring_how_and_why_trees_talk_to_each_other>

——, et al., 'Net Transfer of Carbon between Ectomycorrhizal Tree Species in the Field', *Nature* 388:6642 (1997)

Simpson, Joe, *Touching the Void* (1988; London: Vintage Classic, 2008)

Sleigh-Johnson, Sophie, 'Performance Waves', *Performance Research* 21:2 (2016)

*Smithson, Robert, *The Collected Writings*, ed. Jack Flam (Berkeley: University of California Press, 1996)

Sneider, Noah, 'Cursed Fields: What the Tundra Has in Store for Russia's Reindeer Herders', *Harper's Magazine* (April 2018)

Solnit, Rebecca, *Savage Dreams: A Journey into the Hidden Wars of the American West* (Berkeley: University of California Press, 2014)

Solomon, Andrew, *Far and Away: How Travel Can Change the World* (London: Scribner, 2016)

Sophocles, *Antigone*, ed. and trans. Diane J. Rayor (Cambridge: Cambridge University Press, 2011)

Spirito, Pietro, 'Alla scoperta del Timavo', *Il Piccolo*, 2–23 August 2014

——, 'Nei cantieri sottoterra da anni si dà la caccia al fiume che non c'è', *Il Piccolo*, 23 August 2014

Stokes, Adrian, *Stones of Rimini* (New York: Schocken Books, 1969)

Stone, Alison, 'Adorno and the Disenchantment of Nature', *Philosophy and Social Criticism* 32:2 (2006)

Stranj, Pavel, *The Submerged Community*, trans. Mark Brady (Trieste: Editoriale Stampa, 1992)

Strugatsky, Arkady, and Strugatsky, Boris, *Roadside Picnic* (London: Gollancz, 2012)

Sullivan, John Jeremiah, *Pulphead: Notes from the Other Side of America* (New York: FSG, 2011)

~

Terray, Lionel, *Conquistadors of the Useless: From the Alps to Annapurna*, trans. Geoffrey Sutton (1963; Sheffield: Bâton Wicks, 2000)

Thacker, Eugene, *In the Dust of This Planet* (Alresford: Zero Books, 2011)

Thomas, Edward, 'Chalk Pits', in *Selected Poems and Prose* (1981; London: Penguin, 2012)

Thompson, Mark, *The White War: Life and Death on the Italian Front* (New York: Basic Books, 2009)

Toshihisa, Okamura, *The Cultural History of Matsutake*, trans. Fusako Shimura and Miyaki Inoue (Tokyo: Yama to Keikokusha, 2005)

Trauth, Kathleen M., et al., 'Expert Judgment on Markers to Deter Inadvertent Intrusion into the Waste Isolation Pilot Plant', *Sandia National Laboratories*, SAND 92–1382. UC–721 (1993)

Trentmann, Frank, *Empire of Things: How We Became a World of Consumers, from the Fifteenth Century to the Twenty-First* (New York: HarperCollins, 2016)

Tsing, Anna, 'Arts of Inclusion, or How to Love a Mushroom', *Manoa* 22:2 (2010)

*——, *The Mushroom at the End of the World: On the Possibility of Life in Capitalist Ruins* (Princeton: Princeton University Press, 2017)

——, 'The Politics of the Rhizosphere' (interviewed by Rosetta S. Elkin), *Harvard Design Magazine* 45 (Spring/Summer 2018) <http://www.harvarddesignmagazine.org/issues/45/the-politics-of-the-rhizosphere>

*——, Swanson, Heather, Gan, Elaine, and Buband, Nils, (eds.), *Arts of Living on a Damaged Planet* (Minneapolis: University of Minnesota Press, 2017)

~

Valvasor, Johann von, 'An Extract of a Letter Written to the Royal Society out of Carniola, by Mr John Weichard Valvasor, R. Soc. S. Being a Full and Accurate

Description of the Wonderful Lake of Zirknitz in that Country', in *Philosophical Transactions, Giving Some Accompt of the Present Undertakings, Studies, and Labours, of the Ingenious in Many Considerable Parts of the World*, eds. Henry Oldenburg and Francis Roper, vol. 16 (London: Printed for T.N. by John Martyn, 1687)

Verne, Jules, *Journey to the Centre of the Earth*, trans. Robert Baldick (1864; Harmondsworth: Puffin Books, 1965)

~

Wark, Mackenzie, *Molecular Red: Theory for the Anthropocene* (London: Verso, 2015)

Webb, Dave (dir.), *Fight for Life: The Neil Moss Story* (2006)

Webb, Dave, and Whiteside, Judy, 'Fight for Life: The Neil Moss Story' <www.mountain.rescue.org.uk/assets/files/The Oracle/history and people/NeilMossStory.pdf>

Weir, Andy, 'Deep Decay: Into Diachronic Polychromatic Material Fictions', *PARSE* 4 (2017) <http://parsejournal.com/article/deep-decay-into-diachronic-polychromatic-material-fictions/>

*Weizman, Eyal, *Hollow Land: Israel's Architecture of Occupation* (London: Verso, 2007)

Wells, H. G., *The Time Machine* (1895; Richmond: Alma Classics, 2017)

Williams, Rosalind, *Notes on the Underground: An Essay on Technology, Society, and the Imagination* (London: MIT Press, 2008)

Williams, William Carlos, *The Collected Poems of William Carlos Williams, Volume II 1939–1962*, ed. Christopher MacGowan (New York: New Directions, 1988)

Wilson, Louise K. (ed.), *A Record of Fear* (Salisbury: B.A.S. Printers Ltd, 2005)

Wohlleben, Peter, *The Hidden Life of Trees*, trans. Jane Billinghurst (Vancouver/Berkeley: Greystone Press, 2016)

Wulf, Andrea, *The Invention of Nature: Alexander von Humboldt's New World* (New York: Knopf, 2015)

Wylie, John, 'The Spectral Geographies of W. G. Sebald', *Cultural Geographies* 14 (2007)

Wyndham, John, *The Chrysalids* (1955; London: Penguin, 2018)

~

*Yusoff, Kathryn, 'Geologic Subjects: Nonhuman Origins, Geomorphic Aesthetics, and the Art of Becoming *In*human', *cultural geographies* 22:3 (2015)

~

Zalasiewicz, Jan, et al., 'The Anthropocene: A New Epoch of Geological Time?', *Philosophical Transactions. Series A, Mathematical, Physical, and Engineering Sciences* 369 (2011)

Zhdanova, N. N., et al., 'Ionizing Radiation Attracts Soil Fungi', *Mycological Research* 108:9 (2004)

Zola, Emile, *Germinal*, trans. Havelock Ellis (London: Dent, 1970)

ACKNOWLEDGEMENTS

I thank first those who have shaped this book most as companions, guides and teachers, helping me learn to see in the dark: John Beatty, Hein Bjerck, Sean and Jane Borodale, Bill Carslake, Lucian and Maria Carmen Comoy, Sergio Dambrosi, Steve Dilworth, Bradley Garrett, Meriel Harrison, Lina and Jay, Helen Mort, Robert Mulvaney, Bjørnar Nicolaisen, Þóra Pétursdóttir, Neil Rowley, Merlin Sheldrake, Richard Skelton, Helen and Matt Spenceley, Christopher Toth and Pasi Tuohimaa.

Garnette Cadogan, Walter Donohue, Henry Hitchings, Julith Jedamus, Simon McBurney, Garry Martin, Rob Newton and Jedediah Purdy read all or part of *Underland* in the course of its writing; their responses were invaluable. I hope I have made my profound gratitude apparent to each of you. Several people brought their specialist knowledge to bear on specific sections of the book, correcting and clarifying my work with generous expertise. I am especially grateful to Carolin Crawford (on stars), John MacLennan (on rocks) and Ruth Mottram (on ice). Tanja Trček kindly and bravely translated the *foiba* text for me. Rob Newton was the best research assistant I could have hoped for in the closing months of the book, offering calm counsel and sharp eyes at every turn.

My editor Simon Prosser and my agent Jessica Woollard have been remarkable readers and friends throughout the six and a half years it took me to write *Underland*. At Hamish Hamilton/Penguin I have been exceptionally fortunate to work with Richard Bravery, Dave Cradduck, Caroline Pretty, Anna Ridley, Ellie Smith and Hermione Thompson. At W. W. Norton in America I have benefited immensely from the acuity, support and patience of my editor Matt Weiland and the encouragement of Jim Rutman.

I have learned so much from – and thought much with – my students, especially Jei Degenhardt, Louis Klee, Aron Penczu, Kryštof Vosatka and Lewis Wynn. I thank my close friends for all they have done for me and the book: Julie Brook, Peter

Davidson, Gareth Evans, Nick Hayes, Michael Hrebeniak, Michael Hurley, Raphael Lyne, Finlay Macleod, Leo Mellor, Jackie Morris, Clair Quentin, Corinna Russell, Jan and Chris Schramm, David Trotter, James Wade and Simon Williams. Above all and forever, love and gratitude to Julia, Lily, Tom and Will, and to my parents, Rosamund and John.

Thanks are then also due for many kinds of assistance, information and inspiration over the years: to Glenn Albrecht, Alice and Chris Allan, Tim Allen, Antti Apunen, Marina Ballard, Ariane Bankes, Mattias Bärmann, Ginny Battson, Sharon Blackie, Miguel Angel Blanco, Adam Bobbette, Edward-John Bottomley, James Bradley, Michael Bravo, Julia Brigdale, Julie Brook, Rob Bushby, Jonathan and Keggie Carew, Steve Casimiro, Silvia Ceramicola, Christopher Chippendale, Václav Cílek, Horatio Clare, Erlend Clouston, Michela Coletta, Ray Collins, Adrian Cooper, Holly Corfield-Carr, Nicola Dahrendorf, John Dale, William Dalrymple, Jane Davidson, Jeremy Davies, Tim Dee, Thomas Demarchi, Aly Derby, Hildegard Diemberger, Hunter Dukes, Cody Duncan, Minna Moore Ede, Chris Evans, Garry Fabian-Miller, David Farrier, Kitty Fedorec, Rose Ferraby, Toby Ferris, Johnny Flynn, Xesus Fraga, Robin Friend, Rebecca Giggs, Antony Gormley, Simon Grant, Susan Greaney, Pino Guidi, Beatrice Harding, Kateřina Havlíková, M. John Harrison, Harriet Hawkins, Caspar Henderson, Julia Hoffman, Cymene Howe, Robert Hyde, Bob Jellicoe, Martin Johnson, Stuart Kelly, Michael Kerr, Patrick Kingsley, Andrew Kötting, Paul Laity, Szabolcs Leél-Őssy, Angela Leighton, Emily Lethbridge, Huw Lewis-Jones, Tim de Lisle, Thelma and Bill Lovell, Borut Lozej, Richard Mabey, Helen Macdonald, Jim Macfarlane, Duncan Mackay, Finlay Macleod, Andrew McNeillie, Geoff Manaugh, Kevan Manwaring, Philip Marsden, Jana Martinčič, Rod Mengham, China Miéville, Alex Moss, Helen Murphy, Victoria Nelson, Kate Norbury, Annie O'Garra Worsley, Bjørnar Olsen, Jay Owens, Francesco Panetta, Fabio Pasini, Donald and Lucy Peck, Sibylle Pein, Borut Peric, Pirhuk, Jonathan Power, Andrew Ray, Lara Reid, Fiona Reynolds, Dan Richards, Autumn Richardson, Darmon Richter, Tim Robinson, David Rose, Giuliana Rossi, Corinna Russell, Stanley Schtinter, Adam Scovell, Geoff Shipp, Robbie Shone, Philip Sidney, Iain Sinclair, Ingrid Skjoldvær, Paul Slovak, Jos Smith, Rebecca Solnit, Emily Stokes, John and Katja Stubbs, Kier Swaffield, Sarah Thomas, Louis Torelli, Michaela Vieser, Marina Warner, Jim Warren, Julianne Warren, Giles Watson, Stephen Watts, Samantha Weinberg, Andy Weir, Deb Wilenski, Christopher Woodward, Geoff Yeadon, Benjamin Zidarich and many correspondents on Twitter.

I am grateful to those photographers and rights-holders who have generously allowed me to use their images here. The photograph prefacing the First Chamber is the image of a hand stencil made in El Castillo cave in northern Spain; the earliest of the El Castillo stencils has been dated to at least 37,300 years old, and it is therefore possible that it was made by a Neanderthal artist. It is reproduced with permission of and © to La Sociedad Regional de Educación, Cultura y Deporte of Cantabria (SRECD). The photograph that prefaces Chapter 1, 'Descending', is by Ivana Cajina (@von_co) and is made available for free use under the @unsplash licence. The photograph of one of the Priddy Nine Barrows that prefaces Chapter 2, 'Burial', is © Richard Scott-Robinson. The photograph that prefaces Chapter 3, 'Dark Matter', is by Alexander Andrews (@alex_andrews) and is made available for free use under the @unsplash licence. The photograph that prefaces Chapter 4, 'The Understorey', is by Johannes Plenio (@jplenio), and is made available for free use under the @ pixabay/CC0 Creative Commons licence. The photograph that prefaces Chapter 5, 'Invisible Cities', is of *Le Passe-Muraille* and is © Laura Brown (fuschiaphoto.com). The image that prefaces Chapter 6, 'Starless Rivers', is of the chamber at the bottom of the Abyss of Trebiciano, through which the Timavo river runs. It is an engraving by Giuseppe Rieger from the mid nineteenth century, and I am grateful to the Biblioteca Civica Attilio Hortis, Trieste, and E. Hapulca for permission to reproduce it here. The image that prefaces Chapter 9, 'The Edge', is Harry Clarke's illustration to accompany Edgar Allan Poe's story 'A Descent into the Maelstrom' when the story was reprinted in a 1919 edition of *Tales of Mystery and Imagination*; it is out of copyright. The photographs prefacing Chapters 10 and 11, 'The Blue of Time' and 'Meltwater', are from our time in East Greenland and are © Helen Spenceley. The photograph prefacing Chapter 12, 'The Hiding Place', is of Onkalo and is © Posiva. The photograph prefacing Chapter 13, 'Surfacing', is of the Cueva de las Manos, the 'Cave of the Hands', in Patagonia in 2005; the hand stencils were made using ochre dust blown through a bone pipe, and have been dated to around 9,300 years ago. The image is generously made available by and is © Mariano Cecowski. The jacket photograph is of me approaching the crevasse labyrinth of the Knud Rasmussen glacier, East Greenland; it is © Helen Spenceley. All other images are my own.

For the granting of textual permissions, I am grateful to James Maynard and the estate of Helen Adam for allowing me to quote from 'Down There in the Dark' as an epigraph. It is © the Poetry Collection of the University Libraries, University at Buffalo, The State University of New York. I am grateful to Alexey Molchanov for permission to publish his mother Natalia's poem 'The Depth' in full. The only significant portion of this book to appear in any form before publication was 'Secrets of the Wood Wide Web', in the *New Yorker* online, edited by Emily Stokes; I am grateful to Emily and the *New Yorker* for permission to reuse sentences from that essay here.

Underland could not have been completed without the support of the British Academy in the form of a Mid-Career Fellowship, for which I am grateful beyond easy expression. I am indebted to several other institutions and colleagues, above all Emmanuel College, Cambridge, where I have been privileged to teach for seventeen years now; also the English Faculty in Cambridge and the English Faculty Library (the best library beyond Babel). Among the music and musicians whose work has kept me company above and below ground, I could not have done without *AR, Bon Iver, the Duke Spirit, Elbow, Johnny Flynn, Grasscut, Willy Mason, the Pixies, Karine Polwart, Schubert, Cosmo Sheldrake and Le Tigre.

The cover image of *Underland* is by my long-time friend and collaborator Stanley Donwood. I first saw his luminous painting *Nether* in 2013, the year after I'd begun work on *Underland*. It amazed me at first sight – the eerie glow of the sun, the curling technicolour fingers of the trees, the sense of looking down into a radiant, dangerous underworld – and I knew immediately that I wanted it to be the cover of my book. It's huge, too: 1.5 square metres. Big enough to fall head first into – or down. At its simplest, indeed, the word 'nether' just means 'down', 'downwards'. It more fully suggests, the *OED* records, 'what lies, or is imagined as lying, beneath the earth; of, belonging to, or native to hell or the underworld'. Whenever I felt exhausted or unsettled by work on *Underland* – which was often – I would think of *Nether*. It lit the way.

Except that while *Nether* looks like a vast sun rising at the end of a sunken lane, it's not. I remember asking Stanley about the image when we were together one day on Orford Ness, the shingle spit off the Suffolk coast where nuclear weapons were tested in the decades after the Second World War. '*Nether*,' Stanley said then, 'isn't the sun. It's the last thing you'd ever see. It's the light of a nuclear blast that has just detonated, seen down a holloway. When you look at *Nether*, you've got about 0.001 of a second of life remaining, before the flesh is melted from your bones.' Oh. Lustrous and lethal, fatal and beautiful, nuclear and natural, the image beckons the viewer's eye on and down into the underworld, into its reactor core. As such, it could hardly be truer to the atmospheres of *Underland*.

INDEX